Deterritorializing the Future

Critical Climate Change

SERIES EDITORS: TOM COHEN AND CLAIRE COLEBROOK

The era of climate change involves the mutation of systems beyond 20th century anthropomorphic models and has stood, until recently, outside representation or address. Understood in a broad and critical sense, climate change concerns material agencies that impact on biomass and energy, erased borders and microbial invention, geological and nanographic time, and extinction events. The possibility of extinction has always been a latent figure in textual production and archives; but the current sense of depletion, decay, mutation and exhaustion calls for new modes of address, new styles of publishing and authoring, and new formats and speeds of distribution. As the pressures and realignments of this re-arrangement occur, so must the critical languages and conceptual templates, political premises and definitions of 'life.' There is a particular need to publish in timely fashion experimental monographs that redefine the boundaries of disciplinary fields, rhetorical invasions, the interface of conceptual and scientific languages, and geomorphic and geopolitical interventions. Critical Climate Change is oriented, in this general manner, toward the epistemo-political mutations that correspond to the temporalities of terrestrial mutation.

Deterritorializing the Future

Heritage in, of and after the Anthropocene

Edited by Rodney Harrison and Colin Sterling

O

OPEN HUMANITIES PRESS

London 2020

First edition published by OPEN HUMANITIES PRESS 2020

Text © Contributors, 2020

Images © Contributors and copyright holders named in captions, 2020

Cover Image: Still from Tuguldur Yondonjamts, *An Artificial Nest Captures a King*, 2016, artist film, 25:09 min.

PRINT ISBN 978-1-78542-088-7

PDF ISBN 978-1-78542-087-0

( )

OPEN HUMANITIES PRESS

OPEN HUMANITIES PRESS is an international, scholar-led open access publishing collective whose mission is to make leading works of contemporary critical thought freely available worldwide. More at http://openhumanitiespress.org

Contents

II: Territories

Coda

List of Figures

Preface and Acknowledgements

This book is an outcome of two events which were co-organized as part of our work on the 'Heritage and Posthumanities' subtheme of the UK Arts and Humanities Research Council (AHRC)-funded Heritage Priority Area Leadership Fellowship research project (grant number AH/P009719/1), based at the Institute of Archaeology, University College London. This subtheme aimed to bring together contemporary developments in the posthumanities with the field of critical heritage studies to explore the potential synergies between them. The first of these two events was an extended, whole day conference session on 'Heritage and Posthumanism' which was held at the 4th Biennial Association of Critical Heritage Studies (ACHS) Conference on 'Heritage Across Borders' in Hangzhou, China in early September 2018. The session aimed to explore the emerging contribution of posthumanist thinking to critical heritage studies, and considered a series of interlinked questions: In what ways can concepts in the posthumanities 'animate' debates in critical heritage studies? How does our understanding of heritage shift when considered from the perspective of posthuman futures? Ultimately, if 'heritage' is fundamentally concerned with *human* practices of value generation, is a *post*human philosophy of heritage even possible? Chapters included here by Bohlin, Sterling, Storm and Ugwuanyi were first presented at this conference session alongside several others, and subsequently revised to address the central themes of this volume.

The second was the symposium 'Deterritorializing the Future', which was held at Senate House in London in mid September 2018 following our return from China. The symposium brought together a series of invited scholars across a number of disciplines to explore themes of care, vulnerability and inheritance across human and more-than-human worlds. Again, this symposium aimed to consider a series of linked questions. How can we conceive of memory and the archive beyond the human? What life forms and objects do we inherit *with*? How might scarcity and abundance be reconfigured in the face of environmental catastrophe? In our framing of this event, we suggested that approaching these questions from distinct though interconnected pathways might allow us to 'deterritorialize' the future, picking out moments of solidarity that – in the spirit of Donna Haraway – might provide the basis for possible ongoingness inside what feels to us to be relentlessly diffracting future worlds. Chapters included here by Åsberg & Fredengren, Bohlin, Breithoff & Harrison, Byrne, DeSilvey, Van Allen and Zylinska were presented at the symposium. To these we have added a separate contribution by Venovcevs, who first presented his poem as a spoken performance at the 8th Winter School of the Estonian Graduate School of Culture Studies and Arts in Tallinn in late 2018.

The symposium included a number of interlinked events which significantly helped to frame our thinking around the final set of chapters reproduced here. The first of these was developed as part of an emerging collaboration with Arts Catalyst, a non-profit contemporary arts organization that commissions and produces transdisciplinary art and research. Based at the time of writing in Kings Cross, London, not far from University College London where we are preparing this preface, Arts Catalyst's aims to incubate new ideas, conversations and transformative experiences across science and culture, and to encourage people to engage actively with a changing world, seemed to resonate strongly with our own. It was through Arts Catalyst that we were first introduced to the work of Tuguldur Yondonjamts, a Mongolian artist who draws on symbolic aspects of nomadic cultures of Central Asia in his video, drawing and installation artwork to engage with issues of environmental change and the effects of extractive industries and technologies on marginal landscapes in the Anthropocene. As part of his residency at Arts Catalyst's Centre for Arts, Culture and Society in 2018, we organized a public 'conversation' between Yondonjamts and Denis Byrne, who was visiting us from the University of Western Sydney in Australia, and who is a contributor to this book. Byrne's work, like that of Yondonjamts, draws on aspects of photography, travel writing and autoethnography to engage with questions of globalization, environmental change and their impact on local tradition in Asia and beyond. The public conversation bought together Byrne, Yondonjamts, ourselves and Arts Catalyst curator Anna Santamouro to explore how speculative and investigative artistic practices like those of Yondonjamts might inform the approaches of archaeology and critical heritage studies to the investigation of history, memory and environmental futures, and conversely, how archaeology and heritage studies might be understood to constitute speculative or artistic practices in their own distinctive ways.

The cover of this book features a still from Yondonjamts' film *An Artificial Nest Captures a King* (2016). In the film,

> ... the artist travels from artificial falcons' nests on the Mongolian steppes to the Gobi Desert, where he discovers a fossil crocodile, a mythological creature which he enters and animates. Driving a 1980s Russian utility vehicle, this shamanic journey gives the illusion of continuing its progress in linear time along a desert road, yet from above we see the car caught in the folds of looped time (Arts Catalyst 2018).

These interlinked aspects of the Anthropocene – the spatial and the temporal – and the ways in which they challenge and trouble the categories of 'human', 'non-human' and 'more-than-human' form the two main themes around which this book is organized. We thank the artist for allowing us to use this screenshot from his work as an invitation to think both with and against the grain of the Anthropocene and its material and discursive legacies.

Claire Colebrook's contribution to this volume was originally planned as a separate public keynote lecture to open the symposium, however circumstances (themselves related to territorialization and contemporary geopolitics) meant that Claire was unable to travel to London to participate. Nonetheless, precirculating her paper meant that her arguments about the ways in which the future is already deterritorialized formed a touchstone for participants in the symposium and helped us significantly in developing the arguments we present in the introductory chapter. As such, this revised version of her keynote lecture reproduced here provides a fitting concluding piece to the book, which provocatively and helpfully provides a critical exploration of different ways of viewing the deterritorialization of the future(s) which authors in the volume argue for.

We thank contributors and audience members at each of these events for their comments and insights which have helped us to shape the final content of the present volume. We particularly acknowledge the support of our host institution, the Institute of Archaeology at University College London, and our funder, the UK Arts and Humanities Research Council (AHRC), in helping make each of these linked events possible. Our work has been practically and intellectually supported by other members of the AHRC Heritage Priority Area team, including Hana Morel, Hannah Williams and Susan Sandford-Smith, and enrichened by work undertaken across that project's other subthemes (see further information at www.heritage-research.org). We have also drawn inspiration from the work of collaborators on the Heritage Futures research programme (www.heritage-futures.org), three members of which have contributed directly to the present volume. Bohlin and Appelgren's participation in the Deterritorializing the Future symposium was made possible as part of their collaborations with RH on the Making Global Heritage Futures research cluster of the joint University College London-University of Gothenburg Centre for Critical Heritage Studies (see www.criticalheritagestudies.gu.se and www.ucl.ac.uk/critical-heritage-studies/).

As we write we are struck by the significant acceleration of public discourse relating to the Anthropocene, the climate emergency and anthropogenically instigated species extinction which has occurred in the year since our original symposium on this topic. At the end of this week, what is predicted to be the largest coordinated global climate change protest is to take place, whilst the work of Greta Thunberg and Extinction Rebellion has significantly raised the profile of these issues. Within this context we remain committed to the substantial and meaningful role of the arts, humanities and social sciences in imagining and realizing more-than-human futures which are radically different to the present, whilst critically uncovering the social, economic, political and ecological 'work' of natural and cultural heritage preservation as a central aim of critical heritage studies. The future is already deterritorializing. But what matters moving forward – to remix and extend Donna Haraway's assertion that what matters is which "worlds world worlds" (2016: 35; see also conclusion to Zylinska, this volume) – is which deterritorializing territories deterritorialize. The chapters assembled here demonstrate the significant possibilities inherent in the arts, humanities and social sciences in collaboratively building alternative futures in, of and after the Anthropocene.

Rodney Harrison & Colin Sterling, London, September 2019.

References

Arts Catalyst 2018. Tuguldur Yondonjamts: An Artificial Nest Captures A King + Investigations into the Darkest Dark. https://www.artscatalyst.org/tuguldur-yondonjamts-residency-and-exhibition-arts-catalyst.

Chapter 1

Introduction: Of Territories and Temporalities

Colin Sterling & Rodney Harrison

Utopia, today, is to believe that current societies will be able to continue along on their merry little way without major upheavals. Social modes of organization that prevail today on earth are not holding up, literally and figuratively. History is gripped by crazy parameters: demography, energy, the technological-scientific explosion, pollution, the arms race... The Earth is deterritorializing itself at top speed. The true utopians are conservatives of all shapes and sizes who would like for this "to hold up all the same", to return to yesterday and the day before yesterday. What is terrifying is our lack of collective imagination in a world that has reached such a boiling point. (Guattari 1983 [2009]: 307)

Félix Guattari did not have the terminology of the Anthropocene at his disposal when he was asked to respond to a survey on the subject of Utopia by *La Quinzaine Littéraire* in 1983, but the ingredients are all there. A history gripped by 'crazy parameters', the failure of traditional social systems and the collective imagination to confront a boiling planet, and the Earth itself 'deterritorialized' to the brink of collapse. Critical theory did not need the Anthropocene to see the interconnections between all of these elements, but we cannot deny the generative qualities of the term. As a newly designated geological time interval the Anthropocene signifies a fundamental change in environmental conditions and processes across the globe, one brought about by human activities on a vast scale. From soil erosion and species loss to the chemical composition of the atmosphere, the magnitude of these transformations can only be understood in a multi-scalar fashion, tacking endlessly between the gods-eye view and the molecular, between the satellite and the microbe. This sense of destabilization and boundary crossing has stimulated novel creative practices and redirected scholarly attention in many areas. No matter what angle we approach it from, however, the geological roots of the Anthropocene foreground certain territorial themes and registers: strata, fossils, emissions, extractions, minerals, the Earth itself. More than simply a temporal threshold, the emergence of the Anthropocene as a socio-material concept and empirical reality is marked by this sense of ongoing and irreversible *territorialization* – 'we' have created a new age for the planet, which 'we' must live with in all its contradictions and vulnerabilities. Whether the Anthropocene ends up being added to the Geological Time Scale as a period, an epoch, an age or a boundary event (the difference between these intervals might be "a few billion human lives", Jan Zalasiewicz reminds us (2008: 157)) the term therefore makes a distinct claim on the present and the future – a claim inscribed to varying degrees in bodies, sediments, historical narratives and social worlds. To what extent the grip of the Anthropocene might be loosened is the core concern of this book, framed here through the reciprocal if sometimes counterintuitive logics of deterritorialization and critical heritage thinking.

In an increasingly interconnected world, deterritorialization has emerged as a key conceptual framing through which to apprehend the flow of people, ideas, artefacts and cultural practices around the globe, whether physically or via a disembedded digital mediascape. Arjan Appadurai for example identifies deterritorialization as a 'central force' in the modern world, paying particular attention to the movement of people – especially "labouring populations" – who are brought into the "lower-class sectors and spaces of relatively wealthy societies" (1990: 11). Deterritorialization and globalization here are mutually reinforcing cultural-spatial processes, characterized by the emergence of new social relations in dispersed yet interconnected geographic contexts. This echoes the use of the term in anthropology (e.g. Tomlinson 1999) and mobility studies (e.g. Sheller and Urry 2006), where a core focus has been the weakening of ties between culture and place in a globalized world. Communication technologies are given a central place in this reading, as the ability to maintain close relationships at considerable distance is a key component in the deterritorialized experience of modernity. As Anthony Giddens argued some time ago now, in the modern world "the very tissue of spatial experience alters, conjoining proximity and distance in ways that have few parallels in prior ages" (1990: 140).

This notion of deterritorialization provides a useful jumping off point for the present volume, but it is not our main focus. The apparently immaterial flows of data, people, ideas and cultures around the globe has encouraged a 'whole Earth' vision that is both fundamental to and inconsistent with the Anthropocene as a spatialized and inherently material phenomenon. This contradiction surfaces in well-known projects such as Globaïa's CGI-driven Anthropocene films, which aim to raise awareness of how 'one species changed a planet' (see further discussion in Breithoff and Harrison, this volume). As digital lines representing transport, resource and communication networks connect up towns, cities, countries and continents over the past two centuries – beginning with the Industrial Revolution in England and spreading to every corner of the globe – so the Earth itself fades from view, an invisible territory against which a familiar narrative of globalization and ecological degradation might unfold. While the planetary scale of the Anthropocene is central to its formal designation as a geological time interval (thus underlining the deterritorialized nature of the concept), the legacies and resonances of this global signature are stubbornly territorial, from landfills and plastic islands to polluted cities slowly choking their most vulnerable residents to death. Just as the frontier landscapes of the Western imagination relied on the violent suppression of Indigenous populations, so your ephemeral digital avatar is rooted in poisonous earthly extractions.

It is in this context that deterritorializing the future emerges as a project of urgent theoretical, practical and political concern. While Guattari was right to claim that the Earth has been deterritorializing itself at 'top speed' for some time now, parallel forces and practices of (re)territorialization exert an equally strong pull on the present and the future. Some of these are intentional; driven – as Guattari identifies – by a nostalgic longing to 'return to yesterday'. Others surface in the vast environmental reconfigurations enacted through mining, drilling and land reclamations, as recorded for example by Edward Burtynsky under the banner of The Anthropocene Project (www.theanthropocene.org).

The disorienting scale of Burtynsky's aerial photographs make clear the limitations of familiar representational practices when confronted by this new geological framework. Vast and totalizing, the Anthropocene as seen through Burtynsky's lens reasserts the centrality of the Earth to a supposedly post-industrial and deterritorialized planet. Missing here however are the differential drivers and consequences of such change, at least at the level of human social and political systems. Consequently, the territorializing force of the Anthropocene is universalized and flattened, "obscuring the accountability behind the mounting eco-catastrophe and inadvertently making us all complicit in its destructive project" (Demos 2017: 19).

We might begin to disentangle such universalizing gestures by critically reframing the Anthropocene as a diffuse yet concrete material inheritance; one that requires careful and distinct forms of management in the present, for the future. As Kathryn Yusoff has argued, approaches to the Anthropocene that "flatten agency across different material economies" have little to contribute to the "geological inheritances and forces that are capitalized upon over generations through the vagaries of hominin evolution and deep history" (2013: 791). To help resituate this debate, Yusoff focuses on the human as fossil-to-come – "an ancestral statement" which underlines the "symbolic and imaginative function" of such artefacts, caught up "in the making of stories of history, futurity, and identity" (2013: 793). The framework of inheritance here responds to the multi-temporal nature of the Anthropocene whilst mobilizing a concern for the enduring and shifting qualities of diverse material legacies, questioning "what it is that is taken forward into the future, what is inherited under the concept of the human, and what survives it as excess or exclusion within its formations?" (ibid). This mode of apprehending the Anthropocene recognizes its territorializing qualities without surrendering to these completely: a form of *critical* inheritance that has direct resonances with ongoing work in the rapidly expanding field of critical heritage studies. If this volume can be said to have one aim it would be centring heritage within the Anthropocene debate, not as a nostalgic longing for how things were, but as a means of expanding our collective imagination. This means thinking differently about the temporalities and territories of heritage, which is precisely one of those social modes of organization that Guattari identified as *no longer holding up*.

Critical heritage and Anthropocene futures

A familiar view of heritage – at least in the Western tradition – would evoke themes of continuity and nostalgia, played out through historical consumption and a kind of kitsch romanticism, oriented towards the production of origin myths connecting territory, tradition, citizenship and the nation-state. As a heavily commoditized industry, heritage is closely tied to global tourism and the preservation of 'grand' architecture, but it is also deeply personal and embodied, drawing together both collective and individual genetic, cultural, artistic and economic modes of inheritance. Across these domains, heritage can be seen to intersect with the issues raised by climate change and the Anthropocene in numerous ways. Historic sites around the world are at risk from rising sea levels and melting permafrost; museums have become spaces of protest over sponsorship by big oil companies; biobanks and frozen zoos have been created to house genetic material in danger of becoming extinct; oral history projects have been undertaken to record memories of changed landscapes in an attempt to counteract the 'shifting baseline syndrome'. Custodians of 'natural' and 'cultural' heritage may deal with questions of vulnerability, scarcity, loss and sustainability in different ways, but both are forced to confront lasting and systemic change in the face of climate breakdown. Against this backdrop, exhibitions, museums and heritage sites have emerged as important tools in communicating this threat to the general public (e.g. see Cameron and Neilson 2014; Harvey and Perry 2015), while certain sites have been scrutinized to try and understand how previous civilizations responded to rapid environmental change (e.g. Hambrecht et al. 2018). Case studies in adaptation are not only historical, however. Bringing historic buildings back into use has emerged as a key trend in contemporary architecture, offering an alternative to the damaging ecological impact of new developments. At the other end of the scale, traditional skills have been 'rediscovered' by conservationists and survivalists alike (although with different intentions and motivations). As a sign of their growing interconnectedness, 2018 saw the inaugural Climate Heritage Mobilization meeting at the Global Climate Action Summit in San Francisco – the first time the issue had been given a significant platform at a major climate event. In 2019 the Climate Heritage Network held its launch event in Edinburgh, galvanizing work in this area.

Such activities are an important indication of the different ways in which the *practice* of heritage can overlap with and complement action on climate change, but they are not the focus of this book.[1] There are three main distinctions between the work we want to undertake in this volume and more familiar approaches to heritage and climate change. It is worth introducing these here to help frame subsequent discussions, which in many cases depart significantly from mainstream heritage discourse. This is a reflection of the transdisciplinary approach taken to formulating this collection and – we hope – one of the key strengths of the book.

Perhaps the most obvious point of departure concerns the overarching question of **the Anthropocene**, which we see as *related to* but not *synonymous with* global warming and climate breakdown. Whilst anthropogenic climate change clearly shares many roots and points of origin with the Anthropocene – from rapid industrialization and resource extraction to biodiversity loss and human population increases – the (possible) onset of a new geological timeframe for the Earth does not necessarily follow from changes to climate, no matter how profound these may be. As Lewis and Maslin contend, "people began to change the planet long ago, and these impacts run deeper than just our use of fossil fuels. And so our responses to living in this new epoch will have to be more far-reaching" (2018: 6). The Anthropocene is thus, in the words of Ben Dibley (2012), both *epoch* and *discourse;* a discourse which he notes embodies simultaneous nostalgia and repulsion for the notion of the human and its ending (on these contradictions see Dibley 2015, 2018) and which itself acts as a newly emerging apparatus to direct and determine certain ways of acting in and upon the world.

The emergence of the Anthropocene from this perspective insists on something more than just 'action', as responses to climate change are commonly framed. Indeed, 'action' if tied to endless growth and progress in neoliberal terms is liable to result in even greater environmental degradation. In this sense the Anthropocene represents an opportunity for collective planetary rethinking, not further technocratic solutions. One of the main virtues of the Anthropocene as a geopolitical concept is the fact it *anticipates* our current temporality whilst naming it from within (but see Bastian 2012 and discussion in Ginn et al. 2018). It is both reflective and predictive, which is surely at the root of its take up across the arts and humanities in recent years. A caveat needs to be added here, however. The emergence of a new planet altering species (there have been others previously) is cause enough for contemplation; the fact this transformative potential seems to belong to certain ways of living and not others has prompted an even deeper self-examination. As Christophe Bonneuil and Jean-Baptiste Fressoz put it in a passage that is worth quoting in full:

> The challenges of the Anthropocene demand a differentiated view of humanity, not just for the sake of historical truth, or to assess the responsibilities of the past, but also to pursue future policies that are more effective and more just; to construct a common world in which ordinary people will not be blamed for everything while the ecological crimes of the big corporations are left unpunished; in which the inhabitants of islands threatened by climate change will see their right to live on their territories recognized, without their weak numbers condemning them to statistical and political non-existence; a world in which the 30,000 people who still live as hunter-gatherers and are threatened with extinction by the year 2030 will continue to exist. The wealth of humanity and its capacity for future adaptation come from the diversity of its cultures, which are so many experiments in ways of worthily inhabiting the Earth (2016: 71-2).

It is here that we can begin to locate the second key contribution of this volume in terms of **thinking *with* heritage** in the shadow of the Anthropocene. Following Bonneuil and Fressoz's call for a 'differentiated view of humanity' – one that might bring to the surface marginalized, alternative and experimental ways of inhabiting the Earth – *Deterritorializing the Future* builds on recent scholarship in critical heritage studies that aims to track and stimulate multivocal, heterogeneous and dialogical ways of apprehending the past in the present (see Harrison

2013). Critical heritage studies is an emergent and inherently interdisciplinary field that overlaps considerably with archaeology, anthropology, history, cultural geography, architecture, art and – increasingly – the environmental humanities. Although it has roots in a peculiarly British trend of 'heritage-baiting' (see Lowenthal 1985, 1998; Hewison 1987; Wright 1987; Samuel 1994; Waterton 2010), the scope and target of critique has expanded over the last two decades, with prominent work now carried out in Australia (e.g. Smith 2006; Waterton and Gayo 2018), North and South America (e.g. La Salle and Hutchings 2018; Breithoff 2020), mainland Europe (e.g. Macdonald 2013), Scandinavia (e.g. Storm 2014; Appelgren and Bohlin 2017), Africa (e.g. Meskell 2011; Peterson, Gavua and Rassool 2015; Giblin 2018), the Middle East (e.g. Exxel and Rico 2014) and Asia (e.g. Winter 2011; Byrne 2014; Zhu 2015; Rico 2016), alongside significant multi-regional comparative projects (e.g. Harrison et al. 2020), to name but a few examples. The globalized reach of 'critical' heritage (e.g. Meskell 2015) is testament to the rapid spread of heritage around the world, whether as a set of logics and practices associated with colonization and globalization (Byrne 2014; Harrison and Hughes 2010; Labadi and Long 2010), or as a branch of UNESCO's universalizing agendas and principles (Meskell 2013, 2018). Here it is worth noting that much critical heritage scholarship has focused precisely on the territorializing qualities of these practices, from the insistence on the relationship between culture, history, 'blood', 'soil' and citizenship as part of the logics of the formation of the modern nation state (e.g. Anderson 1983), to the emptying of towns, villages and landscapes in the services of heritage tourism (Winter 2011, 2013, 2019). Pushing back against such developments, critical heritage studies typically seeks to illuminate and examine the socio-material effects of such territorializing practices to encourage a greater awareness of alternative modes of engaging with the past in the present to create more equitable futures. This relies on a nuanced commitment to cultural diversity and the flourishing of lifeways that may challenge universalizing, imperialist and, increasingly, capitalist worldviews – a task that aligns with recent thinking in the Anthropocene debate.

From this perspective we can begin to see how critical heritage studies and critical Anthropocene research might share a common set of interests and underlying impulses that go beyond issues of mitigation, adaptation and sustainability. The central logic of heritage – a cliché paraded on countless reports, tag lines and marketing brochures – is captured in the notion of 'saving the past, for the future' (see Harrison 2013; Harrison et al. 2020). Rather than focus on *what* is being 'passed down' and 'taken forward' in this framework and *how* it might be better protected, critical heritage studies poses a different set of questions that correspond with the geopolitics of climate change and the Anthropocene: Who is involved in decision making processes of inheritance and care for the future? How is this future defined and articulated? What 'pasts' are given priority in the present, and whose histories are obscured through such work? How might alternative and marginalized concepts of nature and culture challenge familiar methods of preservation? What stories are waiting to be told about the past, in the present, and what is their role in shaping future worlds? The historical inequities and present injustices that shadow both heritage and the Anthropocene as universalizing (we might also say territorializing) concepts are brought to the surface through such questions, which provide an important foundation for further transdisciplinary inquiry at the intersection of these fields.

While different strands of research have developed around the micropolitics of heritage as a practice and an industry, a central concern has been with humanizing the discipline (see Smith 2006). By this we mean highlighting social, emotional, affective (e.g. Tolia-Kelly, Waterton and Watson 2016) and cultural factors in the management of the past over and above issues of physical preservation and conservation – an exploration of 'why' people preserve natural and cultural heritage, rather than 'how' they should do it more effectively (c.f. Harrison 2013). Such thinking has been hugely important in driving forwards emancipatory heritage projects that seek to radically subvert the values afforded to people, things, places and cultural practices when it comes to 'saving the past, for the future'. Without denying the impact of this critical agenda, the approach to heritage we foreground in this volume takes the concept beyond familiar notions of social production, commodification and the 'politics of the past' to consider alternative modes of 'taking on' and 'passing down' across human and non-human worlds. Here, we aim to engage with the ways in which heritage and conservation practices, understood broadly, can be seen as practices which actively resource the construction of future worlds (Harrison et al. 2020). This reorientation – the third critical gesture we make in response to the Anthropocene – asks us to rethink contradictory approaches found in natural and cultural heritage management, such as the celebration of existence value in biodiversity conservation and the prioritization of social value in the protection of cultural artefacts (e.g. see Harrison 2015, 2018). The Anthropocene is both a prompt for this reconceptualization and a focal point for assessing the implications of an expanded heritage field (see also Solli et al. 2011; Edgeworth et al. 2014; Harrison 2015; Olsen and Pétursdóttir 2016; Pétursdóttir 2017; Harrison, Appelgren and Bohlin 2018; Saul and Waterton 2019). Our key argument here is that **heritage should not be reduced to a human construct**. Instead we look to apprehend processes of care, inheritance, sustainability and connectivity *in excess of the human*, as a way of thinking through the entangled and dialogical nature of all heritage processes. This is no simple task, but we might find an opening or fissure in the call to reimagine heritage in the wake of the posthumanities (see Fredengren and Åsberg this volume), which aims to dislodge anthropocentric concepts of memory, transmission, precarity and affect, all of which are central to the emergence and ongoing work of heritage across various domains.

The three pathways outlined above – **beyond climate action, thinking with critical heritage studies, more-than-human approaches** – resituate heritage in relation to the Anthropocene. No longer to be seen primarily as a set of places or things to be 'saved' (c.f. DeSilvey 2017; DeSilvey and Harrison 2020) in the present, for the future, heritage as we understand it in this volume is an intersubjective and inherently transdisciplinary space where ongoing concerns over climate breakdown, environmental justice, more-than-human legacies and alternative modes of care and stewardship might be worked through by different actors in different ways. To help explore these overlaps and trajectories, the present volume includes contributions from scholars who are firmly situated in heritage studies alongside essays that may avoid the term completely. It is our contention that the cross-fertilization of geography, media studies, philosophy, archaeology, museum studies and geology provides a more useful grounding for heritage research moving forwards. This line of thinking draws out multiple encounters with the Anthropocene as a concept and as an empirical reality across history, the arts and the social sciences. The territorializing status of the Anthropocene is fragmented through this approach, which begins to imagine alternative futures beyond the destructive legacies of the present.

Deterritorializing what?

By now it has become something of a platitude to suggest that the Anthropocene destabilizes familiar concepts of space and time. In one measure it asks us to look millions of years into the future to consider the human as fossil (Yusoff 2013); in another it seeks to undo taken-for-granted assumptions about the distinction between natural and human history (Chakrabarty 2009). In spatial terms meanwhile the diffuse qualities of the Anthropocene bring distant places into close dialogue. 'The loneliest tree in the world' on a remote New Zealand island is marked by radiation from post-war nuclear tests in Nevada (Turney et al. 2018). Antarctic ice-cores document a short-lived dip in atmospheric carbon-dioxide in the early seventeenth century, the result of huge numbers of people succumbing to disease as Europe colonized the Americas (Lewis and Maslin 2018). There is a material intimacy to the concept when seen from this perspective: a proximity that may appear to contradict the grand sweep of geologic timescales but is in fact densely interwoven with such epic narratives. We see this also in the central conceit of naming the 'Anthropos' as a homogenous geological agent, a discursive gesture that effectively erases historical inequities and present injustices through the figure of a universal human agent. The gravitational pull of the Anthropocene is such that the differentiated spatial and temporal rhythms of contemporary social life collapse in on one another. The Anthropocene as concept and as empirical reality is everywhere and nowhere. It is anchored and free-floating, close and distant. It demands action *now*, yet is only truly legible through the lens of the deep future and the deep past. These paradoxes do not undermine the Anthropocene: they are part of its very fabric.

This nebulous yet grounded character underlines the 'territorializing' dimensions of the Anthropocene. As described above, these are connected to issues of climate breakdown, pollution, biodiversity loss and resource extraction, but also to the adoption (or appropriation) of the term beyond geology and the environmental sciences. In many ways the rapid spread and constant fragmentation of the Anthropocene *as a concept* is a perfect example of how territorialization and deterritorialization work across different spatial, material and discursive contexts. New trajectories of creative practice and critical thinking constantly branch off from and feed back into processes of scientific knowledge production. These operate alongside and often in tandem with other territorializing apparatuses, from data algorithms and digital bubbles to rapid processes of urbanization. As we explore below and throughout this book, the cross-currents between such phenomena are not separate to the Anthropocene, but rather part of its historical formation and anticipatory logics.

Against this backdrop the notion of 'deterritorializing the future' emerges as an important *modus operandi* for critically disentangling the Anthropocene and its effects. First articulated by Deleuze and Guattari in *Anti-Oedipus* (1972), deterritorialization as we understand it here names the movement by which one leaves a territory – a process which simultaneously extends the territory in new ways. Such territories are not solely or even primarily topographic, but instead describe all forms of social, organic and political organization. As Claire Colebrook puts it, "the very connective forces that allow any form of life to become what it *is* (territorialize) can also allow it to become what it *is not* (deterritorialize)" (2002: xxii, emphasis in original). Through the act of deterritorialization a set of relations is undone or decontextualized, allowing new relations and actualizations to occur. The territory of 'the future' can never be reduced to a single space or time, but rather oscillates between a multiplicity of temporalities and potential worlds. In the shadow of the Anthropocene however these worlds seem increasingly narrow, reduced to post-human dystopias or capitalist techno-states. In this reading the very concept of the territory as a thing to hold on to or escape from has been surpassed by a colonizing force that leaves no room for deterritorialization, because the planet cannot become *what it is not already* (i.e. irrevocably altered by humans). Despite its remarkable capacity to generate critical and creative work across the arts and humanities, the geopolitics of the Anthropocene are more despotic than democratic. Put simply, if the Anthropocene can be considered a particular assemblage of past-present-future materialities, practices and legacies, then it is also a territorializing apparatus – not just spatially but discursively and socially. It claims the present and the future as a distinctly human territory. Deterritorialization seeks to undo this, or at least expose its fragilities; somehow making the future less beholden to the present, less dependent on the now.

At this point we need to acknowledge the discursive gap between a present temporality that is viewed from the future and a future reality that is shaped by the present. These are mutually constitutive, for sure, but they point to very different capacities for change and action. From one perspective the present is a thing to be read and interpreted, a dense entanglement of matter and meaning waiting to be deciphered. From the viewpoint of the present however the Anthropocene is a thing to be apprehended and – potentially – (re)directed: a chance to 'take stock' of our impact on the planet and ask what other forms of living with the Earth might be possible. These two outlooks feed into each other in useful ways – highlighting unforeseen material legacies and significant disparities in the (future) geological record, for example – but they can also be counter-productive. Most notably, the first implies a sense of inevitability and temporal distance which may well serve to amplify the sociopolitical inertia of the second. Perhaps this explains the febrile search for a 'golden spike' to help designate a singular moment of origin for the Anthropocene, as if the fluctuating possibilities of the present could be contained in a straightforward genealogy of the future.

Of all the strategies that have emerged to trouble this picture in recent years a key pattern has developed around the morphological transformation of the very term 'Anthropocene.' Neologisms such as Plantationocene (Tsing 2015) and Chthulucene (Haraway 2015) seek to decentre the human from the Anthropocene equation, drawing attention respectively to the specific social formations that have given rise to climate breakdown and the multispecies collaborations that might offer a way out of this predicament. Jason Moore's notion of the Capitalocene (2015, 2017) has gained the most traction in this respect, naming – in the words of Demos – the real culprit behind climate change (2017: 54). Instead of placing the blame for planetary environmental collapse on humanity's 'species being', the Capitalocene thesis emphasizes "complex socio-economic, political, and material operations, involving classes and commodities, imperialisms and empress, and biotechnology and militarism" (2017: 86). As Haraway argues, "If you think the Capitalocene, even in a remotely smart way, you're in a whole different cast of characters compared to the Anthropocene" (2016: 240). While the historiographic possibilities of this concept are immediately apparent, it is less clear how the Capitalocene might help us to imagine alternative futures beyond the more destructive regimes of the present. Worth noting here is the fact that, for Deleuze and Guattari, capitalism in all its fluid, schizophrenic and dissipated states is intimately tied to ongoing processes of territorialization and deterritorialization. As they explain in *Anti-Oedipus,*

> The prime function incumbent upon the socius, has always been to codify the flows of desire, to inscribe them, to record them, to see to it that no flow exists that is not properly damned up, channelled, regulated. When the primitive *territorial machine* proved inadequate to the task, the *despotic machine* set up a kind of overcoding system. But the *capitalist machine*, insofar as it was built on the ruins of a despotic State more or less far removed in time, finds itself in a totally new situation: it is faced with the task of decoding and deterritorializing the flows. Capitalism does not confront this situation from the outside, since it experiences it as the very fabric of its existence, as both its primary determinant and its fundamental raw material, its form and its function, and deliberately perpetuates it, in all its violence, with all the powers at its command. Its sovereign production and repression can be achieved in no other way (1972: 47, original emphasis).

To speak of deterritorializing the future in this context risks maintaining or even celebrating the productive destabilizations of the Anthropocene/Capitalocene. As Colebrook argues in this volume, seen from the perspective of capital and various horizon scanning initiatives, the future is already 'deterritorialized' in ways that many would find profoundly disturbing. But while the capitalist machine may depend on continual processes of territorialization and deterritorialization for its very existence, the Anthropocene seems to ground such flows in environmental degradation, human suffering and species extinction (Jørgensen 2017, 2019). This recognition aligns with Manuel DeLanda's reading of deterritorialization, which builds on Deleuze and Guattari's thinking and forms part of his wider theory of the assemblage (2006, 2016). Assemblages for DeLanda are made up of material and expressive components (things and discourses), which are stabilized or destabilized through processes of territorialization and deterritorialization. Crucially, these concepts are to be understood literally in DeLanda's model, as processes that occur *in a particular place*, from the spatial setting of a conversation through to the architectural manifestations of juridical and bureaucratic organizations. From this starting point – where social relations and human and non-human assemblages are understood in quite concrete terms – deterritorialization is formulated as a process through which change occurs, sometimes causing entirely new assemblages to come into being. Stable entities, concepts and identities are constantly unravelled through such movements, which spatialize change over time through real material connections. There is a dense back-and-forth here between territorial qualities of boundedness and situatedness (however real or imagined) and the flows of deterritorialization *in progress*, which evokes a certain form of liquidity that is easily (too easily?) translatable to the realm of commodity circulation. Deleuze and Guattari would see this as an inescapable component of capitalism, which confronts territorialization and deterritorialization as part of its make-up, rather than a problem to be solved. And yet the fragmentations on which capitalism depends seem to harden in the Anthropocene narrative, which effectively codifies the future – possibly for thousands of years – as a 'human' epoch. Does it help us to label this future as capitalist instead? Probably not. New vocabularies are required to deterritorialize the future in a way that is not beholden either to the human or to capital: a project this book contributes to through the lens of critical heritage thinking.

The varied uses of deterritorialization within anthropology, cultural studies, critical theory and philosophy speaks to the inner vibrancy of the term, and we should not imagine that Deleuze and Guattari's conceptualization marks out an 'original' sense that all subsequent work must follow. By definition it cannot be contained, but neither is it a form of romanticized escape. These are material processes just as much as they are discursive (the two are entangled rather than hierarchical in this reading). While deterritorialization in Deleuze and Guattari's thinking is densely interwoven with the oppressive nature of capitalism, it also names something else: the possibility for branching off and becoming new; the moment of decontextualization that leads to a different state; the uncertain mutations that radically transform a given territory. It is this broader conceptualization that animates our use of the term in this volume, suggesting a fragility and openness that may help to counteract some of the more problematic territorializing gestures of the Anthropocene.

From 'Learning to Die' to 'The Arts of Living': Heritage in, of and after the Anthropocene

The Anthropocene/Capitalocene occupies a central place in what we might describe as **the new inheritance paradigm**. Across science, philosophy, culture and the arts the question of inheritance has been posed anew in various disciplinary contexts, from environmental criticism to biogenetics (van Dooren 2014; Gilbert 2017). There are many branches to this reconceptualization, but a central thread can be located in the slow erosion of boundaries between human and nonhuman, between subject and object, and between 'natural' systems and 'cultural' formations. As Haraway notes, the whole question of nature/cultures is about "the dilemma of inheritance, of what we have inherited, in our flesh" (2016: 221). This 'we' extends beyond the human to consider the diffuse material, chemical and biological residues 'taken on' and 'passed down' in different settings within the Anthropocene matrix. Indeed, in many ways the complexities of the Anthropocene all circle back to this central problem: how to account for and ultimately redirect the entangled inheritances of capital and toxins, of fossil fuels and marginalized groups, of political ideologies and nonhuman genetics. Given that inheritance always points in multiple directions at once – to the deep past and the distant future; to the legacies of yesterday and the relics of tomorrow – these transdisciplinary concerns are also marked by a renewed interest in alternative historiographies and radical futures thinking. It is here that we find a particular role for heritage both as a field of inquiry in and of itself and as a potential mode of critical Anthropocene praxis, focused on the shifting logics, ethics and practices of inheritance. Two contrasting notions of heritage are introduced here to help open up these pathways to further investigation.

Roy Scranton's slight but engaging book *Learning to Die in the Anthropocene* (2015) offers one way of thinking about heritage within this new geological framework. For Scranton the climactic changes wrought by humanity signal the demise of global capitalist civilization: "The sooner we confront this situation," he argues, "the sooner we can get down to the difficult task of adapting, with mortal humility, to our new reality" (2015: 23). Tellingly, Scranton identifies the "variety and richness of our collective cultural heritage" as one of the key facets of this new humility (2015: 24). This leads to a familiar assertion made in the face of the apocalypse: build arks. These would not just be biological but cultural, carrying forward genetic data and 'endangered wisdom' alike: "The library of human cultural technologies that is our archive, the concrete record of human thought in all languages that comprise the entirety of our existence as human beings, is not only the seed stock of our future intellectual growth, but its soil, its source, its womb" (2015: 109).

Such projects are of course already underway. The Memory of Mankind project (www.memory-of-mankind.com) for example aims to store millions of ceramic tablets recording human life in all its banality and diversity deep underground in the mountains of Austria. The Arch Mission (www.archmission.org) meanwhile looks to outer space as a site of preservation, with hi-tech storage devices designed to last billions of years planned for distribution across the solar system and beyond (one such 'Archive of Civilization' was attached to a privately funded lunar lander that crashed into the moon in 2019, another will be orbiting the sun for the next 30 million years in the glove compartment of Elon Musk's Tesla). These join well-known global initiatives such as the Voyager Golden Records and the Svalbard Global Seed Vault (see Breithoff and Harrison, this volume) as premeditated fragments of material, cultural or biological inheritance: a 'gift' from the present, to the future (see discussion in Harrison et al. 2020). What such projects often fail to register however is the fact that – as Scranton admits (echoing arguments in Derrida's *Archive Fever*) – 'the heritage of the dead' always needs nurturing: "This nurturing is a practice not strictly of curation… but of active attention, cultivation, making and remaking. It is not enough for the archive to be stored, mapped, or digitized. It must be *worked*" (2015: 99, emphasis in original).

What are the concepts, practices and methods that will enable heritage to be 'worked' differently in the context of the Anthropocene? To what extent might *doing* and *thinking* heritage in new ways help us to engage with the systemic foundations and (potentially) dire consequences of this new geo-philosophical reality? Can changing the way we approach notions of care and inheritance have a meaningful impact 'at scale,' as the Anthropocene seems to demand? What pasts should be prioritized in this new framework, and what futures might we open up by reconceptualizing heritage as a 'deterritorializing' apparatus?

While *Learning to Die in the Anthropocene* relies on a familiar conception of heritage to take forward certain aspects of the past and the present into the future, other ways of confronting the more-than-human entanglements of the new inheritance paradigm ask fundamental questions about what heritage *is*. Take genealogical research for example – one of the most popular heritage pastimes that has developed into a multinational industry supported by DNA testing, in-depth archival research and popular entertainment (e.g. see Basu 2007; Colimer 2017). Typically framed through human-focused narratives of familial descent, economic inheritance, individual triumph or repressed trauma, the search for 'ancestors' is symptomatic of the free-floating nature of modern life, which searches for roots in historical traces and half-remembered echoes of the past. Such pursuits veer between individual curiosity about lost family members and highly politicized attempts to prove certain connections to history. What these investigations rarely draw attention to however is the fact we are 'multilineage organisms' made up of various human and non-human genomes: "The volume of the microbial organisms in our bodies is about the same as the volume of our brain, and the metabolic activity of those microbes is about equivalent to that of our liver. The microbiome *is* another organ; so we are not *anatomically* individuals at all" (Gilbert 2017: M87-83, emphasis in original). This model of genetic heritage is anathema to a discipline and industry built on the prioritization of human modes of inheritance (whether in cultural, biological or individual form), but it may prove vital if we are to rethink notions of care and vulnerability in the age of the Anthropocene. Just as the Anthropocene destabilizes long-held certainties about the break between human and natural history, so recent work in biology, anthropology and the environmental sciences underlines the co-evolution and embedded entanglement of all life. As Donna Haraway puts it, "beings – human and not – become with each other, compose and decompose each other, in every scale and register of time and stuff in sympoietic tangling, in earthly worlding and unworlding" (2017: M45).

The above quotes are taken from the edited collection *Arts of Living on a Damaged Planet* (Tsing et al. 2017) – a volume which takes the notion of entanglement as a critical point of departure to reconsider the 'monsters' and 'ghosts' of the Anthropocene. Monsters in this reading are held to signify the symbiosis of "enfolding bodies" against the "conceit of the individual," while ghosts act as guides to the "haunted lives and landscapes" of environmental degradation (2017: M3). As the editors note, a major challenge of the Anthropocene is "how to think geological, biological, chemical, and cultural activity together, as a network of interactions with shared histories and unstable futures" (2017: 176). Ghosts and monsters are not fantastical figures from this perspective; they are "observable parts of the world" that we might learn "through multiple practices of knowing" (2017: M3). Arts of living in this context are necessary to counteract threats to our very survival. Crucially, this cuts across technological solutions to ecological collapse, new modes of storytelling and creative practice, and political encounters with diverse forms of oppression and marginalization. "There is something mythlike about this task: we consider anew the living and the dead; the ability to speak with invisible and cosmic beings; and the possibility of the end of the world" (2017: 176).

Working along this grain, we might situate heritage as a vital though often overlooked aspect of the Earth's very 'livability'. There are multiple pathways to think with in this regard. Non-Western practices of care and conservation for example often dissolve the boundaries between natural and cultural heritage through their insistence on the spirituality and enchantment of material things (Byrne 2004; see Ugwuanyi this volume). Alternatively, we might consider Indigenous claims of 'human rights for nonhumans' (Surrallés 2017) as a politically charged mode of heritage protection across natural-cultural worlds, or look to Caitlin DeSilvey's concept of 'curated decay' (2017) to inform new approaches to material and environmental change. Identifying heritage as a key component in the 'arts of living' underlines the need to rethink and redirect notions of care, curation, management and preservation, from museum objects to urban landscapes. These activities draw on and intersect with key questions in geology, biology, history, anthropology and the environmental humanities. Heritage *in* the Anthropocene must embrace this multiplicity to encourage new ways of imagining and engaging with the past in the present to shape alternative futures. There is no single model to adopt in this respect; no 'one-size-fits-all' approach for a radically posthumanist critical heritage practice. Instead we should look to situated and relational forms of knowledge making that transcend human/non-human and nature/culture boundaries, recognizing that such dichotomies are an obstacle to understanding let alone confronting the Anthropocene as a material and conceptual force in the world. This will no doubt require (inter)subjectivities that look beyond liberal humanist ideas of progress and development for critical purchase. Like Anna Tsing (2015) we are not quite sure what form a progressive politics without progress might take, but this does not mean we should not seek it out via new and old ways of *doing* heritage.

An important line of inquiry here concerns the interpretive nature of many heritage 'experiences'. Various storytelling devices are employed by heritage to create links between past, present and future, from audio guides and wall plaques to films and museum displays. As well as constantly rethinking these tools, we need to construct alternative genealogies to populate them. One of the most notable reverberations of the Anthropocene has been a renewed commitment to entangled histories when describing the emergence of the modern world. Such narratives bring together histories of resource extraction and social formations, marginalized voices and non-human agencies. A heritage *of* the Anthropocene will depend on these more-than-human stories and entangled lines of descent. Crucially such accounts also bring to the surface unintended material residues and socio-political legacies. Despite – or perhaps because of – its geological framing, the Anthropocene cannot be divorced from urgent and lingering historical questions surrounding slavery, empire, colonialism and the rise of capital (Ghosh 2016; Moore 2017; Yusoff 2018). Again, in this sense the notion of 'Anthropocene heritage' extends rather than subverts progressive and emancipatory work in critical heritage studies scholarship and related fields. A crucial responsibility here is to constantly differentiate the 'we' of Anthropocenic thinking (c.f. Thomas 2016) – a task that might usefully build on the critiques of universality that characterize critical engagements with 'World Heritage' (Meskell 2018) and the 'endangerment sensibility' (c.f. Vidal and Dias 2016; see also Harrison 2013; Rico 2015) which animates it.

Finally, the possibility of heritage *after* the Anthropocene points in two directions at once. The first concerns the future legacies diligently being produced today (plastic bodies and toxic landscapes, scarred minds and broken climates); the second concerns the critical gesture of *post*-Anthropocene thinking – a peculiar consequence of the rapid take-up of the term in the arts and humanities and the equally swift recognition that it is wholly unsatisfactory as a socio-political diagnostic. What of heritage and the Capitalocene, or the Plantationocene, or the Chthulucene (Haraway 2015)? Such labels ask us to look again at the differential legacies and material disparities of a planet altered by 'humans'. The fossil-to-come is a useful frame of reference for this project, but other modes of post-Anthropocene heritage should also be brought to bear on the subject, from museums and archives to augmented digital experiences. The challenge of the Anthropocene is such that entirely new modes of relating past, present and future are liable to emerge in its shadow, whether as unintended consequences of inheritance and precarity or as subversive strategies of survival and flourishing. Conceiving of heritage after the Anthropocene must remain a speculative gesture at this stage, bound up with the politics of the present and the radical need for new temporal and territorial imaginaries.

Learning to Die and *The Arts of Living* represent two very different ways of thinking about heritage in the context of the Anthropocene. Save, conserve, collect and safeguard, or fundamentally rethink emergent relationalities (see also DeSilvey and Harrison 2020). We might see this as a version of debates already being played out across the academy and wider society. As Guattari warned almost four decades ago now, the environmental impact of capitalist civilization confronts us with stark choices, demanding new modes of social organization to avoid ecological collapse. It hardly needs stating that the current rise of populism across the world, with all its territorializing discourses and agendas, is both a response to this predicament and a doubling down of current global systems. More exploitation, more oppression, more boundaries, more suffering. Against this backdrop the rapid breakdown of environmental conditions is viewed with morbid fascination (see again the work of Edward Burtynsky) or disregarded completely. To imagine heritage after the Anthropocene is really to ask what heritage *sans* capitalism and beyond the confines of the nation-state might look like. Would we still collect and preserve things in the same way? What stories might be told about past, present and future without the buttresses of capitalist modernity? Whether in the form of globalized historical 'assets' or as a component of the reterritorializing discourses of nation, nostalgia and home, heritage is fully immersed in the flows that perpetuate and underpin this system. Despite a superficial concern for the past, it is also inherently future-oriented (this is part of its capitalist formation). Rather than reject the concept outright, however, we want to displace the familiar ontologies and cosmologies on which heritage practices have been built to establish new frames of reference and lines of inquiry. Referencing Tim Morton's call for an 'ecology without nature' (2007), we might think of this new framework as a call for inheritance *without* heritage, recognizing that the *idea* of heritage may well stand in the way of a more meaningful relationship with ongoing and inherently more-than-human concepts and processes of care, transmission and vulnerability. To do this we look to new disciplinary collaborations and practices, as well as alternative and marginalized narratives of life beyond, after and in excess of the Anthropocene.

Heritage unbound

Any story is a form of control, an attempt to wrestle the endless fragmentations of reality into a coherent thread of histories and potentialities. This collection is no different, and may be read as a territorializing apparatus, with all the pitfalls and opportunities this framing implies. However, to borrow another concept from Deleuze and Guattari, the stories told in this volume do offer multiple lines of flight, constantly destabilizing the territories on which our assumptions are based. The mapping we undertake here is transdisciplinary in its composition, drawing on recent and ongoing research across cultural geography, anthropology, literature, philosophy, media studies, archaeology and the arts to inform new theories and practices in and for heritage. At the same time, heritage itself is 'liberated' over the course of this book, with many of the central concerns of the field unsettled through new critical-creative approaches. Loosely assembled around the core themes of time and territory, the chapters gathered here may thus be read individually or sequentially, with each 'unburdening' heritage in different ways.

In their chapter on the waste management plant of Gärstad in Linköping, Sweden, Christina Fredengren and Cecilia Åsberg immediately open up the timescales and materialities of heritage to more-than-human forces and imaginaries. Gärstad is a high-tech garbage disposal plant that turns waste from across Sweden and northern Europe into energy for the local community. It is also the site of an Iron Age sanctuary, where the bodies of the dead were burned with clothes and other personal items. Drawing on feminist and posthumanist perspectives on intragenerational care and cross-species co-becoming, Fredengren and Åsberg place Gärstad at the centre of a broad ecology of material and immaterial inheritances, from prehistoric land clearings that reshaped the environment to the lingering effects of CO_2 in the atmosphere – a by-product of the waste incineration carried out at the plant. Connecting the dots between different "domains of inheritance" – including genetics, pollution, waste, art and heritage conservation – the authors put forward a new model of "equity between non-contemporaries" that does not simply aim at flattening hierarchies, but rather seeks for new companions in 'merriment' and 'awe'. This experimental path is deeply attuned to Gärstad as a 'time-giver' – a place where multiple interventions seen and unforeseen are made across generations, prompting a revised ethics of multi-species ancestry. Extending the logics of preservation, care and inheritance means rethinking such 'temporal relations' across nature-culture boundaries. As Fredengren and Åsberg write, there is no "purity of categories to be had in the Anthropocene, and we cannot afford it anyway".

Staying in Sweden, Anna Bohlin turns our attention to an altogether different domain of inheritance: that of second-hand furniture and the "unfolding [of] human-thing entanglements" that such objects are bound up with. Inspired by vital materialist perspectives and new approaches to ruination across archaeology and heritage, Bohlin investigates the different ways in which "material liveliness" is valorized in the consumption and use of old things. Temporality becomes a key factor in this analysis, as second-hand objects are seen to transform, age and decay over time; they are "porous and leaky things" according to Bohlin – "involved in a form of 'growing' as they accumulate traces and sociality". Here a stark difference emerges with the meaning of time in relation to conventional heritage objects, which are typically 'frozen' at a particular moment. Questioning this "myth of stability and fixity", Bohlin suggests that second-hand objects have a greater freedom to "follow their own trajectories and unfolding" – a realisation that underpins a post-anthropocentric view of sovereignty over things. As Bohlin concludes, this temporal-material shift opens up the possibility of responding differently to the acute Anthropocene challenges of mass-production, over-consumption and waste.

The liveliness of matter is also central to Adrian Van Allen's chapter on museum taxidermy, which brings together themes of care and the more-than-human to investigate the different temporalities associated with animal collections. Drawing on ethnographic and archival research at two natural history museums, Van Allen carefully examines how animal bodies are "made and remade" in relation to shifting logics of ecology, evolution, biodiversity and conservation. Here novel techniques in preparation, storage and analysis sit alongside methods of fixing, preening, dissecting and stuffing that have changed little in over four centuries. Unpacking the simplistic notion that taxidermy animals are "frozen in time", Van Allen explores "the intimate and fluid connections between the minutiae of biological organisms, their tissue samples, their data and their DNA, and the embedded visions for shared human and nonhuman futures". This close reading of a traditional museological environment opens up the future-making practices of heritage to renewed scrutiny. It also helps to unsettle dominant narratives of the Anthropocene by focusing on "specific assemblages" where people, places and things interact. More commonly associated with geological strata and vast extraction sites, Van Allen shows how the Anthropocene is equally made and unmade in the bodies and spaces of the museum and the conservation laboratory.

A similar claim is made by Esther Breithoff and Rodney Harrison in their chapter on biobanks and seed vaults, where the authors ask what it means to conserve 'nature' in the 'post-wild' context of the Anthropocene (Marris 2013; Lorimer 2015). Looking across two sites in particular – the Svalbard Global Seed Vault in Norway, and the UK's Frozen Ark – Breithoff and Harrison identify a shift in the core purpose of such facilities. From an initial role as isolated arks that might "carry endangered DNA into an uncertain future", biobanks are now increasingly valued for their restorative potential, being seen as active players in *current* de-extinction and agricultural renewal programmes. As the authors make clear, the first withdrawal of seeds from the Svalbard vault happened many years earlier than anticipated, as a result of the war in Syria rather than any more widespread climate catastrophe. This unexpected demand acts as a pivot to consider biobanks as a form of *"speculative biocapital accumulation"* wherein new futures are actively shaped as part of the broader bioeconomy of the Anthropocene. Crucially, this economy relies on folding time and nature within the space of the vault, with the seeds themselves characterized as archives of "inter-generational, inter-species, human/plant kinship relations" (van Dooren 2007: 83). A natural companion piece to Van Allen's chapter in many ways, Breithoff and Harrison also push forward the notion that more-than-human heritage is inescapably political, as the things and relations brought together in and through conservation practices enact highly unequal futures.

Drawing together numerous threads from the preceding chapters, Colin Sterling's contribution explores a growing trend in art practice that leverages familiar heritage concepts such as the museum, the ruin and the monument to critique the Anthropocene as a *historical* phenomenon. Here the author focuses on the future anterior temporality of the Anthropocene concept, which implicitly asks us to look forward to view the present as the past. Playing with this notion, projects such as the Museum of Capitalism, the Museum of Nonhumanity and the Anthropocene Monument aim to defamiliarize the present to better understand its underlying tensions and occlusions. Thinking *with* heritage in this way is both a satirical gesture and a form of critical practice, where the Anthropocene is historicized and provincialized to highlight alternative ways of living and acting across human and non-human worlds (Jørgensen 2018). Building on some of the questions around multi-species care and equitable futures outlined in preceding chapters, Sterling suggests that such work not only helps to demonstrate where heritage *might* be heading, but also questions the Anthropocene "as a totalizing concept and inescapable reality". This mode of deterritorialization is played out through curatorial experiments and creative interventions, from floating museums to fossilized iPhones.

While the expanded temporalities of the Anthropocene open up questions of care, inheritance, memory and preservation to renewed critical scrutiny, the territorial dimensions of the concept challenge familiar notions of place, matter, belonging and boundedness. To help explore the place of heritage in this new spatiotemporal frame, the second part of the volume opens with two chapters that deal in very different ways with water as a liquid territory. Joanna Zylinska offers us a way of thinking with the fluid ontology of water in relation to media and mediation, which emerge here as complex, hybrid processes that humans partake in alongside other organisms. Taking two recent films on water – *The Pearl Button* and *Even the Rain* – as critical points of departure, Zylinska looks to build a "water-rich picture of the world, in all its entanglements, spillages and overflows". Water in this sense emerges as an "ethical medium" for the way it foregrounds a lack of enclosure in the definition of any life or being. Drawing on recent media theory that emphasizes the embeddedness of all forms of mediation with infrastructures, elements, atmospheres and bodies, Zylinska outlines a form of "geo-history as heritage" built on the flows and cascades of water rather than the stability of land.

The constant commingling of water with other things, bodies, spaces and environments is also a key concern of Denis Byrne, only here it is the attempt by humans to impose hard boundaries between water and land in the form of coastal reclamations that acts as a springboard to reconsider the territories of the Anthropocene. As Byrne argues, there has been a rapid increase in coastal reclamations for agricultural, industrial, infrastructural and residential purposes over the last two to three centuries, and these waterlines are now a key site of "nervousness and stress" in an era of climate breakdown. Made possible by fossil fuel driven development on a vast scale, such spaces tend to be hard-edged and hostile to non-human life, becoming in the words of Byrne "a signature landform of the Anthropocene". Rather than see these coastal reclamations as part of a progressive heritage of human ingenuity (a familiar narrative in relation to industrial heritage), Byrne asks that we 'unwind' such ecological interventions to give them a history; this being a first step towards understanding how the world was, and how it might be. Deterritorialization in this context implies making the Anthropocene visible and tangible to help inspire "widespread popular mobilization against the dark future which it portends".

The impossibility of drawing boundaries between human and non-human worlds is also central to J. Kelechi Ugwuanyi's investigation of the trees that play such an important role in village life among the Igbo of Nigeria. Drawing on ethnographic fieldwork and his own experiences as a member of this culture, Ugwuanyi asks how human existence and recreation are made possible in the Igbo cosmology through an intimate connection with the territory of trees. Here the author puts forward a novel conception of heritage as "alive", not through human consciousness or society, but as a manifestation of the "utilitarian" provision all things afford in the "community of life". Stitching together animist and posthumanist philosophies, Ugwuanyi focuses on the key question of survival across human and non-human species in the shadow of the Anthropocene, emphasizing a form of territoriality and belongingness in which human beings share life with *Ala* (the Earth). Heritage in this reading is "of the Earth, living among the community of beings, and should belong to all".

Caitlin DeSilvey's photo essay also takes a site-based approach to question and redirect notions of transformation and loss in different Anthropocene territories. Drawing on ethnographic fieldwork carried out at a former mining site in Cornwall, an abandoned military complex on the east coast of Britain and a valley in Portugal identified as a potential rewilding location, DeSilvey suggests that ecological disturbance is now the norm in most parts of the Earth, and heritage agencies must acknowledge that "strategies for survival will depend on making alliances with more-than-human entities and agencies". Such contexts force us to engage with what DeSilvey calls "ruderal heritage" – a term that references opportunistic plant species that are adapted to take root in disturbed environments. Ruderal heritage then "is orientated to ongoing instances of both destruction and renewal, and focused on the opportunities that emerge from inhabiting disturbed substrates and sensibilities". Through images and stories DeSilvey shows how ruderal thinking may offer a productive conceptual tool for heritage practice, which is too often focused on stability and the possibility of returning to an original time or state of being. Such a shift seems vital in the face of ongoing Anthropocene transformations, which emphasize uncertainty as a condition for history and memory across human and non-human worlds.

Anatolijs Venovcev's brief illustrated slam poem – a provocative "call from the North" – continues in this vein of thinking by exploring the uneven impacts of the extractive industries and technologies which have supported the development of a global Capitalocene on geopolitically marginalized landscapes and their inhabitants. His lyrical critique and the accompanying photographs remind us that "New ways of understanding humanity need to be rooted in the real material costs and consequences of our new and future technologies". Urging us to "remember the waste as we venture forth", he engages with one of the key leitmotifs of Anthropocene studies (e.g. see Morton 2013; Bastian and van Dooren 2017) whilst picking up on points made by DeSilvey in the previous chapter. In doing so, he gestures towards new ways of thinking across critical heritage studies which emphasize the relationship between heritage and waste (e.g. Storm 2014, this volume; Holtorf and Högberg 2015; Olsen and Pétursdóttir 2016; DeSilvey 2017; Harrison et al. 2020) and the productive ways of engaging with anthropogenic material and discursive legacies which might emerge from such comparisons.

Anna Storm's chapter ends Part II and brings us back to Sweden by way of the United States and Belarus. Looking across three sites where nuclear power stations have been decommissioned, Storm asks how certain processes of withdrawal and restoration effectively render history and memory invisible, deterritorializing toxic legacies through the production of supposedly "controlled environments". Such human legacies are counteracted and sustained by non-human forces, with animals, vege-tation, bedrock and clay all "attributed the role of guardians of radioactive remains". In these quasi-mythical landscapes, waste and wildlife collide to unsettle narratives of future progress. As Storm makes clear, "it will take several decades, if ever, before children will dig and play in sandboxes on the former nuclear territory".

Finally, in her provocative coda to the volume, Claire Colebrook both challenges and expands the sense of deterritorialization developed over the preceding pages. Here two distinct forms of deterritorialization are identified and critiqued. The first is linked to post-apocalyptic narra-tives and existential threats, where the Anthropos of the Anthropocene is held up as that which must be protected and preserved against all threats. As Colebrook explains, "it is deterritorialization that enables the Anthropocene, both geologically and conceptually; a potentiality of the species reaches such an intensity that it generates a whole new scale and range of relations. A part overtakes the whole; humanity, man, or Anthropos comes to appear as the ground and organizing whole". The future in this sense is *already* deterritorialized, as a "detached fragment" of humanity has "generated a distinct temporality and modality of the imagination" that effectively shuts down other futures. Building on a specific critique of Oxford University's Future of Humanity Institute, Colebrook suggests that what is needed is an altogether different mode of deterritorialization, one that might "expand the range of the problem of the human". Through an engagement with Karen Barad that implic-itly links back to Fredengren and Åsberg's opening chapter, Colebrook stresses the relationality and impurity of life as one way in which deterritorialization may be 'decolonized' to generate new forms of liv-ing in and with the world. As a conclusion of sorts to the volume, this re-theorisation helpfully captures and pushes forward one of the key messages of the book; namely that heritage in all its complexities and contradictions might provide a grounding to imagine ways out of the Anthropocene – or at least that version of the Anthropocene in which humans can think of no future other than their own demise. This open-ing up of the human is intimately bound to a revised conception of the territories and temporalities of a radically posthuman critical heri-tage studies.

Notes

1.   Readers interested in such issues should consult the 2019 ICOMOS report *Engaging Cultural Heritage in Climate and Action*, and the work of David Harvey and Jim Perry (2015).

References

Anderson, B. 1983. *Imagined Communities: Reflections on the Origin and Spread of Nationalism*. London and New York: Verso.

Appadurai, A. 1990. Disjuncture and Difference in the Global Cultural Economy. *Theory, Culture & Society* 7(2-3): 295-310.

Appelgren, S. and A. Bohlin. 2017. Second-hand as Living Heritage: Intangible Dimensions of Things with History. In *The Routledge Companion to Intangible Cultural Heritage*, edited by P. Davis and M. S. Stefano , 240-250. Abingdon and New York: Routledge.

Bastian, M. 2012. Fatally Confused: Telling the Time in the Midst of Ecological Crises. *Journal of Environmental Philosophy* 9(1): 23–48.

Bastian, M. and T. van Dooren. 2017. The New Immortals: Immortality and Infinitude in the Anthropocene. *Environmental Philosophy* 14(1): 1–9.

Basu, P. 2007. *Highland Homecomings: Genealogy and Heritage Tourism in the Scottish Highland Diaspora*. New York: Routledge.

Bonneuil, C. and J. Fressoz. 2016. *The Shock of the Anthropocene*. London and New York: Verso.

Breithoff, E. 2020. *Conflict, Heritage and World-Making in the Chaco: War at the end of the Worlds?* London: UCL Press.

Byrne, D. 2014. *Counterheritage: Critical Perspectives on Heritage Conservation in Asia*. New York: Routledge.

Cameron, F. and B. Neilson. 2014. (eds). *Climate Change and Museum Futures*. New York and London: Routledge.

Chakrabarty, D. 2009. The Climate of History: Four Theses. *Critical Inquiry* 35: 197–222.

Colebrook, C. 2002. *Understanding Deleuze*. Crows Nest, NSW: Allen & Unwin.

Colimer, L. 2017. Heritage on the move: Cross-cultural heritage as a response to globalisation, mobilities and multiple migrations. *International Journal of Heritage Studies* 23(10): 913–927.

DeLanda, M. 2006. *A New Philosophy of Society: Assemblage Theory and Social Complexity*. London: Continuum.

DeLanda, M. 2016. *Assemblage Theory*. Edinburgh: Edinburgh University Press.

Demos, T. J. 2017. *Against the Anthropocene: Visual Culture and Environment Today*. Berlin: Sternberg.

Deleuze, G. and F. Guattari. 1972. *Anti-Oedipus*. London, Oxford, New York: Bloomsbury.

DeSilvey, C. 2017. *Curated Decay: Heritage Beyond Saving*. Minneapolis and London: University of Minnesota Press.

DeSilvey, C. and R. Harrison. 2020. Anticipating Loss: Rethinking Endangerment in Heritage Futures. *International Journal of Heritage Studies* 26(1): 1–7.

Dibley, B. 2012. 'The Shape of Things to Come': Seven Theses on the Anthropocene and Attachment. *Australian Humanities Review* 52: 139–158.

Dibley, B. 2015. Anthropocene: The Enigma of 'The Geomorphic Fold'. In *Animals in the Anthropocene: Critical Perspectives on Non-Human Futures*, edited by HARN Editorial Collective, 36–48. Sydney: Sydney University Press.

Dibley, B. 2018. The Technofossil: A *Memento Mori*. *Journal of Contemporary Archaeology* 5(1): 44–52.

Edgeworth, M., J. Benjamin, B. Clarke, Z. Crossland, E. Domanska, A. C. Gorman, P. Graves-Brown, E. C. Harris, M. J. Hudson, J. M. Kelly, V. J. Paz, M. A. Salerno, C. Witmore and A. Zarankin 2014. Archaeology of the Anthropocene. *Journal of Contemporary Archaeology* 1(1): 73–132.

Exxell, K. and T. Rico. (eds). 2014. *Cultural Heritage in the Arabian Peninsula: Debates, Discourses and Practices*. Farnham: Ashgate.

Ghosh, A. 2016. *The Great Derangement: Climate Change and the Unthinkable*. Chicago and London: The University of Chicago Press.

Giblin, J. 2018. Heritage and the Use of the Past in East Africa. In *Oxford Research Encyclopedia of African History*, edited by T. Spear. DOI: 10.1093/acrefore/9780190277734.013.135

Giddens, A. 1990. *The Consequences of Modernity*. Cambridge: Polity.

Gilbert, S. F. 2017. Holobiont by birth: Multilineal individuals as the concretion of cooperative processes. In *Arts of Living on a Damaged Planet*, edited by A. Tsing, H. Swanson, E. Gan and N. Bubandt, M37–M89. Minneapolis and London: University of Minnesota Press.

Ginn, F., M. Bastian, D. Farrier and J. Kidwell 2018. Introduction: Unexpected Encounters with Deep Time. *Environmental Humanities* 10(1): 213–225.

Guattari, F. 1983 [2009]. Utopia Today. In *Soft Subversions: Texts and Interviews 1977-1985*, edited by S. Lotringer, 307. Los Angeles: Semiotext(e)

Hambrecht, G., C. Anderung, S. Brewington, A. Dugmore, R. Edvardsson, F. Feeley, K. Gibbons, R. Harrison, M. Hicks, R. Jackson, G.A. Ólafsdóttir, M. Rockman, K. Smiarowski, R. Streeter, V. Szabo and T. McGovern. 2018. Archaeological sites as Distributed Longterm Observing Networks of the Past (DONOP), *Quaternary International*. Advanced Online Access https://doi.org/10.1016/j.quaint.2018.04.016

Haraway, D. 2015. Anthropocene, Capitalocene, Plantationocene, Chthulucene: Making Kin. *Environmental Humanities* 6(1): 159–165.

Haraway, D. 2016. *Manifestly Haraway*. Minneapolis and London: University of Minnesota Press.

Harrison, R. 2013. *Heritage: Critical Approaches*. London and New York: Routledge.

Harrison, R. 2015. Beyond 'Natural' and 'Cultural' Heritage: Toward an Ontological Politics of Heritage in the Age of Anthropocene. *Heritage and Society* 8(1): 24–42.

Harrison, R. 2018. On Heritage Ontologies: Rethinking the Material Worlds of Heritage. *Anthropological Quarterly* 91(4): 1365–138.

Harrison, R. and L. Hughes. 2010. Heritage, Colonialism and Postcolonialism. In *Understanding the Politics of Heritage*, edited by R. Harrison, 234–69. Manchester: Manchester University Press.

Harrison, R., S. Appelgren and A. Bohlin. 2018. Belonging and Belongings: On Migrant and Nomadic Heritages *in* and *for* the Anthropocene. In *The New Nomadic Age: Archaeologies of Forced and Undocumented Migration*, edited by Y. Hamilakis, 209–220. Sheffield: Equinox Publishing.

Harrison, R., C. DeSilvey, C. Holtorf, S. Macdonald, N. Bartolini, E. Breithoff, L.H. Fredheim, A. Lyons, S. May, J. Morgan and S. Penrose. 2020. *Heritage Futures: Comparative Approaches to Natural and Cultural Heritage Practices*. London: UCL Press.

Harvey, D. and J. Perry. (eds). 2015. *The Future of Heritage as Climates Change: Loss, Adaptation and Creativity*. London: Routledge.

Hewison, R., 1987. *The Heritage Industry: Britain in a Climate of Decline*. London: Methuen.

Holtorf, C. and A. Högberg. 2015. Archaeology and the Future: Managing Nuclear Waste as a Living Heritage. Radioactive Waste Management and Constructing

Memory for Future Generations. *Nuclear Energy Agency, no. 7259*, edited by NEA, 97–101. Paris: OECD.

Jørgensen, D. 2017. Endling, the power of the last in an extinction-prone world. *Environmental Philosophy* 14: 119–138.

Jørgensen, D. 2018. After none: Memorialising animal species extinction through monuments. In *Animals Count: How Population Size Matters in Animal-Human Relations*, edited by N. Cushing and J. Frawley, 183–199. London: Routledge.

Jørgensen, D. 2019. *Recovering Lost Species in the Modern Age: Histories of Longing and Belonging.* Cambridge, MA: MIT Press.

La Salle, M. and R. M. Hutchings. 2018. Is Canadian Heritage Studies Critical? *Journal of Canadian Studies* 52(1): 342–360.

Labadi, S. and C. Long 2010. *Heritage and Globalisation*. New York: Routledge.

Lewis, S. L and M. A. Maslin. 2018. *The Human Planet: How We Created the Anthropocene*. London: Penguin.

Lorimer, J. 2015. *Wildlife in the Anthropocene: Conservation after Nature.* Minneapolis, MN: University of Minnesota Press.

Lowenthal, D. 1985. *The Past is a Foreign Country*. Cambridge: Cambridge University Press.

Lowenthal, D. 1998. *The Heritage Crusade and the Spoils of History*. Cambridge: Cambridge University Press.

Macdonald, S. 2013. *Memorylands: Heritage and Identity in Europe Today*. London and New York: Routledge.

Marris, E. 2013. *Rambunctious Garden: Saving Nature in a Post-Wild World*. New York: Bloomsbury.

Meskell, L. 2011. *The Nature of Heritage: The New South Africa*. Wiley: London and New York.

Meskell, L. 2013. UNESCO's World Heritage Convention at 40: Challenging the Economic and Political Order of International Heritage Conservation. *Current Anthropology* 54(4): 483–494.

Meskell, L. (ed). 2015. *Global Heritage: A Reader*. New York: Wiley-Blackwell.

Meskell, L. 2018. *A Future in Ruins: UNESCO, World Heritage, and the Dream of Peace*. Oxford and New York: Oxford University Press.

Moore, J. 2015. *Capitalism in the Web of Life: Ecology and the Accumulation of Capital.* London and New York: Verso.

Moore, J. W. 2017. The Capitalocene, Part I: on the nature and origins of our ecological crisis. *The Journal of Peasant Studies* 44(3): 594–630.

Morton, T. 2007. *Ecology without Nature: Rethinking Environmental Aesthetics.* Cambridge, Mass. and London: Harvard University Press.

Morton, T. 2013. *Hyperobjects: Philosophy and Ecology after the End of the World.* Minneapolis: University of Minnesota Press.

Olsen, B. and Þ. Pétursdóttir. 2016. Unruly Heritage: Tracing Legacies in the Anthropocene. *Arkæologisk Forum* 35: 38–46.

Peterson, D. R., K. Gavua and C. Rassool. 2015. *The Politics of Heritage in Africa. Economies, Histories, and Infrastructures.* Cambridge: Cambridge University Press.

Pétursdóttir, Þ. 2017. Climate Change? Archaeology and Anthropocene. *Archaeological Dialogues* 24(2): 175–205.

Rico, T. 2015. Heritage at Risk: The Authority and Autonomy of a Dominant Preservation Framework. In *Heritage Keywords: Rhetoric and Redescription in Cultural Heritage,* edited by K. Lafrenz Samuels and T. Rico, 147–162. Boulder: University Press of Colorado.

Rico, T. 2016. *The Making of Islamic Heritage: Muslim Pasts and Heritage Presents.* Singapore: Palgrave Macmillan.

Samuel, R. 1994. *Theatres of Memory: Volume 1, Past and Present in Contemporary Culture.* Verso: London and New York.

Saul, H. and E. Waterton. 2019. Anthropocene landscapes. In *The Routledge Companion to Landscape Studies,* edited by P. Howard, I. Thompson, E. Waterton and M. Atha, 139–151. Abingdon and New York: Routledge.

Scranton, R. 2015. *Learning to Die in the Anthropocene.* San Francisco: City Light.

Sheller, M. and J. Urry. 2006. The new mobilities paradigm. *Environment and Planning A.* 38(2): 207–226.

Smith, L. 2006. *Uses of Heritage.* London: Routledge.

Solli, B., M. Burström, E. Domanska, M. Edgeworth, A. González-Ruibal, C. Holtorf, G. Lucas, T. Oestigaard, L. Smith and C. Witmore. 2011. Some Reflections on Heritage and Archaeology in the Anthropocene. *Norwegian Archaeological Review* 44(1): 40–88.

Storm, A. 2014. *Post-Industrial Landscape Scars.* Basingstoke, New York: Palgrave Macmillan.

Surrallés, A. 2017. Human rights for nonhumans? *HAU: Journal of Ethnographic Theory* 7(3): 211-235

Thomas, J.A. 2016. Coda: Who is the 'we' endangered by climate change? In *Endangerment, Biodiversity and Culture*, edited by F. Vidal and N. Dias, 242–260. Abingdon and New York: Routledge.

Tomlinson, J. 1999. *Globalisation and Culture*. Chicago: University of Chicago Press.

Tolia-Kelly, D.P., E. Waterton and S. Watson (eds). 2016. *Heritage, Affect and Emotion: Politics, practices and infrastructures*. Abingdon and New York: Routledge.

Tsing, A. 2015. *The Mushroom at the End of the World: On the Possibility of Life in Capitalist Ruins*. Princeton and Oxford: Princeton University Press.

Tsing, A, H. Swanson, E. Gan and N. Bubandt (eds). 2017. *Arts of Living on a Damaged Planet*. Minneapolis and London: University of Minnesota Press.

Turney, C., J. Palmer and M. Maslin. 2018. Anthropocene Began in 1965, According to Signs Left in the World's 'Loneliest Tree'. *The Conversation* February 19.

van Dooren, T. 2007. Terminated Seed: Death, Proprietary Kinship and the Production of (Bio)Wealth. *Science as Culture* 16(1): 71–94.

van Dooren, T. 2014. Life at the Edge of Extinction: Spectral Crows, Haunted Landscapes and the Environmental Humanities. *Humanities Australia* 5: 8–22.

Vidal, F. and N. Dias. 2016. Introduction: The Endangerment Sensibility. In *Endangerment, Biodiversity and Culture*, edited by F. Vidal and N. Dias, 1–38. Abingdon and New York: Routledge.

Waterton, E. 2010. *Politics, Policy and the Discourses of Heritage in Britain*. Palgrave.

Waterton, E. and M. Gayo. 2018. For All Australians? An Analysis of the Heritage Field. *Continuum* 32(3): 269–81.

Winter, T. 2011. *Post-Conflict Heritage, Postcolonial Tourism: Tourism, Politics and Development at Angkor*. London: Routledge.

Winter, T. 2013. Clarifying the Critical in Critical Heritage Studies. *International Journal of Heritage Studies* 19(6): 532–545.

Winter, T. 2019. *Geocultural Power: China's Quest to Revive the Silk Roads for the Twenty-First Century*. Chicago: Chicago University Press.

Wright, P. 1985. *On Living in an Old Country: The National Past in Contemporary Britain*. London: Verso.

Yusoff, K. 2013. Geologic Life: Prehistory, climate, futures in the Anthropocene. *Environment and Planning D: Society and Space* 31: 779–795.

Yusoff, K. 2018. *A Billion Black Anthropocenes or None.* Minneapolis and London: University of Minnesota Press.

Zalasiewicz, J. 2008. *The Earth After Us: What legacy will humans leave in the rocks?* Oxford: Oxford University Press.

Zhu, Y. 2015. Cultural Effects of Authenticity: Contested Heritage Practices in China. *International Journal of Heritage Studies* 21(6): 594–608.

I
Times

Chapter 2

Checking in with Deep Time: Intragenerational Care in Registers of Feminist Posthumanities, the Case of Gärstadsverken

CHRISTINA FREDENGREN & CECILIA ÅSBERG

The generations of men run on in the tide of Time / But leave their destin'd lineaments permanent for ever and ever. (Blake, Milton, f. 20, 11. 24-25)

Although I have for some time accepted the force of Fredric Jameson's dictum that "we cannot, not periodize," until very recently it would not have occurred to me that postcolonial study, critical theory, or the humanities disciplines in general needed to periodize in relation not only to capital but to carbon, not only in modernities and post-modernities but in parts-per-million, not only in dates but in degrees Celsius. (Baucom 2014: 125)

Introduction

In the face of planetary environmental degradation, most agree that societal transformation is necessary. Yet it seems that it is easier today to imagine the end of the world, and even the end of capitalism (see Jameson 2003), than it is to imagine the end of the universal human that would enforce such magnificent societal transformations. Gendered, racialized, fully cognizant and safely zipped up in a skin of his (sic) own, this impossibly unchanging human figuration looms large over debates on the futures of climate change, environmental degradation and heritage alike. He stands there seemingly untouched by the world as it changes its material-temporal transformations, without contingency or bodily subjection to the synergistic dynamics of evolution, mass species extinction, or toxic legacies. Starting from the assumption that such human exceptionalism is disingenuous to the work needed to meet the major challenges of today (Haraway 2008; Braidotti 2013; Colebrook, this volume), we suggest instead to front an (posthuman) analytics of more-than-human relationality and sociability in regard to the co-becoming of bodies and places over time. Such analytics are familiar to many scholars in the environmental humanities (Bastian 2012; Bastian and van Dooren 2017; Åsberg 2018; Fredengren 2018a & b) and to feminist theorists of technoscience, the posthuman and postnatural conditions of the Anthropocene. While differing in emphasis and terminology, we subsume such diverse work under the heading of feminist posthumanities. We opt for this open-ended and inclusive term as we see it as a way of enlivening the human of the humanities (and critical heritage studies) with a mix of relational, technological, nonhuman and more-than-human conditions of co-existence.

Starting from the assumptions of feminist perspectives from such forms of re-invented humanities, this chapter approaches the major research question of *how better to re-tie the material and immaterial knots between past, present and future generations.* This is a research question guiding us in our project on deep-time interventions and intragenerational care that we explore here through the multi-temporal site of the Gärstad waste-to-energy plant. This plant resides just outside the town of Linköping in south-east Sweden, a site we often pass by on our way home or to the university. The over-arching intent of our research is to contribute to the sociocultural and material transformations needed for us all to become more gracious ancestors for multispecies generations to come. More specifically to this chapter, we hope to explore the affordances of such feminist posthumanities analytics and show why such posthuman/postnatural concepts work as thinking technologies to augment research within critical heritage studies. This research field stands out for its critical contributions to a politics of representation in heritage work, but it also carries an inherent anthropocentrism built around questions of the attribution of heritage values by present humans to be passed onto future humans. There is a need for heritage studies to move beyond this to acknowledge and deal with the variety of inherited material workings that are passed on to human *and* more-than-human generations. In pondering the age-old humanities question of what it means to live in the end of times, with our contemporary new ways of knowing change, we contend that new temporal notions, considerate of the multispecies futures we may never know the shape of, are needed to reformat our histories for more diverse futures.

Towards intragenerational justice and care

Intragenerational justice and care are terms that emerge from our post-disciplinary commitments to feminist environmental humanities, and to natural as well as cultural heritage. Here, the related concept of inter-generational justice is commonly defined as how equity is transacted between non-contemporaries and captures temporal commitments in policy texts and law (see Brown-Weiss 1992; Rawls 1977). Hence, this is an institutionalization of how to formally handle and structure various territorializations of the future. However, we emphasize the Karen Baradian 'intra-' (Barad 2010) as a way of evoking co-consti-tutive togetherness and conviviality over time. We deploy the notion of intragenerational justice and care for a more-than-human ethics of coming together as companions for merriment, awe and inheritances across generations, where these generations are entangled with each other in intricate and situated ways. Intragenerational care is something that the more moderate concept of sustainability, used in relation to sustainable development in various ways since the Brundtland Report for the World Commission on Environment and Development in 1987, has difficulties contending with in detail. Famously, this report defines 'sustainable development' as development that meets the needs of the present without compromising the ability of future generations to meet their own needs. Hence sustainability in this version is per-forming an analytic separation of the two, in efforts that could be per-ceived as a move towards deterritorializing the future. However, in the Anthropocene there are not that many places that anthropogenic action has not yet territorialized, and such untainted futures might be hard to come by. Furthermore, as many environmental scholars have pointed out, the very modern idea of sustainable development also reverberates with capitalist assumptions of economic growth without consideration for planetary boundaries, and continues in the tradition of treating the nonhuman world as a resource and not a receiver of inheritances and care across generations. We believe sustainability and heritage concerns need to challenge their anthropocentrism and have a lot to gain from a more-than-human and postnatural take on the links between generations.

To take an obvious example of how inequities may travel between generations and impact futures to come, and that ought to be a matter for deep-time heritage politics, is how we store nuclear waste from our present energy-intensive heydays, or how to handle climate change and environmental degradation itself or how we engage in the handling of garbage (see Fredengren 2015, 2018a). Such legacies will have physical repercussions for hundreds of generations of humans and nonhumans to come. The pharmaceutical drugs we ingest, the make-up we wear, the sofas we sit on or the refrigerators that hold the food we eat: every day we interact with untested chemical cocktails and different compounds that will be part of the genetic and transcorporeal heritage of futures to come (Alaimo 2010; Åsberg et al. 2011; Cielemecka and Åsberg 2019). Changes enacted today have repercussions for the generations that will live and die with them, and are set in motion as inheritances that entangle across generations.

Another example of multigenerational heritages set in action today: the recent jumps and advances in synthetic biology enabled two human baby girls to be born in 2018 without receptors for HIV by way of recent CRISPR technologies. Various critics, such as environmentalist Bill McKibben, and Gene Corea, Vandana Shiva, Rayna Rapp, as well as many other feminist scholars of new reproductive technologies and seed modification in the 1980s and 1990s, have long argued that genetic engineering infringes on informed consent and constitutes an unwarranted imposition of the past on the future. The gene-editing CRISPR techniques, the epitome of the Frankenstein technologies they warned about, have today already been tried on many nonhuman organisms in laboratories worldwide. These and other more mundane but equally transformative genetic modifications are commonplace in laboratories and agricultural practices today, where a wide range of species have been modified for human medical or domestic purposes. It might only be a matter of time before the science-fiction imaginaries of rogue mutations, militarized viral zoonosis or designer babies emerge as CRISPR ecologies of the Anthropocene. Considering the advanced and far-gone reproductive interventions in live-stock breeding, wildlife conservation and long available, even naturalized, reproductive modifications amongst people, perhaps we need to consider ourselves as already living among/with/as hybrid ecologies. Sudden mass mutations in the evolutionary past have after all been drivers of change throughout the planet's history, and nothing points to advanced contemporary science standing outside such previous planetary *modus operandi* even if human scientists are only unintentionally assisting. Yet, all this is of course just speculation in the public face of vigilant science communities and laboratory safety precautions. It stands to reason, however, that transformations of today have enviro-corporeal effects, some unforeseen. They linger, like plastics.

The new immortals, as Michelle Bastian and Thom van Dooren (2017) call the plastics and synthetic materials that already litter landfills, shore lands, cities, ocean streams and seafloors, are also ubiquitous to our own bodies, part of food-chains and other permeable environs. Waste and waste management are therefore crucial societal pivots of contemporary engineering and social planning, as are toxicities inherent in conservation work and cultural heritage. However, these domains of inheritance (genetics, pollution, waste, art and heritage conservation) are seldom connected, mirroring the division of labor between nature (for science) and culture (for humanities). Obviously entangled, waste cultures connect humanistic norms with ecological responses to toxicity, the provinces of the humanities with those of the sciences, and need to be dynamically approached as such. Translations between waste and heritage, toxic embodiment and impure legacy are certainly an urgent sustainability-in-the-Anthropocene concern that we approach as nature-cultural heritage phenomena, and that also extend way beyond any research domains or human control. To tentatively approach these human and more-than-human matters, we ask in our project how intragenerational communities become forged and undone in the toxicities and energy flows at Gärstadverken, a waste-to-energy plant, across the transits of deep time.

Intragenerational care as heuristic prism of analysis

To do more-than-human humanities, we rely in particular on a notion of *intragenerational care* amalgamated from the *anything but* postbiological, postcritical or postfeminist registers of humanities. In fact, we build our understanding of intragenerational care with a synthesis of bio-curious and corporeal feminist philosophy, STS and critical cultural research. Intragenerational care is thus a term that has emerged from our postdisciplinary commitments to feminist environmental posthumanities. We define it as a respectful consideration of always already existing ties between noncontemporaries. Or rather, between co-contemporaries of *a shared time* as the material ties are what exist across shared pasts, presents and futures.

With this analytical device, we emphasize the 'intra-' of intragenerational ethics as a way of evoking the co-constitutive togetherness, 'ongoingness' and conviviality across time and place for which there is always accountability. This is much in the vein of Karen Barad's bringing together of ontology and hauntology (Barad 2010). Inspired by Maria Puig de la Bellacasa's (2017) take on feminist care ethics, it is a more-than-human *considerational ethics* of coming together of companions for merriment, loss, awe and heritages *across generations*. Donna Haraway's multispecies approach to evolutionary conviviality and 'response-ability' in *When Species Meet*, with its emphasis on common grounds, has been trailblazing for us as well (Haraway 2008). In our research project on environmental waste, conservation and heritage issues in present-day Swedish green policy practices and national heritage imaginary, we start from such emerging multispecies and intragenerational approaches to conviviality *and comorbidity* over the long arch of deep time. This offers us a way of moving beyond thinking about gender, race, sexuality, dis/ability and feminist analytics in narrow terms of identity politics, and in terms of who gets represented or not in expressions of heritage. Instead, we explore forces of formation and ask who and what within these formative forces get to live, thrive, suffer or die. *Cui bono?* It moves us analytically and critically in terms of open encounters, response-abilities and reciprocal responses for the deep-time event. We also try to cultivate such sensibilities in the encounter between heritage research and environmental humanities within the heuristic frameworks of conceptual innovation we call feminist posthumanities.

An Anthropocene challenge to the humanities:
How feminist posthumanities gives shape to our analysis

Contemporary philosophers, many outside the discipline of philosophy itself, have taken on the challenge of thinking dangerously and letting themselves be taken hostage by contemporary 'hypercomplexities' such as climate change or mass species extinction. As Peter Sloterdijk and Hans-Jürgen Heinrichs (2011) suggest in the introduction to *Neither Sun nor Death*, such thinkers opt boldly to forsake the present humanist and nationalist world for a wider horizon *at once ecological and global*.

Even though humans certainly impacted their environments prior to this post-Holocene era, the notion of the 'Anthropocene' indicates such a 'hypercomplexity', at once ecological and global, that by now has captured the imagination of many contemporary humanities and social science scholars. Famously, the Anthropocene has been suggested by natural scientists as that epoch we now live in that is defined by human disturbances of ecosystems and climates. As the 'Anthropos' of the Anthropocene, humans have, in a sense, become a 'force of nature', signaling that nature in its classical sense is over (Chakrabarty 2009). This also has some serious philosophical consequences for research in the humanities and social sciences. Consequently, we cannot separate humanity from nature: the environment is in us, and we humans are fully in the environment (Åsberg 2018). Thus, there is a call for an ecological or environmental (post-)humanities that refuses the divide between nature and culture in analytical and disciplinary terms (see also Harrison 2015, Fredengren 2015). However, the study of anthropogenic impact in a range of emerging ecologies is still of utmost concern.

In academic cultures, these shifts in our understanding have meant that humanities research, which previously tended to prioritize culture over nature, minds over bodies and words over things, is shifting rapidly. As 'nature' becomes cultured, the humanities – and contemporary culture at large have become increasingly 'natured' (Alaimo 2010). Here we might consider the ubiquitous climate awareness and influence of public recycling imperatives or the effects of common popular science reporting on our social imaginaries. Such examples point to the values of reinvented forms of humanities research. As Deborah Bird Rose et al. (2012) have argued, environmental humanities redeploy humanities' modes of inquiry such as 'meaning, value, responsibility and purpose' to real-world environmental problem-solving. The environmental humanities in their plurality of longstanding traditions and new transformations (Neimanis, Åsberg and Hedrén 2015; Oppermann and Iovino 2016; Emmett and Nye 2017) are already well poised to take up the challenges of the Anthropocene. Today, new generations of ecofeminist scholars, speculative scientists and practice-oriented philosophers would like to push the humanities' critiques further. These theory-practitioners, like ourselves, claim that the humanities have become all too 'human', and reclaim the notion of 'posthumanities' (Halberstam and Livingston 1995; Wolfe 2003; Åsberg 2009, 2013) for a novel set of postdisciplinary approaches based on taking the ongoing entanglements of nature and culture very seriously.

Grounded in both science-infused archaeology and contemporary eco-gender studies of the bio-curious kind, our research here situates itself within these contexts in general (environmental humanities and feminist posthumanities) and within the 'posthuman turn' to ethics, and the notion of intragenerational care, in particular. The 'posthuman humanities' or, more often, the 'posthumanities', has created a paradigm shift (Åsberg and Braidotti 2018) in humanities scholarship by focusing on embodied subjectivities *after* the singular idea of 'Man' and subsequent debates on who gets to count as human, as the 'Anthropos' of the Anthropocene (Latour 2004; Haraway 2008; Grusin 2017). Instead, posthumanities imagine humans as co-constituted within multispecies relations, thereby decentring humanity from the humanities and making room for human *and more-than-human* ethics. In doing so, it stands on the shoulders of longstanding anti-humanist or anti-anthropocentric claims within continental philosophy, as well as postdisciplinary commitments to historical and emerging medicine, technology and natural science (Haraway 2008; Franklin et al. 2000; Bryld and Lykke 2000).

Taking science, biology and the corporeal seriously is what defines such bio-curious feminist scholarship (Åsberg and Birke 2010; Radomska 2016), indicating its multivalent inheritance as neither postfeminist, postcritical nor postbiological. Rather, it is kin to the science fiction and cyborg approaches of Donna Haraway (2016), and the affirmative ethics of embodied subjectivity of Rosi Braidotti (1993, 2006).

Besides rethinking human subjectivity and ethics, the posthumanities rely also upon another concept with a queer feminist legacy, namely the 'postnatural' (Åsberg 2018). The Anthropocene forces us to recognize the queer (non-normative) situation we find ourselves in and make us question (in a gesture borrowed from queer theory) taken-for-granted assumptions about life as we thought it was (Radomska 2016). For feminist posthumanities, this boils down to the idea that nature was never natural to start with, and that any kind of purism amongst our analytical categories (be it gender or nature) would not serve analysis well. Because the categories of 'nature' or the 'natural' have been used with detrimental effects upon real bodies and ecologies, even 'naturalizing' discrimination and power asymmetries, diverse scholars interested in *just sustainability for the many* have long problematized these categories altogether. Haraway's early work on the technoembodied figure of the cyborg springs to mind as a starting point for such postnatural conceptions. In this vein, many feminist, queer and decolonial scholars assume that 'natural' is a largely ideological space (Butler 1993; Wynter 2003). Nonetheless they see the world as material and real, and subjectivity as corporeal (Braidotti 1993; Haraway 1991), matter as agential and formative, even ethical (Barad 2010; Alaimo 2010), demanding an analytics of material-semiotics for its ongoing materializations forged by forces, flows and abilities of attachment and affirmation (Haraway 1997; Braidotti 1994; Åsberg, Thiele and van der Tuin 2015; Alaimo 2016). In relation to critical heritage studies, such an approach offers ways of thinking about heritage in terms other than those of a narrowly defined representationalism and identity politics. Heritage can through these advances be analyzed as material discursive phenomena composed by a variety of material/immaterial agencies, processes and apparatuses (Fredengren 2015). This avenue enables a meeting between heritage studies, new materialism and feminist posthumanism, allowing for critical heritage to continue as critical and engage with a materializing world beyond the human. Furthermore, it also allows for experimental and affirmative work that views heritagization alongside other worlding practices. As Fredengren (2012) notes, whilst biases and injustices based on identity in heritage selection need to be recognized and critiqued, they do not necessarily need to be perpetuated. In addition, such injustices need to meet up with the bias of human supremacism in times of great injustice to vast ecologies of nonhumans. Such deterritorialization (and reterritorialization) of categories such as gender, race and class, and of queries in terms of what human gets represented or not in humanistic expressions of heritage, move us instead, critically, creatively and tentatively, in the multispecies directions of posthuman and postnatural ethics.

Identity categories, like for instance those of sex or species, are thus not to be taken as either natural or mere social constructions of human ideation. As material-semiotic phenomena they are actively made, maintained and politically charged to serve some interests more than others. In the planetary age of the Anthropocene, human activities clearly link to mass species extinction rates and environmental destruction. This makes dominant human thinking and acting regimens a lot more radically open and suitable for feminist, decolonial or Deleuzian re-inventions in terms of, for instance, 'becoming imperceptible'. Moreover, it makes inquiries into our relationships with the nonhuman animal (also as it relates to those humans deemed less-than-human or even inhuman) seem increasingly necessary.

With intragenerational care as our multispecies prism, we ask what relationships of mutual dependence sustain ecological survival on this planet, and on this specific multi-temporal site (Gärstadsverken) in particular? How can we enable specific relations of care, or a sense of communality, between species that from an evolutionary point of view must be regarded as co-constitutive? Haraway uses the term 'companion species' to describe such mutual dependencies, highlighting the extent to which our human ongoingness is embedded in symbiotic relationship with bacteria and other microbes, with trees, plants and other nonhuman animals. In the vein of outlier biologists such as Lynn Margulis, Carl Sagan and perhaps Haraway, developmental biologist Scott F Gilbert (2017) deconstructs individualist notions of embodiment. He does this in terms of symbiosis, symbionts and holobiont evolution providing postdisciplinary Anthropocene scholars with arguments for avoiding 'human exceptionalism' (Haraway 2008).

Indicative of how feminist posthumanities draw on both arts and sciences, the proto-science of evolutionary development along with a well-established feminist care ethic (Tronto 1993), enables us to explore the affordances of a multispecies care ethic, or at least more-than-human ethics, for deep-time futures. Care, in the words of feminist ethicist Joan Tronto (1993: 103), defines "a species activity that includes everything that we do to maintain, continue and repair our 'world' so that we can live in it as well as possible". A more-than-human approach to intragenerational care, as we will explore it, thus finds footing in both critical and biological literature. For us this means asking questions about what heritages are needed to be able to repair the world for future multispecies uses.

How relations come to matter is a core political concern for feminist theory today. It is also the pivot of evolutionary development theories of relationality. The starting point for both sets of theories is that no organism lives out its life in isolation or lives only to fulfil its own needs during its lifetime. Biologically speaking, symbiont relations play out, for instance, as *mutualism* (mutually beneficial), as *commensalism* (beneficial to one, negligent effect to the other), as *parasitism* (to the detriment of one), as *competition* (detrimental to both), as *mimicry* (masquerading as the other to mutual benefit or as exploitation of the other), as *amensualism* (asymmetric inhibition or annihilation of one whereas the other stays unaffected), and as *co-evolution* (mutually dependent evolution as in the case of infectious disease, ecological communities at large, and famously, flowering plants and associated pollinators). While competition has been given priority in the more commonplace ideas of evolutionary biology, it is by no means the only way, or even the most successful mode, of multispecies relationality. A sophisticated extension of such evolutionary biopower relationship theory is found in Haraway's multispecies approach to conviviality, 'companion species', and the ethics of 'response-ability' (Haraway 2003, 2008). Put to use in the feminist environmental humanities (or, more aptly, posthumanities) registers of Stacey Alaimo (2016: 175) and others, this leads us to maintain a vigilance in relation to environmentalism and sustainability efforts too, as we, working from within such efforts, need also to ask 'What is it that sustainability seeks to sustain'? A similar question needs to be asked in these cases when heritage is used for sustainable development and care between generations, as the concept is far from innocent. For example, are care for future generations only about commensalism (if such relations are at all still possible in the present environmental predicament?). Again, we cannot care for relations with future humans alone, but need to be concerned about all the ranges from mutualisms, to parasitism etc across generations. There is *no purity of categories* to be had in the Anthropocene, and we cannot afford it anyway. A caring notion of sustainability over intragenerational time would need to combine environmental quality with human equality (social justice), extending care without knowing whom the environed subjects of the future would be. This is how feminist posthumanities works to deterritorialize taken-for-granted notions without losing its critical edge. The deep time concept in our more-than-human heritage research provides us with another such notion that wreaks havoc on human-centred conceptions of time and history.

Deep time haunting the Anthropocene

The concept of 'deep time' is rooted in European, early modern challenges to biblical narratives wherein the planet was seen as created in 4004 BC for humans to rule. Scottish sea-cliff strata revealed to early geologist James Hutton that the Earth had to be much, much older, and that humans, in turn, had only been around for a relatively short period (McPhee 1981: 20). Later, and now quite famously for this audience of readers, Chakrabarty (2009: 213-20) argued that in order to better understand the contemporary world with its climate and environmental problems, there was a great need to question the barriers between histories of geological deep time, natural history and that of the human. This has major implications for how history can be studied/comprehended, and how to engage with inheritances of all types and kinds. Such a union of natural and human history also has the eerie capacity to make us query what a world without humanity might look like. For example, Jussi Parikka (2016: 201) points out how the current situation engages timescales orchestrated by a range of more-than-human others and there is a need to act in response to these others. Here, archaeology, as an early form of posthumanities, *avant la lettre*, taking on board human and nonhuman temporalities, provides useful reference points for heritage studies. Such more-than-human heritage accounts would of course contribute to wider debates within the environmental humanities in terms of memory and time, but also with regards 'archaeological' deep-time materialization processes. The research of Lucas (2015), that archaeology troubles the temporalities of the contemporary, is of particular interest so as to avoid the pitfalls of *presentism*, making everything about the now, which is rampant in Anthropocene discourse. In the age of the 'great acceleration' (Steffen et al. 2015), it is important to slow thinking down so we can acknowledge properly a much denser, layered sense of all the materializing multi-temporalities at play in particular locations.

As succinctly framed by Bird Rose (2013: 1), "Time and agency are troubled, relationality is troubled, situatedness is troubled. We are tangled up in trouble". Moreover, the term *chrono-normativity* comes into play here, as it has been used by queer theorist Elizabeth Freeman to describe "the interlocking temporal schemes necessary for genealogies of descent and for the mundane workings of domestic life" or "the use of time to organize individual human bodies toward maximum productivity" (2010: 3) through calendars and schedules. This notion, however, can favourably be adapted for a critique of the temporalities of modernity, namely, clock time, factory time and time-management that so many modern institutions rely upon. As Michelle Bastian (2012) distinctly articulates, such calendars and clocks also structure power relations, where your time-slot directs my life-choreography, and where the temporalities of a range of different, yet relational, nonhuman others are simply ignored. Bastian asks, are there ways in which time and calendars could be designed otherwise, for us to rebuild more sustainable relations amongst human and nonhuman chronologies?

This stands out as an apt question for heritage research with an outlook to the future, where time may not primarily be a noun, but a verb: *how could material processes, times and temporalities and relations be knotted together and made elsewise and in less damaging ways?* How can our temporal appreciation and language be improved and stretched to better encompass deep-time realities of the past and the future? Would it benefit us today, in daily life, to have our deep-time interventions marked out for us, and recognized as such, through exhibitions, apps or ceremonies? Against this background of how we could be 'checking in with deep time' it seems important to focus on how, in place-specific detail, temporal relations could and need to be designed differently, given the challenges we face in terms of social justice, climate and environmental change. Deep-time concerns *haunt* the Anthropocene, and so do a range of slowly moving and shifting materialization processes.

In the onto-ethical takes of her agential realism (and posthumanist performativity), Barad (2010: 266) addresses in concise philosophical terms how ongoing processes of materialization are already inherently enmeshed with ethical issues. Past violence shapes futures. To elaborate on this, Barad makes interesting use of Derrida's (1994) concept of hauntology to address relations between past-present-futures (see also Fredengren 2013: 63–64, 2016). As Elaine Gan, Anna Tsing, Heather Swanson and Nils Bubant (2017) describe it in the vernacular of feminist environmental humanities, ecologies are made and unmade by traces of more-than-human past ways of life, still charged in the present. Extinction leaves traces in the landscape. Similar to the political philosophy of death developed by Rosi Braidotti (2006), 'refusing to forget the past' – a plant's co-evolutionary relationship to now extinct pollinators or large mammal herbivores that had previously carried their seeds to new places, for instance – can fruitfully be described as ghosting those now hampered plants (Gan et al. 2017). Gan et al. make a powerful argument for such an ecological *hauntology of the Anthropocene* in which the 'great acceleration' of species extinction rates makes "Every landscape haunted by past ways of life" (2017: G2). In their Anthropocene anthropology *modus operandi*, "Big stories take their form from seemingly minor contingencies, asymmetrical encounters, and moments of indeterminacy" (Gan et al. 2017: G5). Amongst assemblages of ghosts and wounded landscapes we might detect liveability again.

Environmental change affects differently situated ecological subjects to varying degrees. For example, flooding or toxic mudslides from burst dams affect people and animals living near the lake shore or along the river more than people or animals higher up on dry land. Derrida (1994) argued that Marx's theories continue to haunt society in a ghost-like way.

Whether his political objectives were realized or not, Marx's critique instilled in society an awareness of social injustice and a responsibility to create just outcomes, in future worlds. Despite failing to be fully embraced in the capitalist world, this critique is hard to undo, leading to continuous questioning of the present, which needs to be dealt with in part through historical analysis. This 'hauntology', as Derrida (1994) defined it, works by looking back through the past as a ghost-like apparition, whilst remaining connected to hopes about the future (Barad 2010: 266). However, such re-thinking and re-composition of history cannot be completely novel, nor can it completely erase events or cover up injustices in the past. 'History' like spectres of what we used to regard as nature, will continue to disturb us through its present-absence. In short, the Anthropocene is "haunted" by its exclusions (Barad 2010: 253; Gan et al. 2017). As argued in Fredengren (2015 and 2016: 14) heritages can be explored as onto-ethic-epistemological phenomena, in all their excesses, which in turn may imply a disruption of the unilinear past-present-future arrangements. Here, the lavish workings of material and immaterial pasts can instead be traced as enchanting or haunting – presently at work forming unexpected and queerly formed materializing alliances, not always within human capture or control, but certainly with the capacity to disturb and diffract temporal workings and alliances of different kinds.

Offering more politicized accounts than approaches that leave it to a celebration of the object/thing *per se* or that work for the maintenance of nature-cultures, as this idea dies down in the Anthropocene discourses, we turn instead to the situated analytics and postdisciplinary practices emerging out of environmental humanities and feminist posthumanities. To explain, it is not enough to point to flat ontologies or instances of nature-culture entanglements, and it is not enough to parade science reports on the deteriorating state of nature. Instead, the Anthropocene calls on us to find ways of, in the words of Cate Sandilands, "seeing beauty in the wounds of the world and taking responsibility to care for the world as it is" (Mortimer-Sandilands 2005: 24). In this chapter, we join company with such environmental scholars who try to formulate a more sustainable and equitable ethics of human diversity and multispecies co-existence in dire times.

In this chapter and our continued research we tentatively discuss the politicization of the long term within the natural/cultural heritage sectors and the layers of vernacular temporalities that meet and transform on a particular site of present contestation, namely a high-tech garbage disposal site that is situated on an Iron Age archaeological sanctuary in the city of Linköping, Sweden. At this site high tech energy economies meet up with ancient heritage environments, resulting in a significant deep-time metabolization process where some material features have been removed and others added to facilitate a large modern waste treatment plant that churns over and transforms a range of different left-overs from geographically extensive locations.

Curating energy over time at Gärstadsverken: Organic temporalities

One could look at the waste-to-energy plant as a fine example of futuristic architecture for the Anthropocene imaginary (cf. Turpin 2013). Towering over the East Sweden flatlands right beside the European Highway 4, the Gärstad plant is a steel and glass composition consisting of a district heating plant and a newer building next to it, an incineration plant designed by Winell & Jern Architects in 2016. The openness of the newer, taller building contrasts with the other lower and more closed-off buildings and the surrounding small, fenced-off, sewerage pools, which are teeming with bird life. Perhaps the city of Linköping's most visible building, the high-profile location presented the architects with significant demands in terms of architectural design. Exposition is key to the building's pedagogical aesthetics. The architects describe how they built the framework largely out of glass in order to expose the primary processes of the furnace and the gas purification plant, partly as part of the architectural expression, and partly as a pedagogical element. Tekniska Verken, the municipal company running this waste-to-energy and district heating plant, have indeed a strong pedagogical mission. All the schoolchildren in the town get to visit the sewer and waste operations and learn about the plant processes, and at least monthly the households of Linköping take part in 'green' or 'climate smart' programmes, depositing household waste in regular plastic bags and organic waste in specially designed green bags. The pedagogical mission of exposing the 'green' process of waste incineration is thus strongly connected to both public

discourse and the architectural expression of the building itself. In visual
rhetoric, advertisements, billboards around town, websites and pam-
phlets, the citizens of Linköping are called upon as part of a climate and
energy-efficient community and Gärstadverken is presented as the green
temple, accessible (made for us all) and at the same time inaccessible in
terms of the technological sublime.

All household waste collected in Linköping (together with waste from
another 20 municipalities and from other countries) is transported to the
Gärstad waste-to-energy plant where the garbage is incinerated, and the
district's heating and electricity are produced. In Sweden, more than 99%
of all household waste is recycled through material recycling, incinera-
tion/energy recovery or biological treatment, and less than 1% goes into
landfills.[1] The waste-to-energy plant consists of several facilities that are
themselves fuelled mainly by the burning of garbage, a technique patented

Figure 2.1 — The Gärstad plant at night. (Photograph by Cecilia Åsberg).

and proudly branded 'climate smart' due to its energy efficiency compared to adding garbage to landfills. With an enormous capacity to burn 88 tons of garbage per hour, the three older-type furnaces in combination with the new ones in the glass house (highly visible from the motorways intersecting at the location outside Linköping) generate as much electricity as 293 GWh/year. In 2006 the Gärstad plant was modernized with two newer furnaces, in a project entitled *Future Gärstad*. Now we live with it, and the extensive CO_2 emissions of waste incineration, for centuries to come.

The Gärstad plant now supplies practically the whole town of Linköping (160,000 inhabitants) with its energy, central heating and electricity. Tekniska Verken in Linköping AB, the company running Gärstadverket, is also owned by the Linköping municipality and employs around 500 people, making it one of the largest employers in the region. This large-scale energy business provides, besides energy, services and service operations for electricity, lighting, water, district heating, district cooling, waste management, broadband, biogas and efficient energy solutions. Not to mention more CO_2 per megawatt of energy than burning coal, at least in the short run.[2] In fact, the CO_2 emissions that are produced from the waste itself over the long period of natural decomposition are in the waste-to-energy incineration process speeded-up, condensed into the present and not spread out over a long duration.

As such Gärstadsverket could be said to represent the built heritage of the Anthropocene epoch, but also its folding of time. In a classical heritage sense, this plant stories the green arts of engineering and the burning of garbage from all over Europe. As an imploded node of material-semiotics with future repercussions, it condenses the infrastructures and politics of heating and purifying the town of Linköping, to include extensive European trade-links and interconnectivities across national boundaries. As such, the plant stands as an important industrial heritage. While this may be a classical way of arguing for making heritage, it would fail to pay attention to the various other, unwanted effects of the site. It would be as if the Anthropocene was something still ahead of us and not something processing us now. There is a certain sense of modern denial involved. And in addition, such accounts of the plant as heritage would fail to consider it as *an artificial time folding agent*, postnaturally taking on the brunt of CO_2 emissions from waste now instead of later in futures to come.

In reports from the Swedish Environmental Protection Agency (*Naturvårdsverket*), waste-to-energy incineration of plastics are troubled by uneven combustion and subsequent emissions of dioxins and furans as cancerous agents present in the remaining ashes and are spread via air. Waste disposal sites and treatment facilities, like the plant at Gärstad, provide water ponds and waste areas that provide important sanctuary, resting and feeding sites for migrating and resident birds, such as jackdaws, ravens and gulls. Bird species have been known to recover from PCP and DDT pollution after such substances affected the eggshells. However, the precise levels of sustained dioxins and furans emissions from waste incineration, accumulating over a long time period into the environment and into environed bodies, are presently unknown. And it is not an issue for birds alone, as effects of dioxins and furans are detrimental to all vertebrate organisms. As it appears when searching for literature on the topic, it remains quite unknown how such recent increases in emissions from burning waste affect the gatherings of birds, and the teeming bird life in the pond waters of Gärstadsverken. The town of Linköping has hosted large flocks of jackdaws and other birds of the crow species, and during the daytime Gärstadverken is their main habitat. Amongst the air-born ashes from waste incineration not caught in the filters, and amongst the waste deposits, many scavenger birds such as gulls and ravens have made Gärstad their main site of attraction. What postnatural heritage of persistant organic pollutants for multispecies generations to come are accumulating in these avian bodies? Such bodily accumulations are indexing time in the Anthropocene, layering bio-temporalities for future generations.

It would be highly cynical to discuss the Gärstad plant and its haunted past-present as merely conglomerates of nature and culture, and simply celebrate its underpinning of nondualistic theories or flat ontologies. After all, some pasts clearly matter more than others, and some futures will materialize more than others in the deep sense of posthuman politics. We would suggest a revisioning of the plant through the environmental posthumanities heuristics we have delineated above. This cannot fail to take into account the multivalent intragenerational comparison and choice between, on the one hand, dispersal of large amounts of *slow* CO_2 emissions from 'naturally' decaying waste, and, on the other, gifting the near-present with massive and rapid emissions immediately. The uneven folding of toxic bio-accumulation and CO_2-time at Gärstadsverken begs consideration beyond conventional heritage discussions.

Sweden likes to regard itself as a forerunner in terms of green waste management. The Gärstad plant also functions in more self-satisfied heritage narratives about national green pride, as Swedish recycling policy has been described in international media as so revolutionary that the country has run out of rubbish. Gärstadsverket has been discussed in the *New York Times* as part of a Swedish success story – being one of 34 plants in the country where garbage is turned into energy. In governmental discourse and media, the engineering of Swedish waste-to-energy politics is articulated as no less than 'the Swedish Recycling Revolution'.[3] However, other voices have also been raised; according to Larsink's waste hierarchy (Figure 2.2), energy recovery is the second least wanted option, with only disposals in landfills being worse. Burning waste is, after all, hardly the same as recycling it. The incineration of garbage, although its dioxin-rich fumes are filtered, contributes to environmental pollution and has adverse climate effects. Hence, Owen Gaffney at the Stockholm Resilience Centre describes it as a short-term solution and environmental watchdog organizations, such as the Environmental Integrity Project, even see incineration/waste-to-energy options as green-washing (Yee 2018). Other critics have taken this stance, pointing out that while import-incineration practices *pass as recycling* (bringing down the costs

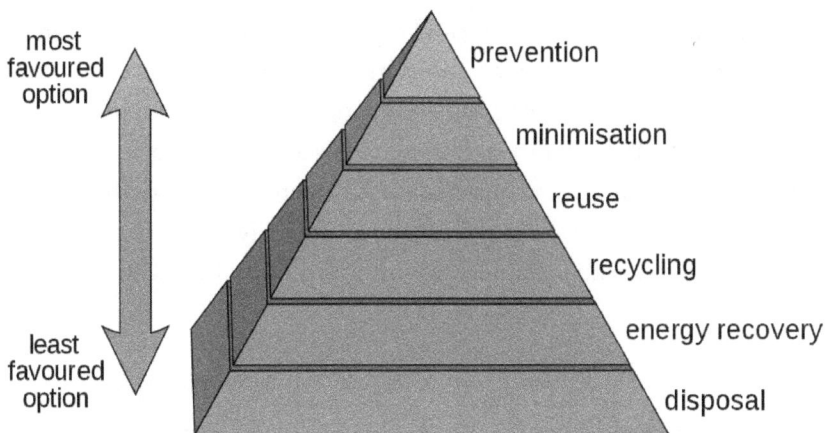

Figure 2.2 — Larsink's waste hierarchy. (Drawn by Drstuey at the English language Wikipedia, CC BY-SA 3.0).

and fines for landfill), it is not recycling, thereby creating a false sense of 'garbage Nirvana' (Hogg 2016). To use the same vocabulary, perhaps this insistence on incineration instead creates negative karma for generations to come. This is also because the waste-to-energy plants of today are so efficient, they have created *a commercial demand for waste*, its main fuel source. Such market demands for more waste to burn hardly creates an incentive for either waste prevention or actual, practical recycling of refuse materials. The Swedish Environmental Protection Agency (*Naturvårdsverket*) therefore proposes a higher levy on waste collection.

On the other hand, landfills – also an integral part of the waste plant at Gärstad – could, as suggested by some more visionary researchers, be turned into veritable treasure troves. Tanha and Zarate (2012) at Linköping University have calculated that there might be some 100,000 tons of aluminium, but also significant amounts of expensive minerals such as copper and zinc, to be retrieved from the depot at Gärstad. To the extent that these landfills have the potential to become future garbage mining sites, they also evoke the latent societal and corporeal toxicities of other forms of mining politics. Attending to the deep infrastructures of valuable metals sitting on site awaiting extraction, other Linköping researchers have pointed to the dangers involved in digging into decades of toxic landfill materials. The jury is out, when it comes to this waste-to-energy plant and to what hypercomplex heritages it may be a part of, with its impact on climates, environments and temporalities. Clearly, the congregated levels of toxicity associated with the untested practice of landfill extractions in Sweden begs the question, when we turn matter into refuse and throw it out of sight, what are we trying to forget? Picking at the surface of deep landfills that we might come to see as barely covered human-induced earth wounds, alive with a toxic agency of their own, we ask what different, more safe, possibilities could there be for recuperation, inhabitation and healing at the Anthropocene site of the Gärstad plant?

Multi-temporal clock sites

As discussed in Michelle Bastian's (2012) essay, time and agency are deeply interconnected, not only in linear narratives of cause and effect, but also when figuring out action capacities and relationalities in the

Anthropocene. If heritages are understood as mere social constructions in the present, they become void of time, they become timeless, and hence without any agency or material effects. As evidenced in the previous sections, Gärstad stands out to us as anything but void of temporality and nonhuman agency. It is teeming with interlinked and embedded human and nonhuman activities and with multiple temporalities, unfolding as the story develops. One can argue with Bergson (1907: 14) that "The present contains nothing more than the past, and what is found in the effect was already in the cause"; however, this would, in our posthuman take on time, be a present that reaches across the entire arch of deep time, including deep futures. Of concern for us is how a particular place, the site of Gärstad, re-territorializes and acts-out past/present/futures in situated ways. Gärstad clearly holds many temporalities in place. Essentially, we care about how places and practices with varying temporal rhythms and durability interfere with intragenerational matters in sometimes unexpected, and not always linear ways.

Following Bastian's reasoning, we ask in particular how this site is telling time. We consider the way human and more-than-human forces and flows of the Gärstad plant are dragging multispecies beings into future existence, or not. Organisms-in-place become multi-temporal clock sites by way of embodiment. They become indicative of past intercessions as woven together with present interventions, for futures to come.

In the periodic sense of temporality, time comes with an order, a past precedes its present and its future. This can be measured, clocked and performed culturally as aesthetic or artificial devices for understanding the passing of time. In evolutionary terms of multispecies liveability, time accounts for degrees of adaptation and extinction in organisms. Large herbivores, like aurox and tarpan, now extinct, would have been the major environmental engineers of the post-ice age site we now call Gärstad, influencing its vegetation and the composition of woodland species. In an anachronic sense, some herbs (like sorrel) that they munched are still prepared as food (herb soups) by contemporary people while other herbs have lost their most advantageous means of dissemination due to the extinction of these large, grazing animals. Large herbivores, now long vanished, still haunt the present and the possibilities for the spread of plant life on site. Using terms from film studies, the herb sorrel for instance can be said to mediate time, providing us with a 'flashback' (analepsis) of past pastures around Gärstad, and so on (Table 1). Taking into consideration such temporal categorizations and temporal relations possible at the site of Gärstadverken, what time would Gärstadverken tell?

Economical or social time/Time as noun	Ecologizing time/time as verb
Time as history/Periodizing time	History working beyond 'the historian's code'
Past/Contemporary/Future: Extinction/Adaption: Modern/Altermodern: Obsolescence/Innovation: Anticipation/Unexpected	Processes and relations weaving through past/ futures – a thickening of the contemporary – and possibly bursting this time fold
Time as calculation/Measuring time	Intra-historical workings
Clock/Lived: Synchronic/Anachronic: Human/Planetary: Serial/Simultaneous: Emergency/Everyday	The holobiontical labouring of organisms, the kiss of death to certain life-forms by toxicities, environmental damage and climate change
Time as culture/mediating time	Supra-historical workings
Aesthetic/Prosthetic: Analepsis/Prolepsis	The dynamics and cascading effects of CO_2 emissions, isostatic movements, long-term storage of heavy metals

Table 1: Table of humanistic times (after Burges and Elias 2016), adapted from Lucas (2015), Bastian (2012) and Barad (2010) and further inspired by Baucom (2014), Gilbert (2017) and transformed.

As we will continue to explore in the research project, this landscape is interwoven with materially entangled temporalities, perhaps where we might explore something like ecological or even ecologizing time, where it is made through a variety of materializing relationalities, braided together in situated ways. Suffice to say, the high modern wonder of engineering of the Gärstad plant is in fact also located on an archaeological site. Bit by bit, by careful excavation, the archaeological layers have recently had to make way for the expansion of this waste-to-energy plant. A variety of material and temporal processes can still be mapped in the area of Gärstad. Geological, hydrological and archaeological temporal changes perform transformations that are still taking place in the landscape. The land is still slowly recovering from the last ice age, but it is not clear if and for how long the land rise will counter-act increasing water-levels from sea-level rise. Topographically transformed by modern exploitation, the small remaining archaeological area is located on two elevations that lie between Lake Roxen and, to the west, River Mörtlösa (now straightened out, managed and partly laid in culverts). The shorelines in the nearby geography have retracted since prehistory – for instance as a result of ongoing isostatic movement, where land is still recovering from having been under pressure. The small river supplies its own hydrological temporalities, running in a more seasonal register, with overflows in spring after ice-melts, and perhaps droughts in the summer. This seasonal change is adapted and tamed, handled by the culverting. However, with climate change, such volatility also means increasing prospects of future flooding and human emergencies to come, from swamping to deluging and other water related events.

There are remains of Neolithic activities in the area – hut sites and some of the pottery from recent excavations are discerned as being from this period. In fact, the general area of Gärstad is internationally known as one of the major rock art sites in the region of East Sweden (Östergötland). Here the Neolithic layers meet up with pre-Roman Iron Age and Medieval strata. It has been argued that this was a sacred place, an attraction for people on foot, where both the act of carving the stone and the flow of bronzes from Europe were handled by an elite (Tilley 2008 and cited literature therein).

More precisely, there is rock art (both figurative and non-figurative) in the fields between Linköping, Lake Roxen and just north of Gärstad Mansion (a geographical area). A sun horse and a ship carving, together

Figure 2.3 — The sun horse and ship in Gärstad. (After Wikell et al. 2011).

with several cup marks, have been found here (see Figure 2.3). These were times when the sun was understood to be carried over the skies by a horse (Wikell et al. 2011), hence this panel captures the temporalities of day and night. However, the rock art that originated in the Bronze Age also had relevance for futures to come, as argued by Nilsson (2017), as the past was of importance also in the past. Traces of previous rituals including fire sites haunt the location to this day. There is even archaeological evidence for small bonfires from later times, i.e. the Iron Age, telling stories of the interplay of fires and folks. It is likely that these fire sites were connected to cattle herding, as it appears that the sites of rock art, at the time, were resting places for people and animals as they moved across the flat landscape, leaving imperceptible marks (Nilsson 2017: 82–87).

Broadly speaking, the archaeologies of the place run on a variety of temporal registers. Not only does the figure of the sun horse relate to and describe the temporalities of the day according to what may be a Scandinavian Bronze Age cosmology; the rock art and the associated sites materially remind us of alternative ways of forging human-animal relations. Furthermore, the rock art captures the intragenerational, seasonal and daily rhythms of this landscape. This place mixes up the temporalities of synchrony/anachrony, as it has been available and in use during subsequent periods of time, therefore marking out places of return. The rock art also points towards human-animal relations and marks resting and meeting places, along with possible ways to move around in the landscape under other climate regimes than the present.

Looming over this area is the wilder, more extreme, both wetter and dryer Anthropocene climes. This general area of Östergötland was intensively settled during the Bronze Age as people took advantage of the more stable Holocene landscape. Here, material archives of sediments from wetlands and lakes as well as archaeological waste deposits tell stories of land change. During the Middle Bronze Age, the broad covering oak forests (favoured by now extinct large herbivores), with vegetation and undergrowth, were cleared to make room for human settlements and fields. Botanies of more light-loving species, such as salix, rowan and hazel, took their place. Various types of plant such as dandelions, yarrow, coltsfoot and nettles that thrive in the company of humans and animals came into being (see Carlsson 2014 and cited literature therein). The landscape has never fully recovered from these clearings. Such widespread deforestation is still regarded as an archaeologically registered deep-time intervention that affected many futures to come. Materializing at this stage were the more familiar, open and flat farming landscapes of today.

Gärstad is also a place that harbours temporalities of life and death of other cultural kinds. The plant is located near one of the larger excavated burial grounds from the pre-Roman Iron Age in Sweden, with over 500 graves. This too has been removed to make room for the expanding waste plant (Helander 2017: 11–12).

These material and temporal layerings of the site tie together the knots of past and present generations but had to be undone for the power plant to expand. New temporal layers, in completely different registers, have been formed as the burial ground (Figure 2.4) was removed to make way for the large footprint of the waste plant buildings, where artefacts,

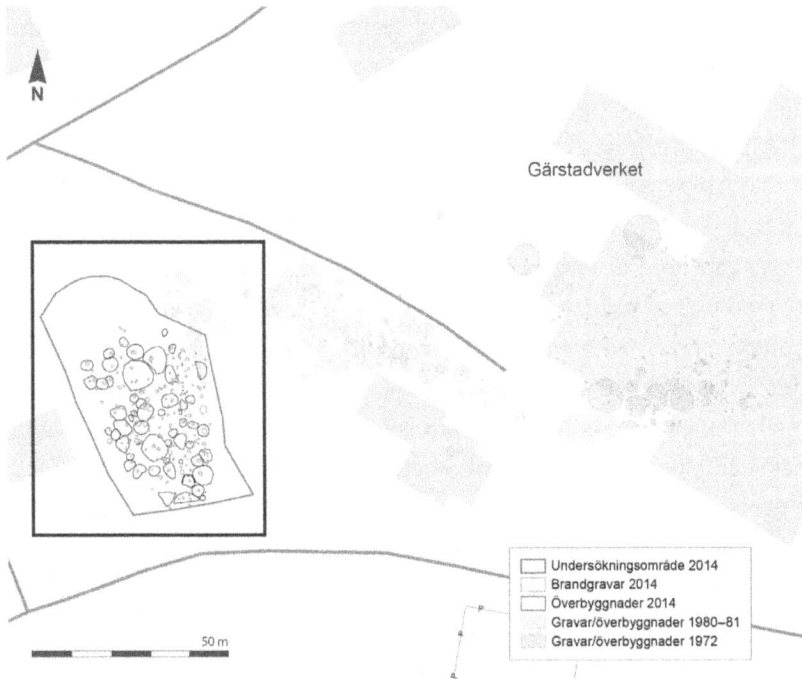

Figure 2.4 — The location of the burial ground in relation to Gärstadverket. (Reproduced from Helander 2017: Figure 3; courtesy of Arkeologerna, National History Museums).

burnt skeletons and soil samples have left some of their relations behind in this place and made their way into and continue their relational working in other contexts. Hence, they are carrying with them some relations, losing others, but are also capable of forming a part of new ones within museum ecologies. In the Iron Age, the bodies of the dead were burned, with clothes and personal items, and when the funeral pyres died out the bones were gathered and placed in the ground with some of the attire. At the plant, clothes and organic materials are burned at a much higher rate today, where consumption/waste materialize at an accelerating speed. The Iron Age funerary rite included meals with the dead, with food placed in the grave (Helander 2017). Today, small green plastic bags containing leftover food get dumped, fermented and eventually burned just behind this location. The earliest Iron Age is often seen as a period when humans had a comparably light footprint on the Earth. It was to some degree a period characterized by today's fashionable *de-materialization movement*, with less prestige-consumption than the Bronze Age. Yet, it was still a period that brought substantial human-induced landscape change, with more organized field systems and farming. In later days it was a *tingsplats*, a site of deliberations and ruling, a place where people met to discuss and pass judgment on current affairs (Helander 2017). As a heritage site, it works both in, through and towards the Anthropocene, as a location where intragenerational justices and judgment are played out today too.

As captured in research (Fredengren 2016 and cited works therein), archaeological remains occasionally appear unexpectedly and disturb modern chains of events, contributing to clashes and disjunctures in time. Encounters with materials from deeper temporal strata can give rise to what is understood as *enchantment effects*. Bennett (2010) has argued that such effects are important when people step up from environmental ethical thinking to real and substantial environmental action. Here, however, the archaeological finds from the excavations interfered with the development of the plant from the 1980s and onwards as it expanded.

Presently, the buildings and asphalted surrounds have their own life cycle and temporality of use. It is often held that buildings have life cycles of about 100 years, and there is no indication that the large glass and iron structures of the plant would be much different.

Figure 2.5 — Archaeological remains, still protruding in the field, with the plant in the background. (Reproduced from Helander 2017: Figure 4; courtesy of Arkeologerna, National History Museums).

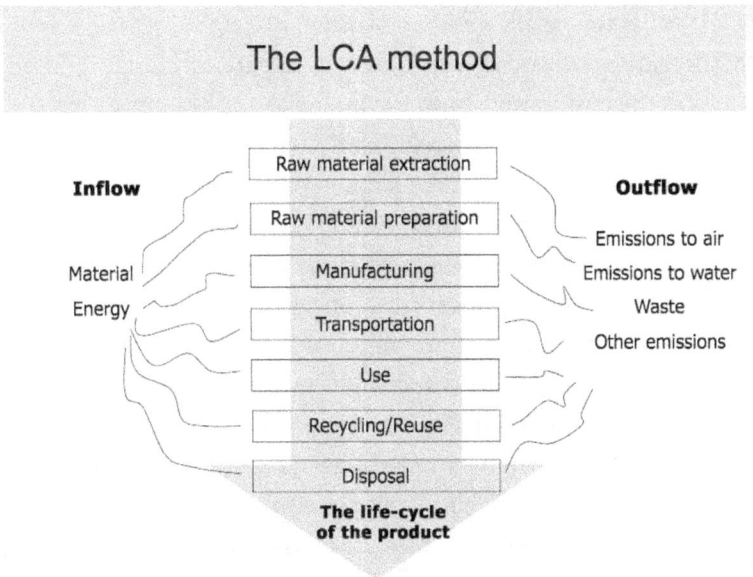

Figure 2.6 — Life-cycle assessment method. (Drawn by Linda Tufvesson, SLU (Swedish University of Agricultural Science)).

Life cycle assessment is a method that tracks the resource use and environmental impact during an item's life, from raw material extraction, through manufacturing, transportation and use (cf. Finnveden et al. 2009). Besides being applied to the garbage plant itself, it can also be used to formulate the garbage content. Garbage can be seen to have reached the end of its life cycle. But there is more to it, this garbage also gets a second life through the process of recycling, or, as in the case of the Gärstad plant, through the waste-to-energy process. It passes into the metabolism of the town, in the form of energy for heating, with the possibilities for future garbage-dump mining. While the model favourably points to all the necessary energies and flows that go into the making of a product, the question is whether the model really knits together a cycle for *all* the garbage here? As most of the waste is incinerated, and not really recycled, and takes on the precious form of energy it seems to join in the more volatile register of the *new immortals*, i.e. less degradable forms of polymers and plastics that defy decay and interfere with future lives in unexpected ways (cf. Bastian and van Dooren 2017). Waste products actually recycled, such as batteries or parts of refrigerators, stay in circulation albeit in various forms. But the waste that flows into the massive incineration process takes on other life forms as toxicities in bodies and CO_2 emissions to air, water and land.

While garbage disposal certainly was a part of human and animal activities in the past as well, both the intensity and volume differ substantially. The present-day waste accumulation and incineration at the waste plant of Gärstad is part of a much larger world system. It plays into transformations on the planetary level with insidious and irregular impacts on climate. These large-scale transformations cascade with extended time lags between causes and observable effects, toxification processes and the invisibility and extent of air pollution. The circulation of toxins to the planet's polar areas, vast dispersal of pollution and other large-scale matters of environmental degradation are scientifically measurable, and part of what some researchers refer to as 'the great acceleration' (Steffen et al. 2015). Compared to the Iron Age, this site has transformed into a massive gathering site for waste from all over the world, following the great acceleration of global consumption. This is the insignia of the Anthropocene where the human impact on planetary ecosystems and geology has increased significantly. The waste plant of Gärstad brings into sharp contrast the intensity and speed induced by the super-efficient incineration process and its circulation of energy, emissions and waste matters. The waste-to-energy plant is part of a transnational network of consumption and waste production, and serves at the furthest end of the life cycle of the product. Waste finds its way to this site from all over the world, as commodity and fuel goods, central to a whole new waste economy. The plant is supported by garbage from the whole of Sweden, but also from Britain and the Continent. It thereby creates spatio-temporal relations between consumption in different countries and depositions in others. It makes us, on the one hand, part of the same Eurocentric garbage community, but on the other, it also keeps us in chains of temporal and material dependencies. The furnace needs waste to support us with energy and heating. As high-energy consumers we depend on these networks. However, these waste networks and consumer desires cannot be sustained over the long term. The activities at Gärstad, through the burning of garbage and emissions into the atmosphere, cause serious *deep-time interventions*. They feed into climate change and set up a relation to future multispecies generations. In the vernacular of Bastian, they shape the action capacities of deep-time futures, unsustainably.

What times does Gärstad tell?

At face value Gärstad appeared very provincial to us. But in more ways than we imagined at the start of this research, it connects us with the world and with the global infrastructures of waste, energy and power. This occurs not only through the creation of elusive sets of new immortals, but also by enticing us into a perceived re-cycling nirvana where, to borrow from Marx, all that is solid melts into polluted air. With scholars like Stacey Alaimo (2012), we stand concerned with what type of sustainability really gets sustained through the hyper consumption enabled by such garbage disposal places. What type of wounds and deep-time interventions, not only at Gärstad, but also around the world, do the underlying desires and consumptions perpetuate? This site not only remembers and tells the time of the sun horses, cattle and their herdspeople, the dead herbivores or the loss of trees, but also makes kin with the gaps and wounds in other landscapes following mining, oil fields, plastic production, chemical cocktails and treasure trove consumption. Hence, tuning in with Gärstad as a time-giver opens up the observation of a range of othered relationalities connected with the materializing practices of international consumption and garbage economies. Parts of the world connect, congeal and transform by incineration at Gärstad, as do a multitude of planetary temporalities.

Our brief survey and analysis of this situated place has started to outline some of the linked times and places that still have a traceable endurance in the landscape. On the one hand, this material could be placed into ordinary linearized time-slots, such as the Holocene, which contains the Neolithic, Bronze Age, Iron Age, Medieval and Modern period. Voices have been raised to also periodize the Anthropocene and make it into a series of consecutive phases, such as early Anthropocene, the first acceleration, the great acceleration (post-1950) and then into the more futuristic good or bad Anthropocene depending on what human actions are taken (Kunnas 2017). One could continue this series with a *baroque Anthropocene* of heavy impact, a *rococo Anthropocene* of asymmetries and emotions – for fun and as an ironic and playful pedagogic contribution. However, the material processes knotted together in this location would have been an interlacing of a variety of sources and forces, where a myriad of materialization processes are in action, that would make such periodization unbound, as they stitch through time in, for us, rather uncanny ways. Here, these protruded time layers make the place around the waste plant of Gärstad a haunted site of Anthropocene refusals, and denials, that time as a noun would hide.

The multi-temporal site of Gärstad has interfered with the present generations, with traces of deaths, depositions from several archaeological periods and comorbidities to come. The archaeological excavations have temporarily hindered the expansion of the garbage site, from time to time, and perhaps have spooked the development (albeit in a rather modest way, with low-key ghosts). Hence, this first mapping exercise suggests that Gärstad is a materializing political ecology of a variety of temporalities coinciding in this place, and a multiplicity of places haunting us in time, a place with futures both territorialized and deterritorialized. We live in haunted landscapes (Gan et al. 2017), where such pressing temporalities can be captured as Derridean hauntologies (Fredengren 2015, 2016). Here we are bothered, and spooked, by both pasts and futures to come, injustices already made, that are stitched into the fabric of the world.

As Bastian (2012) argues, various phenomena 'tell the time' in particular ways, and some of them, those that are connected with pressing environmental challenges, need to be better coordinated with our actions. Such a re-alignment also concerns who we need to care for more in times of environmental upheaval and climatic deep-time changes. When such temporal analysis is applied to Gärstad, it could highlight that we are already in intra-action with a number of past-present-future generations and trajectories. While the standard sustainability endeavour to have a commensurable relationship to future generations i.e. where practices of today have negligent effects on future generations are admirable, the co-constitution over time takes other turns. We are in the process of both becoming-with and becoming undone by our involvements with future generations, in for example parasitic, co-evolutionary, amensualistic or mutually beneficial ways. Bird Rose (2013: 7) alerts us to the fact that "our past is now racing towards us from the future". A variety of deep-time interventions, woven into the fabric of the past with distinct future trajectories, would, if we contemplated them in full, overwhelm us with shadows. Not only past generations, but also present and future ones, are coinciding at the waste plant of Gärstad. This site is telling the time of those intragenerations, reaping what was once sown here, but the processes are far from linear. And this has a bearing on intra-generational ethics and practices.

Ways of living and dying well together

If our species does not survive the ecological crisis, it will probably be due to our failure to imagine and work out new ways to live with the earth, to rework ourselves and our high energy, high consumption, and hyper-instrumental societies adaptively... We will go onwards in a different mode of humanity, or not at all (Plumwood 2007: 1).

With Gärstad as our starting point for checking in with deep time, we aimed for encounters with all sorts of heritages – past, present, future – to probe into what responsibilities 'inheritance' and becoming decent ancestors for future generations place upon us.

In considering this waste plant as Anthropocene heritage, we knew a feminist ethics of intragenerational coexistence would need to take an experimental path. Grosz (2005: 14) would refer to it as an untimely ethics, as it orients itself to the needs of the present and the immediate future, but also accommodates for nicks in time. Haraway (2016: 130) enrols Arendt and Woolf, to cultivate *response-ability* and remind us of "the high stakes of training the mind and imagination to go visiting, to venture off the beaten path to meet unexpected, non-natal kin, and to strike up conversations, to pose and respond to interesting questions, to propose together something unanticipated, to take up unasked-for obligations of having met". So far, we have with this brief set of encounters visited cattle and herders, moved through the land and rested by bonfires next to the rock art at night. We have seen the loss of tree-canopies and herbs, times and places of possibly just and unjust rulings. But we also, possibly, met non-natal generations to come in challenged futures, suffering the bad karma inherited from a variety of consumer/garbage networked relations and materialities. What we are beginning to discuss here is what types of virtualities open up in close encounters with Gärstad, its past, present, future inhabitants and its ongoing energy processes, stretching from before and after. Imagination and sympathy are key to this endeavour. In this, perhaps we find the hope of retying the knots between species and generations over time, in less harmful ways. Yet again, in the words of Donna Haraway, "Outcomes are not guaranteed. There is no teleological warrant here, no assured happy or unhappy ending, sociologically, ecologically, or scientifically. There is only the chance of getting on together with some grace" (Haraway 2008: 15).

Becoming better ancestors:
Re-visiting intragenerational ethics

As explored above, several nonhuman generations (waste burnt into CO_2, jackdaws with survival skills, sun horses next to grazing cattle) already co-habit this landscape of Gärstad, and they also leak into other places and other bio-temporal knottings. By these action capacities, we consider them more than spatially and temporally contained inter-generations, but rather intra-generations. As living things are facilitated by queer temporalities and heritages, where one generation is already related both to past, present and condition future lives, they (we) are as such co-constituted. Here, the question of how to live and die together comes to the fore, as well as how future generations can develop more-than-human kin and affinity. Against this background, the traditional take on intergenerational ethics and law (cf. Brown-Weiss 1992) experiences problems in capturing the queer links between past/present/future generations, primarily because of their focus on bounded human individuals. Entanglements of self and other, cultures within worldly nature, and pasts, presents and futures, have emerged. Barad's work inspires our intragenerational take on multispecies ethics. The entanglements of Barad's 'phenomena' – or 'agencements' or assemblages according to Deleuze – do not come from an interconnectedness of separate entities but are instead "specific material relations of the ongoing differentiation of the world" (Barad 2010: 265). And these entanglements, these onto-epistemological processes of "becoming with" (Haraway 2008: 15) are, we argue, in relation to intragenerationality constantly being both territorialized and deterritorialized. These entanglements are also relations of obligations: they come with a more-than-human ethics. For that reason, both past and future generations are already here to haunt us, and as we have done here, can always be summoned from the shadows.

De la Bellacasa (2017: 4–5), following Tronto (1993: 105–8), teases out the various dimensions of care, and the tensions that there may be in-between those dimensions. Working in a similar vein for *intragenerational* care may mean tracking what affects and affections are linked to engagements over deeper periods of time. In this chapter we have touched upon these sensibilities when they perform as hauntings, or an eerie feeling that they (whoever they are) are here with us. However, there is also the revelation of joy and pleasure in convivial interlinkages over time. We have discovered various types of labour/work/material processes as makers of time and temporalities. They range from local, regional and international disposal and recycling practices to the possibilities of preservation for the multigenerational environments of the Gärstad waste plant. In the super-accelerated waste-to-energy cycle, the grand-scale unwinding of consumption might be out of reach for most people, even if they dutifully sort their garbage in the most appropriate way. The issues at stake are too large for individual interventions. However, working with the waste plant at Gärstad in this way invites contemplation on how to heed the pressing materializing temporalities of this site-in-relation to figure out *how better to re-tie the material and immaterial knots between past, present and future generations.* It opens up a space for mourning, regrouping and alterworlding, for training us in an ethics of becoming better multi-species ancestors. It might also demand of us that we design more clearly when we make deep-time interventions, and to ask what such actions impress onto future generations. We believe it invites a more careful re-stitching of ourselves into futures, a way for heritage to get in touch over time, sustainably.

Notes

1.   An official Swedish government website claimed that 99 per cent of Sweden's waste is 'recycled': https://sweden.se/nature/the-swedish-recycling-revolution/. This claim was picked up in *Global Citizen* https://www.globalcitizen.org/en/content/sweden-garbage-waste-recycling-energy/ and debated, as incineration is not recycling, at Tree Hugger: https://www.treehugger.com/energy-policy/no-sweden-does-not-recycle-99-percent-its-waste.html. After the writing of this chapter, in March 2019 the page was updated, and the claim to recycling moderated.

2.   A 2013 European Environmental Agency report, 'Municipal Waste Management in Sweden', claims that incinerating garbage releases 2,988 pounds of CO_2 per megawatt hour of electricity produced. This compares unfavourably with coal (2,249 pounds/megawatt hour) and natural gas (1,135 pounds/megawatt hour). However, most of the stuff burned in waste-to-energy processes (such as paper, food, wood and other stuff created from biomass) would have released the CO_2 embedded in it over time, as 'part of the Earth's natural carbon cycle'. If adjusted for the slow temporality of the natural carbon cycle, here speeded up by way of incineration, the calculations for CO_2 emissions are on a par with those of energy derived from natural gas, but without the advantage of getting rid of waste in the process.

3.   The Swedish Recycling Revolution: https://sweden.se/nature/the-swedish-recycling-revolution/#start

References

Alaimo, S. 2010. *Bodily Natures: Science, Environment, and the Material Self.* Bloomington: Indiana University Press.

Alaimo, S. 2012. Sustainable This, Sustainable That: New Materialisms, Posthumanism, and Unknown Futures. *PMLA* 127(3): 558–64.

Alaimo, S. 2016. *Exposed: Environmental Politics and Pleasures in Posthuman Times.* Minneapolis: University of Minnesota Press.

Åsberg, C. 2009. Den posthumanistiska utmaningen. *Tidskrift för genusvetenskap* 2009(2–3): 64–68.

Åsberg, C. 2013. The Timely Ethics of Posthumanist Gender Studies. *Feministische studien* 1: 7–12.

Åsberg, C. 2018. Feminist Posthumanities in the Anthropocene: Forays into the Postnatural. *Journal of Posthuman Studies* 1(2): 185–204.

Åsberg, C. and L. Birke. 2010. Biology is a feminist issue. *European Journal of Women's Studies* 17(4): 413–423.

Åsberg, C., R. Koobak and E. Johnson. 2011. Beyond the Humanist Imagination. *NORA – Nordic Journal of Feminist and Gender Research* 19: 218–30.

Åsberg, C., K. Thiele and I. Van der Tuin. 2015. Speculative before the turn: Reintroducing feminist materialist performativity. *Cultural Studies Review* 21(2): 145.

Åsberg, C. and R. Braidotti. 2018. Feminist Posthumanities. An Introduction. *A Feminist Companion to the Posthumanities,* 1–22. Doordrecht: Springer.

Barad, K. 2010. Quantum Entanglements and Hauntological Relations of Inheritance: Dis/continuities, Space Time Enfoldings, and Justice-to-Come. *Derrida Today* 3(2): 240–68.

Bastian, M. 2012. Fatally Confused: Telling the Time in the Midst of Ecological Crises. *Environmental Philosophy* 9(1): 23–48.

Bastian, M. and T. van Dooren. 2017. Editorial Preface: The New Immortals: Immortality and Infinitude in the Anthropocene. *Journal of Environmental Philosophy* 14(1): 1–9.

Baucom, I. 2014. History 4: Postcolonial Method and Anthropocene Time. *Cambridge Journal of Postcolonial Literary Inquiry* 1(1): 123–42.

Bennett, J. 2010. *Vibrant Matter: A Political Ecology of Things.* Durham, NC: Duke University Press.

Bergson, H. 2007 (1907). *Creative Evolution.* London: Palgrave Macmillian UK.

Bird Rose, D. 2013. Anthropocene Noir. *Proceedings of the People and the Planet 2013 Conference: Transforming the Future*. Melbourne: RMIT University.

Bird Rose, D., T. van Dooren, M. Chrulew, S. Cooke, M. Kearnes and E. O'Gorman. 2012. Thinking through the Environment, Unsettling the Humanities. *Environmental Humanities* 1: 1–5.

Blake, W. 1908. *The Poetical Works of William Blake*, ed. John Sampson. London: Oxford University Press.

Braidotti, R. 1993. Embodiment, Sexual Difference, and the Nomadic Subject. *Hypatia: A Journal of Feminist Philosophy* 8(1): 1–13.

Braidotti, R. 1994. *Nomadic subjects: Embodiment and Sexual Difference in Contemporary Feminist Theory*. New York: Columbia University Press.

Braidotti, R. 2006. *Transpositions: On Nomadic Ethics*. Cambridge: Polity Press.

Braidotti, R. 2013. *The Posthuman*. Cambridge: Polity Press.

Brown-Weiss, E. 1992. In Fairness to Future Generations and Sustainable Development. *American University International Law Review* 8(1): 19–26.

Bryld, M. and N. Lykke. 2000. *Cosmodolphins: Feminist Cultural Studies of Technology, Animals and the Sacred*. London and New York: Zed Books.

Burges, J. and A. J. Elias. 2016. *Time: A Vocabulary of the Present*. New York: New York University Press.

Butler, J. 1993. *Bodies that Matter: On the Discursive Limits of "Sex"*. New York: Routledge.

Carlsson, T. 2014. En bronsåldersgård i Sjötorp, Ekängen. Östergötland, Linköpings kommun, Rystad socken, Ekängen, inom Sjötorp 1:1, RAÄ 327. Särskild arkeologisk undersökning. Linköping: Riksantikvarieämbetet, Arkeologiska uppdragsverksamheten, UV Öst,. Linköping.

Chakrabarty, D. 2009. The Climate of History: Four Theses. *Critical Inquiry* 35(2): 197–222.

Cielemęcka, O. and C. Åsberg. 2019: Introduction: Toxic Embodiment and Feminist Environmental Humanities. *Environmental Humanities* 11(1): 101–107.

De la Bellacasa, M. P. 2017. *Matters of Care: Speculative Ethics in More Than Human Worlds*. Minneapolis: University of Minnesota Press.

Derrida, J. 1994. *Spectres of Marx: The State of Debt, the Work of Mourning and the New International*. New York: Routledge.

Emmett, R. S. and D. E. Nye. 2017. *The Environmental Humanities: A Critical Introduction*. Cambridge, MA and London: The MIT Press.

Finnveden, G., M.Z. Hauschild, T. Ekvall, J. Guine, R. Heijungs, S. Hellweg, A. Koehler, D. Pennington and S. Suh. 2009. Recent developments within Life Cycle Analysis. *Journal of Environmental Management* 91: 1–21.

Franklin, S., C. Lury and J. Stacy. 2000. *Global Nature, Global Culture*. London: Sage.

Fredengren, C. 2012. Kulturarvets värde för en hållbar samhällsutveckling. In *I valet och kvalet. Värdering och urval av kulturarv*, edited by C. Fredengren, O. W. Jensen and Å. Wall, 189–223. Stockholm: Riksantikvarieämbetet.

Fredengren, C. 2013. Posthumanism, the Transcorporeal and Biomolecular Archaeology. *Current Swedish Archaeology* 21: 53–71.

Fredengren, C. 2015. Nature: Cultures: Heritage, Sustainability and Feminist Posthumanism. *Current Swedish Archaeology* 23: 109–130.

Fredengren, C. 2016. Unexpected Encounters with Deep Time: Bog Bodies, Crannogs and 'Otherworldly' Sites. The Materializing Powers of Disjunctures in Time. *World Archaeology* 8(4): 482–99.

Fredengren, C. 2018a. Re-wilding the Environmental Humanities. A Deep Time comment. *Current Swedish Archaeology* 26: 50–60.

Fredengren, C. 2018b. Archaeological Posthumanities: Feminist Re-invention of Humanities, Science and Material Pasts. In *Reinventing the Humanities*, edited by R. Braidotti and C. Åsberg, 129–40. New York: Springer.

Freeman, E. 2010. *Time Binds: Queer Temporalities, Queer Histories*. London: Duke University Press.

Gan, E., A. Tsing, H. Swanson and N. Bubandt. 2017. Introduction: Haunted Landscapes of the Anthropocene. In *Arts of Living on a Damaged Planet*, edited by A. Tsing, H. Swanson, E. Gan and N. Bubandt, G1–G14. Minneapolis: University of Minnesota Press.

Gilbert, S. F. 2017. Holobiont by birth. Multilineage individuals as the Concretion of Cooperative Processes. In *Arts of Living on a Damaged Planet*, edited by A. Tsing, H. Swanson, E. Gan and N. Bubandt, M73–M89. Minneapolis: University of Minnesota Press.

Grosz, E. 2005. *Time Travels: Feminism, Nature, Power*. Durham, NC: Duke University Press.

Grusin, R. (ed). 2017. *Anthropocene Feminism*. Minnesota: University of Minnesota Press.

Halberstam, J. M. and I. Livingston (eds). 1995. *Posthuman Bodies*. Bloomington: Indiana University Press.

Haraway, D. 1991. *Simians, Cyborgs, and Women*. London: Routledge.

Haraway, D. 2003. The *Companion Species Manifesto: Dogs, People and Significant Otherness*. Chicago: Prickly Paradigm Press.

Haraway, D. 2008. *When Species Meet*. Minneapolis: University of Minnesota Press.

Haraway, D. 2016. *Staying with the Trouble: Making Kin in the Chthulucene*. London: Duke University Press.

Harrison, R. 2015. Beyond 'Natural' and 'Cultural' Heritage: Towards an Ontological Politics of Heritage in the Age of Anthropocene. *Heritage and Society* 8(1): 24–42.

Helander, A. 2017. *Gravfältet vid Gärstad*. Rapport 2017: 15. Arkeologisk Undersökning. Stockholm: Statens.

Hogg, D. 2016. The dark truth behind Sweden's 'revolutionary' recycling schemes. *The Independent Newspaper*, Tuesday 13 December 2016. https://www.independent.co.uk/voices/sweden-recycling-rates-revolutionary-dark-truth-behind-uk-wales-incineration-a7471861.html.

Jameson, F. 2003. Future City. *New Left Review* 21: 65–79.

Kunnas, J. 2017. Storytelling: From the Early Anthropocene to the Good or the Bad Anthropocene. *The Anthropocene Review* 4(2): 136–150.

Latour, B. 2004. Why Critique has run out of steam? From matters of fact to matters of concern. *Critical Inquiry* 30(2): 225–48.

Lucas, G. 2015. Archaeology and Contemporaneity. *Archaeological Dialogues* 22(1): 1–15.

McPhee, J. 1981. *Basin and Range*. New York: Farrar, Straus & Giroux.

Mortimer-Sandilands, C. 2005. Unnatural passions? Notes toward a queer ecology. *Invisible Culture: An electronic Journal for Visual Cultures*. http://ivc.lib.rochester.edu/unnatural-passions-notes-toward-a-queer-ecology/

Neimanis, A., C. Åsberg and J. Hedrén. 2015. Four Problems, Four Directions for Environmental Humanities: Toward Critical Posthumanities for the Anthropocene. *Ethics and the Environment* 20(1): 67–97.

Nilsson, P. 2017. *Brukade bilder. Södra Skandinaviens hällristningar ur ett historiebruksperspektiv*. Stockholm: Stockholms Universitet.

Opperman, S. and S. Irvino 2016. Introduction: The Environmental Humanities and the Challenges of the Anthropocene. In *Environmental Humanities: Voices from the Anthropocene*, S. Opperman and S. Irviono (eds), 1–21. London: Rowman & Littlefield International.

Parikka, J. 2016. Deep Times of Planetary Trouble. *Cultural Politics* 12(3): 279–292.

Plumwood, V. 2007. A review of Deborah Bird Rose's 'Reports from a Wild Country: Ethics for Decolonisation'. *Australian Humanities Review* 42: 1–4.

Radomska, M. 2016. *Uncontainable Life. A Biophilosophy of Bioart.* Linköping Studies in Arts and Sciences, No 666. Linköping: Linköping University.

Rawls, J. 1971. *A Theory of Justice.* Cambridge, MA: Harvard University Press.

Sloterdijk, P. and Heinrich, H-J. 2011. *Neither Sun nor Death.* Los Angeles: MIT Press.

Steffen, W., W. Broadgate, L. Deutsch, O. Gaffney and C. Ludwig. 2015. The Trajectory of the Anthropocene: The Great Acceleration. *The Anthropocene Review* 2(1): 81–98.

Tanha, A. and D. Zarate 2012. *Landfill Mining: Prospecting metal in Gärstad Landfill.* Examensarbete Linköpings Universitet.

Tilley, C. 2008. *Body and image. Explorations in landscape phenomenology.* Walnut Creek: Left Coast Press.

Tronto, J. 1993. *Moral Boundaries: A Political Argument for an Ethics of Care.* New York: Routledge.

Turpin. E. (ed). 2013. *Architecture in the Anthropocene: Encounters Among Design, Deep Time, Science and Philosophy.* Michigan: Open Humanities Press.

Wikell, R., S.-G. Broström and K. Ihrestam. 2011. Ostkustens första solhäst (The First Sun Horse on Sweden's East Coast). *Fornvännen* 106: 179–88.

Wolfe, C. 2003. *Animal Rites: American Culture, the Discourse of Species, and Posthumanist Theory.* Chicago, IL: University of Chicago Press.

Wynter, S. 2003. Unsettling the Coloniality of Being/Power/Truth/Freedom: Towards the Human, After Man, Its Overrepresentation – An Argument. *CR: The New Centennial Review* 3(3): 257–337.

Yee, A. 2018. In Sweden, Trash Heats Homes, Powers Buses and Fuels Taxi Fleets. *New York Times* 21 September 2018. https://www.nytimes.com/2018/09/21/climate/sweden-garbage-used-for-fuel.html

Chapter 3

The Liveliness of Ordinary Objects: Living with Stuff in the Anthropocene

ANNA BOHLIN

[T]he bodily disciplines through which ethical sensibilities and social relations are formed and reformed are themselves political and constitute a whole (underexplored) field of 'micropolitics' without which any principle or policy risks being just a bunch of words. There will be no greening of the economy, no redistribution of wealth, no enforcement or extension of rights without human dispositions, moods, and cultural ensembles hospitable to these effects. (Bennett 2010: xii)

Introduction

When pondering the future of humanity and the planet there is a tendency to prioritize the 'big' questions such as nuclear waste, artificial intelligence, complex media technologies, human-machine interactions and the like, at the expense of more mundane and humdrum aspects of everyday life. Yet, one of the most pressing issues could well be right under our noses, embedded in the seemingly trivial quotidian practices of living with domestic belongings. The volume of anthropogenic objects on Earth is steadily increasing, having a direct impact on life conditions for a multitude of species on land and in the sea – in the form of waste and toxins – and indirectly affecting the future through the unsustainable use of natural resources required for its production. It is clear that we cannot continue choking the planet with discarded stuff, and that we need to find alternatives (Corvellec 2019; Liboiron, Tironi and Calvillo 2018). However, whenever our relationship to our everyday things is mentioned in public debate, or even when studied academically, a common depiction of contemporary lifestyles is that they are almost completely dominated by fleeting and shallow attachments to consumer goods; that we live in a 'throwaway society', somehow incapable of long-lasting or caring engagements with the things around us (e.g. Bauman 2007; Kennedy 2014; Hulme 2015; see also Gregson, Metcalfe and Crewe 2007 for a critique of the term 'throwaway society').

Drawing on recent debates within critical heritage studies showing how heritage scholarship fruitfully can be applied outside the traditional domain of heritage (Byrne 2014; Harrison 2013, 2015, 2018; Holtorf and Högberg 2014; Olsen and Pétursdóttir 2016), this chapter argues that we need to unpack the abstract and all-encompassing categories of 'consumption' and 'consumer goods' and pay attention to the fine-grained, situated and temporally unfolding human-thing entanglements that they involve. Inspired by vital materialist perspectives (Bennett 2010; Braidotti 2013; Ingold 2012) and scholarship on ruination (DeSilvey 2017; Pétursdóttir 2013) it asks: what is the role of material transformation and decay in such entanglements, and under what circumstances is material liveliness tolerated, or even valorized? When might oldness be preferred over newness, and what might this mean for a post-anthropocentric ethics of expanded responsibility that includes care for nonhuman things and processes (Zylinska 2014)?

The objects in focus in this chapter, domestic items that have been acquired second-hand, were chosen because of their significance in this regard. Deriving their identity from a normative position that privileges things as new and unused, the status of being 'second-hand' in itself draws attention to the temporal and processual nature of objects. The chapter, based on ethnographic fieldwork in Swedish homes, shows that such belongings invited specific forms of usage, and engendered embodied and affective responses that differed from those involved in two other common classes of objects that have received much scholarly attention: newly produced commodities, on the one hand, and conventional heritage objects, on the other.[1] Occupying a loosely defined phase in between these object positions (Rubio 2016), second-hand items are interesting in the way that they induce particular responses to their obvious 'ongoingness' – their material fragility, change and impermanence.

As things already set on a trajectory of being in use, their continued trans-formation is expected. Rather than seeking to control and order emergent and unanticipated properties and events, interlocutors engaged in a somatic 'dialogue' with the things, listening to creaking, or sensing material signs of decay, and responded with a readiness to adjust their behaviour. Responding to this everyday and domestic form of 'ruination' (DeSilvey 2006, 2017) can be seen as a vernacular heritage practice (Appelgren and Bohlin 2017) that entails the capacity to be "attuned to work with, rather than against, processes of change" (Harrison 2018: 1378–9). Significantly, some of these responses involve dynamic, receptive and caring ways of being-with things that contain embryonic, everyday forms of the kind of extended ethics of care that is often called for as a response to critical Anthropocene challenges (e.g. Corvellec 2019; Hawkins 2017).

Anthropological fieldwork, involving interviews and participant observation, was carried out intermittently between 2015 and 2019 in second-hand shops and households in a major Swedish city.[2] The method of ethnographic fieldwork, with its emphasis on 'being there' through immersing oneself into social life, is particularly useful for investigating the entanglement of humans and things. As an open-ended, explorative and embodied form of knowledge production, it engages a broad range of sensory and affective registers and allows for first-hand experiences of the relational affordances of things (Ingold 2011; Pink 2015). In addition to conventional fieldwork, the study also draws on material generated by an experiment in which objects in charity second-hand shops in two major Swedish cities were 'ringed', i.e. equipped with tags, encouraging the new owner to submit information about their involvement with the thing via social media.[3]

The temporality of domestic objects

While various becoming-inspired ontologies (e.g. Bennett 2010; Ingold 2007, 2012) have animated discussions on a broad range of objects and materials, such as waste (Gregson and Crang 2010; Hawkins 2005, 2017; Alexander and Reno 2012), craft (Ingold 2012; Ingold and Hallam 2014), design (Appelgren 2019) and art (Rubio 2016), surprisingly few studies have applied this perspective to the topic of contemporary domestic belongings. Within the field of material culture and consumption studies there is a rich literature on how people in advanced capitalist societies engage with consumer goods, particularly as part of identity processes and as means of negotiating social relationships, status and class (e.g. Belk 1988; Bourdieu 1984; Douglas and Isherwood 1979; McCracken 1988; Miller 1987, 2008, 2012; Shove, Trentmann and Wilk 2009). However, largely missing from this literature is the question of how people relate to such objects over time, and how they perceive and handle their ongoing material transformation, aging and decay (although see Gregson 2007; Gregson, Metcalf and Crewe 2009). In as far as the temporality of domestic belongings has been considered, it has primarily been through a focus on inherited objects, antiques or home interiors (Lee Dawdy 2016; Lipman 2019); on domestic memorializing and curating (Marcoux 2001; Macdonald 2002, 2013); in relation to acts of acquisition and discarding (Gregson, Metcalfe and Crewe 2007; Hetherington 2004; Lastovicka and Fernandez 2005; McCracken 1986) or hoarding and collecting (Bennett 2012; Cherrier 2010; Kilroy-Marac 2016). Another track has been concerned with how objects acquire new meaning and value as they shift between contexts and value regimes (Appadurai 1986; Kopytoff 1986; Hoskins 1998), e.g. as second-hand and vintage objects (Baker 2012; Balthazar 2016; Clarke 2010; Gregson and Crewe 2003; Holland 2018; Jenss 2015; Knowles 2015; for non-European contexts see Tranberg Hansen 2000; Norris 2010). With the noteworthy exceptions of Gregson (2007), who studies the movement of objects within the household, and Gregson, Metcalfe and Crewe (2009), who focus on domestic repair of household belongings (2009), as Domínguez Rubio points out, remarkably few studies have engaged with the question of how people on a daily basis need to deal with the reality that all objects are fragile and unstable material processes. What is missing in contemporary explorations, he argues, is "an account of temporality and change, of the fact that the objects they describe are always being outgrown, betrayed and transformed by the constant unfolding of things" (Rubio 2016: 64).

In order to explore this topic in relation to domestic belongings, the present study focuses on a category of objects that makes up a significant part of an average Swedish household: items that have been acquired second-hand, from charity shops, flea markets and online auctions, such as trinkets, decorative objects, lamps, furniture, books, clothes and similar. The chapter builds on the idea that such belongings are porous and leaky things, involved in a form of 'growing' as they accumulate traces and sociality (Appelgren and Bohlin 2015). While second-hand objects enable domestic and everyday encounters with the past, forming part of a vernacular, 'living' type of heritage practice (Appelgren and Bohlin 2017), the following discussion will focus on a specific dimension of their heritage qualities: those related to their material transformation. For the moment, then, it will set aside consideration of other ways in which such objects are animated, for example the common practice in Sweden of referring to second-hand objects as 'having soul' or some kind of inner being. During fieldwork, people certainly spoke of – and sometimes to – their second-hand objects as if they were somehow living sentient beings, a form of animist relationality that can be compared to the dispersed personhood, or 'dividuals' sedimented semiotically in certain objects observed in a US context (Newell 2014; see also Lee Dawdy 2016 for a discussion of mana in antiques in New Orleans). However, while not distinct from the type of liveliness of materials and things in focus in the present discussion – in fact, they are closely related – a full consideration of them is a topic in its own right, and is beyond the scope of this chapter.

In exploring the material transformation of second-hand objects, this chapter draws on a theoretical perspective proposed by Rubio (2016), informed by vital materialism (e.g. Ingold 2012), but distinct in his conceptualization of objects. While things in this view are material processes unfolding over time, objects refer to particular physical and semiotic positions that things must occupy, and remain within, in order to be legible as specific kinds of objects, and participate in different regimes of value and meaning (Rubio 2016: 62). This perspective takes the fragility and temporality of objects as a starting point, and directs attention to the ongoing work required to reduce the discrepancy between thing and object, i.e. to prevent the thing from breaking out of its object position. While this is a useful starting point, more can be said about the varying degrees to which different types of object positions allow for 'thingness' to be expressed without the identity of the object being threatened. For example, as will be shown, a particular affordance of second-hand objects is that more than most, and certainly more than the world-famous Mona Lisa which Rubio takes as his case study, they are expected to express themselves as unfolding, 'growing' things; this is a quality that forms part of the criteria for the object position of being second-hand.

This takes us to another question: how physical change in itself can be a source of heritage values. Studying processes of ruination and decay, particularly of built structures and landscapes, cultural geographer Caitlin DeSilvey emphasizes the generative and productive aspects of processes of material change and becoming (2006, 2017). She shows how attention to the way natural and cultural substances intermingle in entropic processes can allow for the articulation of new kinds of dynamic heritage stories that map unexpected political, economic and ecological connections. Structural disintegration does not necessarily amount to a loss of cultural meaning but speaks to other ways of experiencing pastness than those typically associated with conventional heritage. In a similar vein, archaeologist Þóra Pétursdóttir writes about abandoned industrial buildings, discussing how the past is differently felt and experienced in them compared with conventional, well-managed heritage sites. While the latter presents heritage objects through circumscribed and predetermined epistemological and interpretative frames, ruins allow for more unmediated and undisciplined encounters with things as they really are, their "unruly thingness" (2013: 46). The key insight for both scholars is that ruins can have heritage values, and be valued in terms of their historical significance, while not being asked to speak to a single, elevated moment of that past. Rather, they suggest a mode of remembrance that is framed around movement, and considers not just past, but also future becoming. While these studies mainly focus on the material decay of abandoned buildings, built structures and landscapes, the following will explore what happens when people live in close bodily proximity to things as they are undergoing change. Thinking of such processes as domestic forms of 'ruination' might help us to critically examine conventions that valorize some temporal states and processes and not others, and to understand the different entanglement of matter, semiotics, desire and affect they involve.

Using

Malin showed me a pair of chairs in her kitchen that she bought from a charity shop. The week before, she had discovered that one of the chairs started creaking because one leg was beginning to break. She said that if it had been a new chair, she would have become very annoyed, or even angry. But because she got it second-hand, she shrugged, and simply rearranged the chairs around the table so that the semi-broken one was standing in a less accessible spot. That way the four chairs still looked nice together, but the fragile one would be used less intensively. "It's just that when sitting on it one needs to be a bit extra careful," she said. As we walked around Malin's home, she pointed to various physical imperfections and quirks in things that she had bought second-hand: threads poking out from the stuffing of an old comfy chair ("a real Ikea from the seventies!"); a table with ring-shaped stains from a hot pot, and curtains that were of a particularly attractive mustard colour that she liked. They were a bit too short, ending a good fifteen centimetres above the floor, but that didn't matter: "These things were made to be used, right? It would be wrong not to continue using them".

This excerpt from fieldnotes describes things and activities that on the surface might seem ordinary to the point of trivial. Yet, by juxtaposing such observations of the way people interacted with their second-hand belongings with reflections on the human-thing entanglement involved in two other well-known categories of objects, new commodities and conventional heritage objects, the following discussion intends to draw attention to some unexpected parallels and differences between them.

A key observation during fieldwork, reflected in the vignette above, was that second-hand things were there to be used. To use Rubio's term, their readiness to be used was part of the implicit and taken-for-granted criteria for the particular 'object position' that second-hand objects occupy. Furthermore, in nearly every household, I encountered variations on the idea that second-hand objects should be used because they have already been in use. In other words, since somebody had already

Figure 3.1 — Malin's kitchen chairs, with the fragile one placed by the window, where it will be used less intensively. (Photograph by Anna Bohlin).

set the object on a trajectory of being entangled in human life, part of its expected 'destiny' as a second-hand object was the continuation of such entanglement. To not allow this entanglement to continue, by refraining from using it, or worse, by throwing it away so that no one else would have the possibility of using it, would be, as Malin pointed out above, 'wrong'. Rather than merely pointing backwards, towards the past, then, typical second-hand physical qualities such as the smell of chlorine from a cabinet that had once stood in a public bath house; the burn marks on a table; or the shortness of the curtains, having been once cut to fit a previous owner's home, simultaneously also pointed towards the future, encouraging and inviting further use.

Usage was also foregrounded in the responses submitted by those who came across items that we have tagged as part of the ringing experiment. One woman emailed the following, drawing attention not just to the way she used her objects, but also to their ability to withstand heavy usage: "All my stainless-steel plates are from second-hand [shops]. They are perfect in the oven, on the grill, to serve on and they can be machine-washed. If the dirt gets too bloody stuck they can even be scrubbed with steel-wool. Survive falls from tall heights (smiley)." Another person posted a photo of a porcelain bowl that she had bought, tagged as part of our experiment, and originally fitted with a clasp for hanging that she had

Figure 3.2 — A photo by an interlocutor, showing a much appreciated quality in second-hand objects: that they can be used intensively, here washed in a dishwasher. (Photograph by Lena Ekelund).

removed. She wrote: "At our place we don't hang it on the wall. Today it was filled with potato-, lentils-, celery- and carrot soup."

Sometimes usage also involved more active interference in the structural integrity of objects, for example through painting or stripping, taking parts from one object to be used for another, or removing, swapping or adding details such as handles or knobs. Creative engagement in objects was particularly visible in responses from the ringing experiment described. In some cases, such engagement would remove an item altogether from one object position and transfer it into another. Examples included one woman who submitted photographs of how she had turned a man's shirt, bought second-hand, into a baby's dress for her grandchild, and a young man who bought a book with Manga cartoons, reminiscent of his childhood, from which he would cut out pages and create a wall decoration.

The active and intensive usage of second-hand items took on a particular salience when contrasted with how people behaved around, and spoke about, items that they had bought new, from 'regular' shops. Such things – clothes, household appliances, china – were often treated with caution, used reverentially and sometimes not at all, stored away in cupboards or other protected places. The reason for this, I was told, was that if such items should break, or get a mark or a scratch, especially soon after they were bought, this would cause intense feelings of disappointment and frustration. Lisa, for example, an engineer in her fifties, and an avid second-hand shopper, out of principle tried to avoid buying newly produced goods, but had made an exception when she bought a new washing machine. She had carefully read up on tests in order to select a good one, and when it broke, she felt "mad with anger". It wasn't just the fact that it had cost a lot of money, she said, but that she felt cheated. Had it been a washing machine bought second-hand that had malfunctioned, she explained that she would simply have become "tired" because of the practical task of needing to repair or replace it, but would have felt none of the anger and resentment.

The potential of newly produced commodities to evoke strong affects and feelings of disappointment and resentment, a recurrent theme during fieldwork, can be understood as directly related to the transaction involved in purchasing first cycle objects (as opposed to second cycle, Gregson and Crewe 2003). Buying a first cycle commodity can be likened to a contract, in which money changes hands, not merely for the exclusive right to the thing, but also for the peace of mind of knowing that the thing, and the materials it is made up of, can be expected to stay stable and unchanging for a certain, often regulated amount of time. In Sweden, for example, it is three years (Ministry of Justice 1990).[4] If, during this time, the thing for some reason expresses its material liveliness in an unexpected manner – a thread poking out, a leg starting to creak – it is not just the integrity of the physical object that is threatened, but also that of the underlying contract that governed the purchase, along with the expectations of material boundedness and stasis that it is premised on.

Second-hand things, in contrast, besides being less expensive, were typically bought with the expectation that they were likely to change and transform. Their status as pristine, new commodities had already been transformed into that of 'second-hand', an object position that in itself entails the temporal notion of becoming or 'growing'; of having a history of being pre-used, and a future of expected further transformation. Changes to the material integrity of the second-hand thing, then, did not necessarily translate into a loss, the way a new object risks quite dramatically changing its position from flawless, valuable and intact to a lesser version, one that is broken, scratched or faded. While a new commodity was of course also designed to be used, its desired stability and fixity, at least for a given amount of time, called for caution, and the ever-present possibility of unexpected and unwanted transformation circumscribed how the object was handled and related to.

A vital characteristic of second-hand objects, then, was that they were easy and undemanding to be with, and invited more relaxed, intimate and less reverential relations than did new commodities. Less was expected of them – for example, the curtains that Malin bought, despite knowing they would be too short, an inadequacy that she would not have tolerated in a new item – but they also demanded less. In particular, they demanded less respect for their present physical states. Rather than trying to prevent or arrest their transformation, people tended to respond to their lively material qualities by adjusting their own behaviour towards them – working with, and actively participating in, their various ways of becoming. Engaging in a somatic 'dialogue' with the items – listening, touching, smelling and sensing their signs of aging – such material changes were experienced as affordances to be responded to, rather than as failures to be kept at bay in order to avoid financial loss and emotional upset. For Malin, for example, the creaking from the chair did not signify the end of a specific stage for the chair, in which it was pristine and unblemished. Listening to the leg creaking she interpreted the sound as a signal that its material properties were undergoing change, as they were expected to do, and this prompted her to adjust her usage and find a solution, temporarily, to its gradual transformation towards disintegration.

Caring

While partly a pragmatic response to the specific temporal and material characteristics of pre-used items – having been in use for longer than new commodities, they tend to fall apart and break sooner – such responsive usage was often animated by a strong ethics of care and responsibility for the longevity of things and materials. The expression such care took, its motivation and the extent to which this was articulated, varied. Many explicitly connected it with concerns for environmental degradation and a sense of urgency in needing to live more sustainably, which included reusing as much as they could, primarily in order to save materials and prevent unnecessary wastage. Others felt more of a responsibility towards the things themselves and were motivated by a protective urge to prolong the objects' lives. One woman described herself as having "hoarding tendencies", and would fit well with Jane Bennett's description of hoarders as endowed with particular sensibilities towards things (2012; see also Kilroy-Marac 2016). Unlike her husband and sons, whom she described as "deaf", she explained that she could hear the calling from objects, an ability that could sometimes be tiring, as well as impractical as she tended to collect too many objects. She described how she used to visit her local second-hand shop on the way home from work, but tried to resist the urge to take things home: "I actively train myself to think 'I don't need to rescue it. It can receive the same care in someone else's home.'"

This woman also happened to be particularly fond of small bowls and dishes in pressed glass, a quaint category of kitchenware that had been out of fashion for some time. For this reason, she used to be able to find it for very little money in charity shops in the relatively low-income area of town where she lived. She believed that most of it had come to the shops after the passing away of elderly people, who often had only washed these items by hand, a procedure that did not quite get rid of the dirt. Hence, the dishes would typically be opaque and dull when she found them in the shop, and she took immense pleasure from putting them in her dishwasher to reveal their sparkling facets. Once, she had bought a number of pressed glass dishes for a few kronor, taken them home, washed them in the machine and returned them to the shop. "It was a way of showing them: this is how nice they can become." Being familiar with the way that such objects tended to be physically transformed by their history, she actively participated in their becoming by revealing their latent material qualities. The example also shows that responsibility for objects was not necessarily connected to personal ownership, but could take the form of care distributed serially, across a succession of imagined future owners (Bohlin 2019).

Figure 3.3 — Pressed glass dishes, washed to reveal their sparkling facets.
(Photograph by Anna Bohlin).

As a mode of being with second-hand things, then, care tended to involve attention to things' qualities as unfolding material processes and directed towards becoming and the future. It typically involved close physical proximity and intimacy, and was characterized by a permissiveness and an openness to unanticipated incidents and surprises. Borrowing terms from Tim Ingold, originally used to describe human-animal relations, care for second-hand things could be said to be characterized by 'relations of trust' as opposed to 'relations of dominance'. Rather than seeking to order, dominate and control their emergent and unanticipated properties, people responded to second-hand objects in ways that allowed for emergent and serendipitous properties and events (Ingold, quoted in Lorimer and Driessen 2014: 174-5).

In contrast, care extended towards new commodities had a different directionality and entailed very different affective responses and dispositions. Whereas second-hand objects were given more scope to affect their own trajectories, the care extended to new commodities was typically directed towards containing, stabilizing and controlling their becoming. The goal of such care was stasis and stability, or even to preserve a specific past stage in the object's trajectory: ideally, the commodity should remain exactly as it had come out of its packaging. Caution, avoidance, distance and control of external factors and potential threats characterized such care, and the associated affect and emotions involved tension and even anxiety.

Heritage values and human-thing entanglements

A curious aspect of the care expressed for new commodities is its resemblance to that for conventional heritage objects, where things are valued for their historic or representative qualities and subjected to various formal regimes of preservation and management. Although the two object positions are located at very different points in the chronology of a thing's being, one at the beginning of its commodity phase – or so it is often imagined, although of course its becoming started long before, in the extraction and production phases – and the other at the end, their respective positions in this process have surprisingly similar implications for how their material liveliness is handled. Both the new commodity and the heritage object require being in a fixed state, with material activity contained, stabilized and, ideally, kept to a minimum. For both, the aim is to prolong this phase for as long as possible, although with the new commodity, due to usage, this phase is expected to be transitional and only last a certain time.

Compared to these two categories, second-hand objects occupy an interesting position in between. They have moved out of the pristine phase, no longer bound by the contract of stability and fixity that the first-cycle purchase entailed, but have not yet entered a phase of formally recognized and regulated heritage value in the conventional sense (a stage that most objects, moreover, never will) when material liveliness must again be strictly controlled. The differing expectations and requirements with respect to material becoming have significant consequences for how human-thing entanglement is handled within each of the object categories. The conventional heritage object is valorized precisely because of its temporal entanglement with human beings, but traces and evidence of this entanglement are typically preserved and frozen in time, rather than allowed to evolve (Jones 2006; Harrison 2013). For the new commodity, in contrast, its value resides in a conventionalized notion or fantasy that human entanglement has not yet begun; all use value of the object remains to be capitalized on. In comparison with both these, second-hand objects allowed for human-thing entanglements that were freer to follow their own trajectories and unfolding, and invited ways of being with things that involve physical proximity and intimacy in the form of bodily, sensory and affective engagement and attunement.

One aspect of such close and bodily involvement with things is that it allows for them to act back on us, enabling us to take in their presence and temporal being. Greg Kennedy, focusing on disposable consumables such as polystyrene mugs and packaging, argues that it is precisely the lack of sensory engagement with such consumables that allows the illusion that they are evanescent and instantaneous. Already cast out of the future the moment we grab them, they are 'imprisoned in immediacy' and deprived of all temporality (Kennedy 2014: 138, see also Hawkins 2018 for a discussion of how, until recently, the liveliness of plastic has been obscured by presentism). In contrast, second-hand objects can be seen as unconventional and unregulated kinds of heritage objects that draw attention precisely to issues of temporality and unfolding. Their heritage value exists in a mixture of the sensory, affective and aesthetic qualities related to their age and ongoing trajectory, and to their particular affordances not just as things that are continually becoming, and are expected to transform, but that also invite us as co-creators of their becoming. As such, they fit well with Harrison's reconceptualization of heritage, not as preservation of that which is about to be lost, but as "collaborative, dialogical and interactive, a material-discursive process in which past and future arise out of dialogue and encounter between multiple embodied subjects in (and with) the present" (Harrison 2015: 27).

As a mode of caring for things and materials, living with second-hand objects also has ethical dimensions. Not forgetting that the circulation of things on second-hand markets may well be implicated in intensifying loops of consumption, offering convenient conduits of disposal of commodities that make room for new purchases (Appelgren and Bohlin 2015; Bohlin 2019), this chapter has tried to highlight some of its other, more hopeful, dimensions. Besides overt ethical commitments to saving resources by ensuring longevity of objects and materials, the examples discussed here also involved more subtle and unspoken forms of responsive, receptive and caring ways of being with things. This resonates with calls for a post-anthropocentric ethics of care that extends responsibility towards nonhuman things and processes (e.g. Hawkins 2017; Zylinska 2014), in particular speaking to the need for an ethics grounded in "curiosity and openness to things' being, how they affect us upon encounter, and which, importantly, takes seriously how they persist, gather and outlive us" (Olsen and Pétursdóttir 2016: 40). To live with pre-used items entails being prepared to adjust one's behaviour to the unpredictable unfolding of things, accept imperfection and to let go of ambitions of control and sovereignty over things – to recognize that "we do not act over a world but exist within it together" (Beacham 2018: 544).

Conclusion

This chapter has highlighted temporal dimensions of living with domestic objects that are significant not because they stand out, but precisely because they are ordinary and often unnoticed, existing alongside other dimensions of consumption that have received more scholarly attention. Yet, despite having been relatively invisible in the academic literature, they contain seeds of understanding ways of being with things that may prove important when responding to acute Anthropocene challenges related to continued mass production, over-consumption and waste. In particular, they complicate the dominant mode of capitalist storytelling that claims the superiority of new over old, and that presents deterioration and change as always negative and undesirable. The examples of second-hand objects show that besides the more traditionally valorized things with history, such as conventional heritage objects, antiques or precious collectables, there is also a broad range of more mundane and ordinary objects that are appreciated partly for their age, but also for other kinds of material-temporal dimensions. Allowing human-thing entanglements that involve freer, less referential and more intimate forms of interaction, pre-used objects have affordances that new commodities do not. While the former evoke sensory engagement, affective attunement and an openness to unanticipated events and surprises, the latter demand a circumscribed engagement that was characterized by control and avoidance.

Finally, some remarks will be offered on the relationship between the broadly posthumanist perspectives adopted here, and the issue of critique. It would be easy to criticize the examples discussed above for failing to adopt what Jane Bennett has called a 'demystifying' perspective, in other words one that aims to reveal for example the webs of class or unequal global power relations that they form part of, or to expose them as practices that obscure and divert attention from supposedly more real or worthy political causes (2010: xiv). Indeed, this kind of criticism has been levelled at posthumanism more generally in a review of, among other texts, the books *Staying with the Trouble* (Haraway 2016) and *The Mushroom at the End of the World* (Tsing 2015) (Hornborg 2017). In his review, Alf Hornborg expresses frustration with what he considers the authors' failure to provide rigorous analyses of what he calls the money-energy-technology complex, and, therefore, to adequately critique capitalism. In his words, "the promotion of posthumanist discourse is ultimately tantamount to looking away while neoliberal capitalism continues to destroy the planet. In other words, it can only serve as a convenient accomplice of neoliberalism" (2017: 67). At heart, this critique gestures towards something Claire Colebrook highlights in her contribution to this volume; namely, at what level in the emergence of life one chooses to situate what counts as the properly political. Do we locate 'politics' primarily in large-scale entities such as class formations, global economic forces, multinational corporations or technologies, the way Hornborg seems to imply we should? Or do we also seek it in the myriad of ways that our own becoming and embodied needs of care depend on a multitude of relations with similarly conditioned things and beings? Through its focus on commodities, consumption and consumer goods, phenomena that are often missing in discussions on the Anthropocene, this chapter has tried to highlight how the intimate, the lived and the ordinary constitute key aspects to explore if we are to more fully comprehend the workings of the money-energy-technology complex that Hornborg rightfully draws attention to. In other words, far from failing to critique our current state of affairs, the kinds of posthumanist insights and perspectives that have inspired this chapter provide us with tools to at least partially start mapping out the connections and formations that are necessary for any critique of neoliberal capitalism to be effective. It is also in this regard that posthumanism opens exciting possibilities for critical heritage studies, which, through its focus on the temporal aspects of any entanglement of the social and material, can help broaden and refine our view of 'the political' even further.

One of the strengths of posthumanism, then, is its openness to the possibilities of going beyond a critical mode, to formulate new, or search for other, unrealized insights, alternatives and ways of being. Inspired by Braidotti's mode of affirmative critique that is both critical and creative (2018, see also Fredengren 2015), this chapter has suggested that if we are to find alternatives to current scales of generating objects, and prevent further damage to the planet, we need to begin by focusing on the embodied and sensory practices of the everyday. Besides being significant in their own right – as sites of micropolitics – it is only through these that more abstract political principles and goals go from being mere ideas to being enacted, as argued by both Bennett (2001, 2010) and Hawkins (2005). Without properly understanding affective dispositions and practices that are open to, and resonate with, the 'right' of things and materials to enter into caring and responsible human-thing entanglements, any calls for an end to wasteful consumerism will remain ineffective. Yet the challenge remains to engage in such affirmative critique while keeping an eye on scalar proportions and providing 'just right' assessments of universality (Zylinska 2014).

Notes

1. In line with Jansen (2016) I do not follow the 'affective turn' in the sense of resisting attempts to find 'closure' or search for patterns, but rather investigate affect in the more modest sense of using traditional tools of hermeneutic interpretation and conceptualization.

2. In total, some sixty interviews were conducted with women and men, from young adults to pensioners. All were initially encountered in connection with second-hand activities, indicating that they had a particular interest in the topic. Yet, they were fairly representative of Swedish urban households in terms of socio-economic background and lifestyle, with a tendency towards middle-class. Interviews were recorded and partly transcribed.

3. In the spring of 2018, about 200 hundred objects (e.g. clothes, porcelain, lamps, decorative objects, toys, shoes, trinkets, and furniture) in second-hand stores run by charity organizations in Gothenburg and Stockholm were "ringed" using two types of tags, made from cloth or pieces of thin, scrap chipboard, laser-printed with a text in a personalized style. The tags were co-designed with then Masters design student Ingrid Furuta, and attached to objects using string or elastic bands. The text encouraged the new owner to send information (picture and text) to social media platforms. At the time of writing, about twenty individual owners had submitted information, the most recent one in May 2019, fifteen months after the ringing. This method was motivated by the insight that experiments can be used to complicate the notion of a neat divide between the controlled environment of the lab, on the one hand, and the unadulterated reality of "the field," on the other, allowing for unexpected epistemological connections and insights (Lorimer and Driessen 2014: 170).

4. Goods can always be reclaimed if 'the fault' is reported within two months. After three years the right to reclamation is lost. Note the essentialist use of the term 'fault' to cover a range of unintended and unexpected material activity of a given object.

References

Alexander, C. and J. Reno. 2012. Introduction. In *Economies of Recycling: The Global Transformation of Materials, Values and Social Relations*, edited by C. Alexander and J. Reno, 1–34. London and New York: Zed Books Ltd.

Appadurai, A. 1986. Introduction: Commodities and the Politics of Value. In *The Social Life of Things: Commodities in Cultural Perspective*, edited by A. Appadurai, 3–63. Cambridge: Cambridge University Press.

Appelgren, S. 2019. Building Castles out of Debris: Reuse Interior Design as a 'Design of the Concrete'. *Worldwide Waste Journal of Interdisciplinary Studies* 2(1): 1–10.

Appelgren, S. and A. Bohlin. 2015. Growing in Motion: The Circulation of Used Things on Second-hand Markets. *Culture Unbound: Journal of Current Cultural Research* 7(1): 143–68.

Appelgren, S. and A. Bohlin. 2017. Second-hand as 'Living' Heritage: Intangible Dimensions of Things with History. In *Routledge Companion to Intangible Cultural Heritage*, edited by P. Davis and M. L. Stefano, 240–50. London: Routledge.

Baker, S. E. 2012. Retailing Retro: Class, Cultural Capital and the Material Practices of the (Re)valuation of Style. *European Journal of Cultural Studies* 15(5): 621–41.

Balthazar, A. C. 2016. Old Things with Character: The Fetishization of Objects in Margate, UK. *Journal of Material Culture* 21(4): 448–64.

Bauman, Z. 2007. *Consuming Life*. Cambridge: Polity Press.

Beacham, J. 2018. Organising Food Differently: Towards a More-than-Human Ethics of Care for the Anthropocene. *Organization* 25: 533–49.

Belk, R. W. 1988. Possession and the Extended Self. *Journal of Consumer Research* 15: 139–68.

Bennett, J. 2001. *The Enchantment of Modern Life: Attachments, Crossings, and Ethics*. Princeton, NJ: Princeton University Press.

Bennett, J. 2010. *Vibrant Matter*. Durham, NC and London: Duke University Press.

Bennett, J. 2012. Powers of the Hoard: Further Notes on Material Agency. In *Animal, Vegetable, Mineral: Ethics and Objects*, edited by J. Cohen, 237–69. Washington, DC: Oliphaunt Books.

Bohlin, A. 2019. 'It will keep circulating': Loving and Letting Go of Things in Swedish Second-hand Markets. *Worldwide Waste Journal of Interdisciplinary Studies* 2(1): 1–11.

Bourdieu, P. 1984. *Distinction: A Social Critique of the Judgement of Taste*. Cambridge: Harvard University Press.

Braidotti, R. 2013. *The Posthuman*. Cambridge: Polity Press.

Braidotti, R. 2018. A Theoretical Framework for the Critical Posthumanities. Special Issue: Transversal Posthumanities. *Theory, Culture, Society* 36(6): 31–61.

Byrne, D. 2014. *Counterheritage. Critical Perspectives on Heritage Conservation in Asia*. New York and London: Routledge.

Cherrier, H. 2010. Custodian Behaviour: A Material Expression of Anti-Consumerism. *Consumption Markets & Culture* 13(3): 259–72.

Clarke, A. 2010. The Second-hand Brand: Liquid Assets and Borrowed Goods. In *Cultures of Commodity Branding*, edited by A. Bevan and D. Wengrow, 235–54. New York: Routledge.

Corvellec, H. 2019. Waste as Scats. For an Organizational Engagement with Waste. *Organization* 26(2): 217–35.

DeSilvey, C. 2006. Observed Decay: Telling Stories with Mutable Things. *Journal of Material Culture* 11(3): 318–38.

DeSilvey, C. 2017. *Curated Decay: Heritage Beyond Saving*. Minneapolis: University of Minnesota Press.

Douglas, M. and B. Isherwood. 1979. *The World of Goods: Towards and Anthropology of Consumption*. New York: Basic Books.

Fredengren, C. 2015. Nature: Cultures. Heritage, Sustainability and Feminist Posthumanism. *Current Swedish Archaeology* 23: 109–30.

Gregson, N. 2007. *Living with Things: Ridding, Accommodation, Dwelling*. Wantage, UK: Sean Kingston Publishing.

Gregson, N. and L. Crewe. 2003. *Second-hand Cultures*. New York: Berg.

Gregson, N., A. Metcalfe and L. Crewe. 2007. Identity, Mobility and the Throwaway Society. *Environment and Planning D: Society and Space* 25: 682–700.

Gregson, N., A. Metcalfe and L. Crewe. 2009. Practices of Object Maintenance and Repair. *Journal of Consumer Culture* 9(2): 248–272.

Gregson, N. and M. Crang. 2010. Materiality and Waste: Inorganic Vitality in a Networked World. *Environment and Planning A* 42: 1026–32.

Haraway, D. J. 2016. *Staying with the Trouble: Making Kin in the Chthulucene*. Durham, NC: Duke University Press.

Harrison, R. 2013. *Heritage: Critical Approaches*. London: Routledge.

Harrison, R. 2015. Beyond 'Natural' and 'Cultural' Heritage: Toward an Ontological Politics of Heritage in the Age of Anthropocene. *Heritage & Society* 8(1): 24–42.

Harrison, R. 2018. On Heritage Ontologies: Rethinking the Material Worlds of Heritage. *Anthropological Quarterly* 91(4): 1365–83.

Hawkins, G. 2005. *The Ethics of Waste: How We Relate to Rubbish*. Oxford: Rowman & Littlefield.

Hawkins, G. 2017. Ethical Blindness: Plastics, Disposability and the Art of Not Caring. In *Living Ethics in a More-Than-Human-World*, edited by V. Kinnunen and A. Valtonen, 15–28. Rovaniemi: University of Lapland.

Hawkins, G. 2018. Plastic and Presentism: The Time of Disposability. *Journal of Contemporary Archaeology* 5(1): 91–102.

Hetherington, K. 2004. Secondhandedness: Consumption, Disposal, and Absent Presence. *Environment and Planning D: Society and Space* 22: 157–73.

Holland, S. 2018. *Modern Vintage Homes and Leisure Lives: Ghosts and Glamour*. London: Palgrave Macmillan.

Holtorf, C. and A. Högberg. 2014. Communicating with Future Generations: What Are the Benefits of Preserving for Future Generations? Nuclear Power and Beyond. *European Journal of Post-Classical Archaeologies* 4: 315–30.

Hornborg, A. 2017. Dithering while the Planet Burns: Anthropologists' Approaches to the Anthropocene. *Reviews in Anthropology* 46(2–3): 61–77.

Hoskins, J. 1998. *Biographical Objects: How Things Tell the Stories of People's Lives*. London: Routledge.

Hulme, A. 2015. *On the Commodity Trail: The Journey of a Bargain Store Product from East to West*. London and New York: Bloomsbury Academic.

Ingold, T. 2007. Materials Against Materiality. *Archaeological Dialogues* 14: 1–16.

Ingold, T. 2011. *Being Alive: Essays on Movement, Knowledge, and Description*. London: Routledge.

Ingold, T. 2012. Toward an Ecology of Materials. *Annual Review of Anthropology* 41: 427–42.

Ingold, T and E. Hallam. 2014. Making and Growing: An Introduction. In *Making and Growing: Anthropological Studies of Organisms and Artefacts*, edited by E. Hallam and T. Ingold, 1–24. Farnham: Ashgate.

Jansen, S. 2016. Ethnography and the Choices Posed by the Affective Turn. In *Sensitive Objects: Affect and Material Culture*, edited by J. Frykman and M. Povrzanovic Frykman, 55–77. Lund: Nordic Academic Press.

Jenss, H. 2015. *Fashioning Memory: Vintage Style and Youth Culture*. London: Bloomsbury Academic.

Jones, S. 2006. 'They made it a Living Thing, didn't they': The Growth of Things and the Fossilization of Heritage. In *A Future for Archaeology: The Past in the Present*, edited by R. Layton, S. Shennan and P. Stone, 107–26. London: UCL Press.

Kennedy, G. 2014. The Ontology of Trash: The Disposable and its Problematic Nature. PhD diss., University of Ottawa.

Kilroy-Marac, K. 2016. A Magical Reorientation of the Modern: Professional Organizers and Thingly Care in Contemporary North America. *Cultural Anthropology* 31(3): 438–57.

Knowles, K. 2015. Locating Vintage. *Necsus: European Journal of Cultures Studies* 4(2): 73–84.

Kopytoff, I. 1986. The Cultural Biography of Things: Commoditization as Process. In *The Social Life of Things: Commodities in Cultural Perspective*, edited by A. Appadurai, 64–94. Cambridge: Cambridge University Press.

Lastovicka, J. L. and K. V. Fernandez. 2005. Three Paths to Disposition: The Movement of Meaningful Possessions to Strangers. *Journal of Consumer Research* 31(4): 813–23.

Lee Dawdy, S. 2016. *Patina: A Profane Archaeology*. Chicago and London: University of Chicago Press.

Liboiron, M., M. Tironi and N. Calvillo. 2018. Toxic Politics: Acting in a Permanently Polluted World. *Social Studies of Science* 48(3): 331–49.

Lipman, C. 2019. Living with the Past at Home: The Afterlife of Inherited Domestic Objects. *Journal of Material Culture* 24(1): 83–100.

Lorimer, J. and C. Driessen. 2014. Wild Experiments at the Oostvaardersplassen: Rethinking Environmentalism in the Anthropocene. *Transactions of the Institute of British Geography* 3: 169–81.

Macdonald, S. 2002. *Behind the Scenes at the Science Museum*. Oxford: Berg.

Macdonald, S. 2013. *Memorylands: Heritage and Identity in Europe Today*. London: Routledge.

Marcoux, J-S. 2001. The Refurbishment of Memory. In *Home Possessions: Material Culture behind Closed Doors*, edited by D. Miller, 69–86. Oxford and New York: Berg.

McCracken, G. 1986. Culture and Consumption: A Theoretical Account of the Structure and Movement of the Cultural Meaning of Consumer Goods. *Journal of Consumer Research* 13: 71–84.

McCracken, G. 1988. *Culture and Consumption: New Approaches to the Symbolic Character of Consumer Goods and Activities*. Bloomington, IN: Indiana University Press.

Miller, D. 1987. *Material Culture and Mass Consumption*. Oxford: Blackwell.

Miller, D. 2008. *The Comfort of Things*. Cambridge: Polity Press.

Miller, D. 2012. *Consumption and its Consequences*. Cambridge and Malden, MA: Polity Press.

Ministry of Justice. 1990. *Consumer Sales Act* (Konsumentköplag). Swedish Code of Statutes 1990: 932.

Newell, S. 2014. The Matter of the Unfetish: Hoarding and the Spirit of Possessions. *Hau: Journal of Ethnographic Theory* 4(3): 185–213.

Norris, L. 2010. *Recycling Indian Clothing: Global Contexts of Reuse and Value*. Bloomington: Indiana University Press.

Olsen, B. and Þ. Pétursdóttir. 2016. Unruly Heritage: Tracing Legacies in the Anthropocene. *Arkæologisk Forum* 35: 38-45.

Pétursdóttir, Þ. 2013. Concrete Matters: Ruins of Modernity and the Things Called Heritage. *Journal of Social Archaeology* 13(1): 31–53.

Pink, S. 2015. *Doing Sensory Ethnography*. London: Sage.

Rubio, F. D. 2016. On the Discrepancy between Objects and Things: An Ecological Approach. *Journal of Material Culture* 21(1): 59–86.

Shove, E., F. Trentmann and R. Wilk (eds). 2009. *Time, Consumption and Everyday Life: Practice, Materiality and Culture*. Oxford and New York: Berg.

Tranberg Hansen, K. 2000. *Salaula: The World of Secondhand Clothing and Zambia*. Chicago and London: University of Chicago Press.

Tsing, A. 2015. *The Mushroom at the End of the World: On the Possibility of Life in Capitalist Ruins*. Princeton, NJ: Princeton University Press.

Zylinska, J. 2014. *Minimal Ethics for the Anthropocene*. Ann Arbor, MI: Michigan Library.

Chapter 4

Folding Time: Practices of Preservation, Temporality and Care in Making Bird Specimens

ADRIAN VAN ALLEN

Introduction

Through crafting specimens and corresponding categories of life, natural history museums have been apparatuses for articulating knowledges, power and natures into an ordered whole, practices that have extended through to contemporary natural history museums and their genetic collecting programmes. This chapter considers the material and semiotic practices of making futures through making bird specimens, drawing on ethnographic and archival research at two national museums of natural history: the Smithsonian National Museum of Natural History (NMNH) in Washington D.C., USA and the Museum National d'Histoire Naturelle (MNHN) in Paris, France.[1] Examining the ways that animal bodies are made and remade at these two sites I explore how they are configured into specific representations of types of time – as windows into ecological pasts, markers of deep evolutionary time or as instruments for future biodiversity conservation policies. Following scientists and their specimens into the workrooms, laboratories and biorepositories of these museums, I learned to stuff bird skins, take tissue samples, extract DNA and assemble genomic data. I suggest that the practices of integrating new technologies into historical techniques is a form of 'folding time'. That is, new modes of making do not merely replace old ones, but instead encapsulate and transform them, folding them into the details of practice. In doing so scientists incorporate not only new materials into specimen preparation methods, they also incorporate new perceptions of preservation, endangerment and care – all oriented towards charting the unified genomic biodiversity of life and preserving it for uncertain futures.

In this comparative study, I examine the origins and implications of preparing bird specimens for natural history collections in the shifting domain of a genomic age. Through making bird specimens and preserving pieces of them in specific ways, museum scientists construct specific types of time that re-inscribe their disciplinary and institutional histories on the one hand, while also constructing imagined future uses for the increasingly abstracted animal-objects they create on the other. This brings to the fore questions of how specific types of time and different temporalities are created, modified, maintained and reproduced in the museum, requiring particular kinds of labour.

Attending to the materials themselves – from feathers and cotton thread, to tissue samples and DNA extracts – highlights the vital role of the materials themselves in the current project of 'archiving' life through biobanking tissue samples and genomic data[2] – projects that frame an understanding of our current ecological crises (Rose and van Dooren 2011; Waterton, Ellis and Wynne 2013) shaping potential futures preserved and understood through frozen materials such as ice cores, seed banks and blood samples (Antonello and Carey 2017; Harrison 2017; Radin and Kowal 2017; Breithoff and Harrison 2020). I suggest that one method for accessing these processes of making time and constructing types of care is through examining the histories, materials and techniques that constitute them, analyzing the negotiations between these different aspects as time is 'folded' together in changing practices on the lab bench.

Both museum research sites – the Smithsonian NMNH and Paris MNHN – have distinct histories that shape their contemporary research agendas, collecting expeditions and specimen preparation practices. Following scientists and their specimens through these two museums, I ask: How are specimen collections made in response to projected futures of extinction, based in the details of material practices? How are these practices then transformed as biotechnology moves into the museum and redefines what is preserved, by what methods and for what intended purposes? Finally, how are specimens used to perform types of time as they move across boundaries, from field to freezer, from lab to collection, from database to scientific publics?

To answer these questions, I focus on two sets of objects used in bird specimen preparation, as a method to unfold the narratives of temporality, preservation and care at work in the contemporary natural history museum. First, I take up the ideas of *preservation* and *care* through the set of tools used in bird specimen preparation, from scalpels and thread to handwritten specimen catalogues. I compare my own experience in the Paris MNHN workrooms stuffing birds with the centuries of bird preservation history at the same institution, within histories that stretch back to the origins of ornithology and to specimen preparation manuals from the sixteenth century. Next, I explore the concept of *temporality* through a survey of the kits each specimen preparator at the Smithsonian NMNH has assembled. Each container holds many of the same items for cutting, cleaning and sewing bird skins, bones and feathers, but also new additions such as cryovials and superglue. Through a close attention to these new materials I examine how types of time – narratives of reconstructed pasts and imagined futures – are shaped by individual makers through the materials themselves. These include birds' skins but also biobanked tissue samples that function not only as indexes to the bird collections, but also as an index of the current crisis to preserve biodiversity in the face of mass extinctions. This ethnographic study examines how scientists work to integrate emerging technological structures, such as genomic collections and globally dispersed data, while also maintaining continuity with disciplinary pasts.

Through learning to use these sets of tools to skin and stuff bird skins, take tissue samples and log specimen data I examine the material practices used by scientists to craft specimens, reify histories and construct futures. Using specimen preservation techniques that have changed little over centuries, I explore how scientists are working to integrate new biotechnologies into existing practices. Within the context of these longer histories of specimen preparation, I argue that as birds are taken apart and reassembled in the museum, they articulate different purposes with different pieces. Further, I suggest that the capacities or limitations of the materials themselves are a vital component of these future-making practices – what materials are used to construct specimens' shapes not only the form of the stuffed bird, but its perceived potential and imagined future utility. The body of the preserved bird is then assembled not only from biological materials such as feathers and bones, cotton and thread – but also from the invested interests of the makers and their visions of caring for the future that they fold into the making of their specimens.

Practices, care and preservation,
or how to fold time in the museum

Linear, accelerating time has been a marker of conventional Western views of temporality, which in turn has shaped both cultural and natural heritage conservation efforts (Fabian 1983; Harrison 2015; Radin and Kowal 2017). In this research I suggest that the different dispositions towards time that each specimen preparator brings to the worktable shapes the specimens they are creating. They describe the specimens they prepare as needing to persist through "long and uncertain futures" (Van Allen 2017: 534). Intent on making something that will last for centuries, the move towards freezing tissue samples of birds to preserve them echoes previous salvational efforts, such as Joanna Radin's (2013) work on the 'latent futures' conserved in human blood samples taken from Indigenous populations. Frozen in time, the collections of frozen tissues carefully created and preserved in museum biorepositories do much work. These include orienting ideas about what constitutes proper methods of preservation for specimens – that is, what should be kept and for what (imagined) future uses – as well as what these specimens stand for in both taxonomic communities and for larger publics.

The current epoch of the Anthropocene serves as a frame in which human beings have become a primary force in shaping ecological worlds on a geologic time scale, stakeholders in what Deborah Bird Rose has called the 'ethical time' of human decisions shaping the fate of multispecies ecosystems (Rose 2012). This shift in time, bringing humans into contact with immeasurably long expanses of geologic 'deep time' (Ginn et al. 2018) on the one hand has also made immanent futures of cataclysm and loss starkly visible (Tsing et al. 2017). This framework of both deep time and immanent loss has shaped practices of care for those in peril, from endangered species, to ecologies, to optimistic futures. If we take the various discourses on the Anthropocene and its iteration of -cenes (e.g. Crutzen and Stoermer 2000; Haraway 2016; Harrison 2015; Lorimer 2015; Neimanis, Åsberg and Hedrén 2015; Smithsonian Institution, Living in the Anthropocene Consortia 2015; Tsing et al. 2017) to all, in some way, centre on the subject of marking time, then I suggest that museum specimens – and specifically museum collections documenting historical sequences – can be conceived as a form of 'core sample' through these narratives of (albeit, linear modernist) time. They do so as biological samples collected over time, as was their intended purpose for the scientists who originally collected them, yet importantly they are also a chronology of the *techniques* used for 'archiving' life. Further, they function as an archive of the orienting concepts of preservation and care bundled within those techniques.

The importance of materials and an attention to techniques has long been a part of Science and Technology Studies (cf. Law 2010; Latour 1988) as well as anthropology, particularly in the following of the *chaîne opératoire* (or 'operational chain') of how materials and ideas are wound together into processes (cf. Dobres 1999; Lemonnier 1986). However, I suggest that while it is important to attend to what people say versus what they do with materials, it is equally important to go further and examine what the materials themselves can communicate when engaged firsthand (Mol 2003).

For the scope of this chapter I focus on the practices used for producing bird-objects in the museum, unravelling their multiple biographical trajectories and their roles in producing multiple kinds of knowledge (Knorr-Cetina 1999; Rheinberger 1997). Moving into the behind-the-scenes museum workspaces where matter and meaning were woven together in the daily routines and techniques, I explored the intimate and fluid connections between the minutiae of biological organisms, their tissue samples, their data, their DNA and the embedded visions for shared human and non-human futures. I observed the ways in which the taxonomic community in the museum, inheritor of several centuries of specimen collecting bound up with centuries of colonialism, now finds itself caught up in the changing landscapes of several wider intersecting domains. These include the genomic life sciences, biodiversity policy and the increasingly fraught activity of collecting and transporting specimens (now categorized as 'national biowealth' after the ratification in 1992 of the Convention on Biological Diversity and Nagoya Protocol in 2014) across international borders.

Over the course of several years (since 2014 at the Smithsonian and 2017 at the MNHN in Paris) I interviewed and worked alongside curators, conservationists, collections technicians and visiting researchers. We carefully skinned and stuffed birds, took tissue samples, extracted

DNA, sorted genomic data and carried taxidermy through narrow corridors. This provided me with 'a view from below' (Harding 2008), that is, access to the wealth of mundane details and occasional moments of epiphany that constitute the ongoing labour of museum work.

While an attention has long been put on practices – what people do, and how they speak about those actions – I advocate for a more articulated and embodied engagement with precisely how 'matter comes to matter' (Barad 2003) not just by observing it, but by transforming it with one's own efforts. Doing so hopefully transforms one's understanding of how the various and shifting worlds we inhabit are continually made and remade. The materials themselves can be a subject as they offer up their own narratives, with the qualities of materials either bending to the task at hand or fighting every step of the process. Materials have much to tell, not only about their life histories, their capacities, their limitations, but also about one's own physical and psychological engagement with the research site. As fingers slide over blood, bone and feathered skin, triggering a tinge of nausea or feeling the ache in your wrist from pipetting extracted DNA – these actions have the potential to shift one's understanding of the maker, the animal-turned-object, and their co-constituted process of coming into being.

Previously I have argued that scientific objects such as genomic tissue collections and museum specimens were far from neutrally composed (Van Allen 2017, 2018, 2019a, 2019b). This builds on scholarship that suggests the complexities of the biographies of scientific objects could contribute to the value they accumulate (Bowker 2000; Ellis 2008; Sunder Rajan and Leonelli 2013), taking on and performing multiple layers of meaning and value as they travel in and between different sites, communities of practice and epistemic expectations. This transformation of practices, integrating new materials and techniques and transforming others, is what I call 'folding time' as histories fold into the present, as imagined futures shape the human and non-human subjects of the natural history museum.

Orienting museum collections towards the future

By the end of the twentieth century, taxonomy, like most areas of the life sciences, was being reframed by state-of-the-art genomic approaches, increasingly relying on DNA and molecular techniques and powerful computer technologies to 'split' or 'lump' the natural world. Taxonomists themselves were classified as either 'splitters' or 'lumpers' depending on their propensity to either divide species into an increasing number of sub-species or mass them together into one species 'lump'. These methods of analysis built on the 1990s 'genome revolution', reifying an underlying assumption that molecular genetic sequences were the code of life, deterministic of all that was imagined to follow (Fujimura 2003; Parry 2004; Waterton 2010; on museum genomics see: Murray et al. 2011; Nachman 2013). This genomic reframing enabled an acceleration of taxonomic practice and allowed the discipline to expand its scale and speed of knowledge production, to collect, categorize and build a map of all life on Earth. As one curator put it, "We have to know what there is before we know what we have to save" (Van Allen 2016: 237).

With the influx of biotechnology into museum practices, museum collections are currently being reframed as a resource now available for big data science. Natural history collections have been accumulating for hundreds of years, with the amount of "untapped biodiversity resources" (Kress 2014: 1310) compressed into museum collections, botanical gardens and university collections estimated to be as high as three billion specimens (Bi et al. 2013; Hykin, Bi and McGuire 2015) – which "suggest the magnitude of this storehouse of information about the natural world" (Kress 2014: 1310). However, I would argue that this 'storehouse of information' has been configured in a specific way, based on the specific cultural histories that formed it, which in turn have shaped the kinds of information it can then offer up. Or, more precisely, that it can be *conceived* of offering up. "Our predecessors in [the Division of] Birds collected these specimens, they had a very specific idea of what they were going to be used for," a curator in the Smithsonian NMNH Division of Birds told me, "Now we use them for things they never could have imagined."[3] When I ask her what future uses she can imagine the collection being put to, she turns over the bird skin she's holding in her hands before replying, "We can't know, of course, what direction technology will go. But we can prepare things in different ways – like pickling the specimen [preserving in alcohol] so the entire organism stays intact, making sure we don't lose anything, or at least we will lose less... For the future, we just have to be very detailed in the data, make sure we keep it all connected, record everything... You never know what might end up being relevant."[4]

Much of the current scientific understanding of several recently extinct species – such as the Tasmanian tiger or Thylacine (*Thylacinus cynocephalus*), the Caribbean monk seal (*Neomonachus tropicalis*) and the passenger pigeon (*Ectopistes migratorius*), to name but a few – have directly resulted from genomic information extracted from museum collections (Miller et al. 2008; Schipper et al. 2008). This includes not only genomics but other kinds of extractions and abstractions of bio-materials: "combining DNA-, amino acid- and isotope-based analyses of a few grams of bone from a historical specimen of an endangered Pacific seabird, the Hawaiian petrel, has illuminated aspects of the bird's diet, past population demographics, food chain dynamics and the deleterious impacts of industrial fishing on this oceanic predator" (Rocha et al. 2014: 814). From this perspective, museums are being recast as unparalleled – and largely untapped – resources for creating genetic collections of extinct species, part of large-scale genomic studies of animals and plants.

Potential future needs, as imagined by curators and preparators, compel museums to continue collecting and preserving for as-yet-unknown uses. As the "common language of the biological sciences" (Kress 2014: 1310), collections not only speak for the past, but must be maintained and added to with new specimens to speak for the future as well. Although most museum specimens were not originally collected for the purposes for which they are now used, new technologies will "continue to reveal new information previously unanticipated in scientific specimens" (Hykin, Bi and McGuire 2015: e0141579). According to many at the Smithsonian (Rocha et al. 2014; Sholts, Bell and Rick 2016) and beyond (Droege et al. 2013; GGBN 2015) the collections need to be added to – 'extended' with genomic samples that are tied to the bird body they came from – to maintain their value and 'keep in time' with the time series already marked out by the existing collections. For instance, DNA extracted from the toe pad of a bird skin collected a century ago can now be sequenced and compared with one collected last year, or one living in a zoo (Grealy et al. 2017).

As scholarship in both the biological sciences (Pyke and Ehrlich 2010; Suarez and Tsutsui 2004; Winker 2004), history of science (Daston 2004; Strasser 2010) and in science studies (Fujimura 1996; Kohlstedt 2005) have shown, many scientists continue to use collections to discover, describe and document plants and animals with traditional methods – such as the stuffed bird skins I learned to produce. However, the application of new technologies to study specimens is expanding, becoming integrated into the traditional practices, or in some cases disrupting them. In the following section I take up the idea of preserving specimens as a method of care for both endangered species and as a way to care for uncertain futures. I do so through a set of tools used to make bird specimens, comparing my own experience making birds in the context of techniques, tools and histories that stretch back to the origins of ornithology.

Making birds, preserving histories, constructing futures

Paris, February 2018. Sitting at a large communal table, with tools and materials piled in the centre (cotton, sawdust, scissors, scalpels, tissue tubes) I skinned and stuffed birds with a group during weekly sessions at the Muséum National d'Historie Naturelle (MNHN).[5] We sat next to a freezer where the tissue samples we cut from the birds were stored. Liver samples were being kept from the falcons for a study on heavy metal contaminants, feathers were pulled from others for a plumage study and a combination of liver-heart-muscle were carefully snipped and stuffed into 2ml cryovials for as yet unknown future uses, an expansion of the collection of stuffed skins and taxidermy mounts into the realm of the molecular (Figure 4.1).

The process of crafting a bird study skin broke down into a series of processes of measuring, skinning, sampling, washing, drying, stuffing, sewing and pinning out to dry. The third of these – sampling – was the most recent addition to the workflow, while most of the process remained little changed from the procedures set out in historic manuals centuries earlier (Belon 1555; see also: Farber 1977). As I stepped through these processes, I felt a certain kind of vertigo as my hands went numb in the thawing bird blood – a visceral reminder of the formerly living thing I was slowly dismembering. Past and future both seemed very present: as my hands pinned out feathers which looked like a vignette out of one of the historic specimen preparation manuals, I experienced a link to the past and a continuity with histories of collecting (Kalshoven 2018). Various futures also seemed immanent through the potential uses for the tissue samples I was taking, coupled with the idea that I was making something that would be kept in perpetuity, tended to for unknown decades or centuries.

As I prepared a different bird each week – weighing and measuring, pulling the feathered skin off of the body, washing and drying it with a hairdryer, stuffing it with cotton, sewing it closed, pinning it to dry – I

Figure 4.1 — Preparing study skins, Paris MNHN Department of Birds, 2018. (Photograph by Adrian Van Allen).

recognized the same techniques as described in the sixteenth- to nine-teenth-century French taxidermy manuals I had been reading in the museum's archives. Printing out pages from these manuals, I brought them to a weekly session and asked the bird preparators about their sense of time, history, heritage and tradition. "I did not realize how little things had changed," one said as she pulled a page towards her, showing a taxidermist's tools from an 1853 manual. "These look so familiar, just like I use now,"[6] she said as she sorted through the tools on the table in front of her – a pair of scissors, a scalpel, tweezers, linen thread and wads of cotton – aligning them with the image from the antique taxidermy manual (Figures 4.2–3) (Brown 1853: 71).

As we compare the tools in front of us to the image from the book, she describes how she still uses each of these tools in her process of making bird specimens, and why she spends so much time on the details. She explains there are many moments in the process of specimen preparation when one can "go quickly and the bird skin is stuffed and pinned to dry, very fast," she said, but when going fast "there are consequences."[7] These

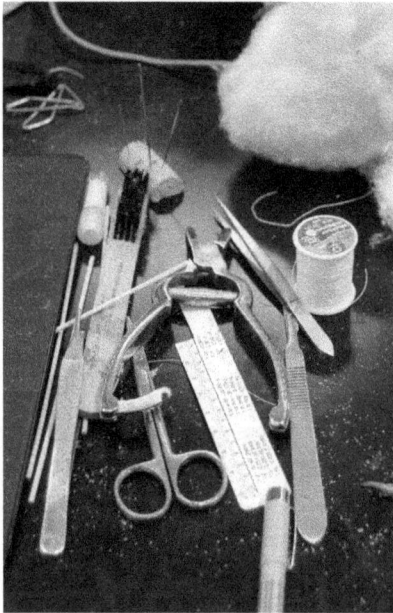

Figure 4.2 — Preparators' tools, circa 2018. (Photograph by Adrian Van Allen).

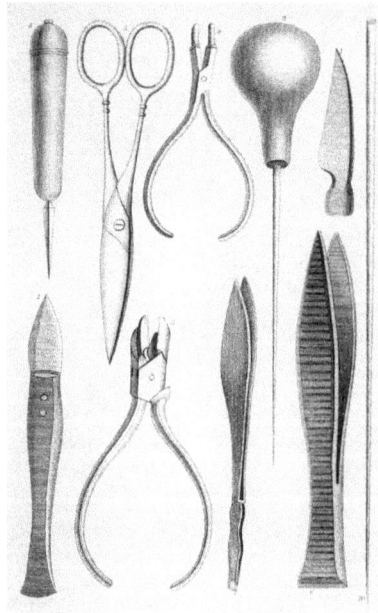

Figure 4.3 — Preparators' tools, circa 1853. (Brown 1853: 27).

consequences are not only for the longevity of the individual specimen, but also have the potential for pest problems that can destroy entire collections.

To 'make specimens last as long as possible' and not provide a food source for insects involves a series of time-consuming tasks. These include using a needle to clean marrow from leg and wing bones, puncturing toe pads with a small needle to drain the fat, cutting the delicate skull open with scissors, slicing out the brain with a scalpel and using bits of cotton to carefully clean out any remaining brain matter. "The insects they love to eat fat, and the brain – it is all fat – so it must go."[8] The tools used for each of these processes have persisted through time, as they align with the continuing intent to make specimens that will last for a century or more. Taking the tissue sample for the frozen tissue collections also uses the same tools: a scalpel to cut off a section of muscle and tweezers to slide it into a small plastic tube – the intent is still to make a (tissue) specimen that will last indefinitely. While these tools are used to preserve specimens for unknown futures, they also preserve parts of the past.

Continuity with the past then stretched into the future. I kept this sense of long histories and longer futures in mind as I learned to skin, measure, sample, wash, dry, stuff, sew and pin my bird study skin. Sewing up the bird, I thought about the other kinds of 'conservation' done in the museum with needle and thread, and what versions of time they create or recreate. How is sewing up a bird skin like conserving other sewn objects in collections?

As I sew the bird skin closed, I think about a conservationist at the Victoria and Albert Museum in London who analyzed her process of repairing a seventeenth-century Spanish Mantuan gown – the tailored bodice modified over generations to fit the bodies of new owners and new styles. She questioned which version of stitches was the 'authentic' one, arguing that each series of actions she took in 'conserving' the object was also an ethical choice and construction of a specific 'truth' made in relation to that object (Malkogeorgou 2011: 442). She sees each stitch she saved or removed as an ethical choice about how to preserve different "life histories" (2011: 443) of the garment and its wearers.

I would argue this can be carried over to the bird skins and tissue samples on the lab bench, where each decision about producing a specimen – be it morphological or molecular – can be thought of as the same process of ethical choices made in relation to constructing 'authentic' life histories for these animal-objects. The choices made determine (and are determined by) contemporary concepts of value and use – what is removed, what is saved and how what remains is classified. The museum scientists also reflect on the changing values of specimens, as mentioned earlier, where they contemplate that their scientific predecessors could never have imagined what use they would be making of specimens collected a hundred years ago, such as sampling the study skin of an extinct species for DNA.

Continuities with past practices

Paris, October 2018. Re-inscribing continuity with past practices, as well as constructing predictive futures for objects, is part of the daily work of museum staff. At the MNHN large paper books are still used to catalogue their specimens, marking down DNA samples in the same margins that once noted seizures from the Versailles Menagerie during the French Revolution. In recent years this data has also been duplicated on databases, but the entries are still meticulously entered on paper in black ink (Figure 4.4). The nineteenth-century catalogues record the same data as

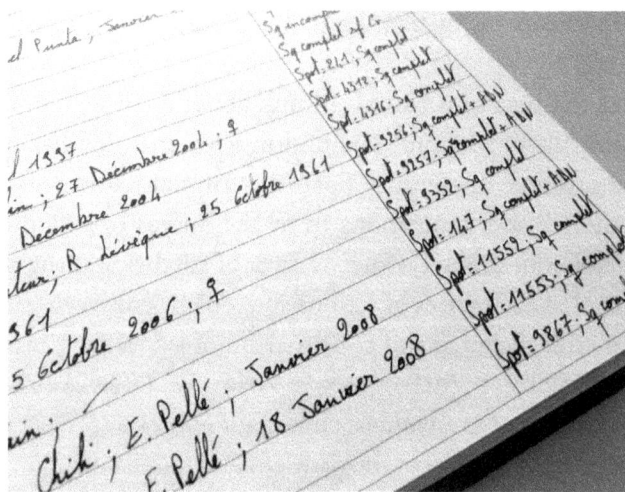

Figure 4.4 — Paper catalog books at the MNHN Department of Birds, 2018. The notes for DNA ('ADN' in French) are visible in the margin. (Photograph by Adrian Van Allen).

2018 entries: name and date collected, locality, but the type of preparations from the specimen now include not just a skin, a skeleton, but also a
DNA sample, carefully penned into the narrow margin. Different curators' lifetimes are spanned across these pages, the collections they created
marked in their own handwritten script.

"It gives me a sense of tradition, part of the history of this place," as
one curator described his use of the catalogues, "and it will always be
there… just in case something happens to the database."[9] Like the taxidermy manuals from the sixteenth century with illustrations of tools
identical to those on the workbench in 2018, these paper catalogues offer
a view into a continuity with past practices – a record of how museum
staff find ways to weave new types of data into existing infrastructures
even as those objects shift status within the museum.

Tracing the continuities and changes for preserving birds in the taxidermy manuals and the specimen catalogues traces the shifting value of
a bird carcass. Once used to map the diversity of living things, preserved
birds have now become sites for care in the face of mass extinction. The
museum practices of collecting and cataloguing individuals of a species,
I suggest, are conceived as a form of caring for the species as a whole. For
example, one scientist articulated the act of collecting and sampling an
individual bird for the heavy metals in its liver tissue as useful for shaping
environmental policies, which in turn could potentially help preserve the
species (Berny et al. 2015). Museum collections have also been used to
compare historical and contemporary biodiversity surveys, with the collections mined for past ecological changes and the ensuing impacts on
endangered species, and the data then used to inform policy decisions on
conservation strategies (Hanner, Corthals and DeSalle 2009). In doing
so, scientists articulate their specimen collecting and preserving as salvational tools, that is, the collected dead in museum collections are being
used to preserve their living kin in the wild.

Preserving birds and shaping ornithology

Pierre Belon's *L'histoire de la nature des oyseaux*, published in Paris in
1555, is one of the earliest known examples of preservation techniques
for making bird skins into objects for scientific study, and includes

examples of exotic birds such as parrots collected during French expeditions (Belon 1555; Schulze-Hagen et al. 2003) – birds that still cluster on the shelves of the collections (Figures 4.5–6). Most of the bird specimens prepared for early cabinets of curiosity were of articulated bird skeletons, eggs, feathers or dried wings – parts of birds that were less prone to sunlight, humidity and insects (Farber 1982; Pomian 1990). As techniques for preserving the delicate materials of skin and feathers were developed, whole bird skins were preserved in more life-like poses, wings spread as if in flight or perched on wooden stands. Walking past the taxidermy workrooms at the MNHN in 2018, I glimpsed a scene that could have been from 1718 or 1918 – workbenches filled with bones and pliers, paintbrushes and glue, feathers and glass eyes. 'Nature' in the process of being crafted, in intricate detail.

The history of taxidermy – which comes from the Greek words *taxis* ('arrangement') and *derma* ('skin') – used complex techniques to preserve a life-like appearance of animals, evolving slowly over centuries of experiments with mummification, drying and chemical preservatives (Péquignot 2002; Wonders 1993). Yet the process in the European tradition for preserving a bird skin has remained essentially unchanged for the past 400 years. The instructions for making a bird skin I read in Belon (1555) or Buffon (1800) were almost identical to those I learned at the workbench in the MNHN in 2018, from using the same materials (salt, sawdust, cotton) to the tools (knife, tweezers, needle and thread).

In brief, the process involves: (1) opening the bird body and removing the flesh, fat and viscera, including the eyes and brain – anything that would attract mould or insects, (2) rubbing the insides with salt or another preservative, historically ashes or pepper were used to deter insects before the invention of arsenic soap in the eighteenth century, but this toxic substance is no longer used in current practice, (3) stuffing the bird skin with cotton, straw or wood wool (shredded wood fibres) wrapped around an armature of wire, wood or plaster, (4) sewing it closed and arranging the body and feathers into the desired shape, and (5) setting it to dry, sometimes pinned in position or bound with thread to hold the feathers in place during the drying process.

However, these life-like poses slowly gave way to a different form, what is called a study skin or round skin (Figure 4.7). In this form, the

Figure 4.5 — Green parrots in Pierre Belon, *L'histoire de la nature des oyseaux* (1555: 298–99).

Figure 4.6 — Taxidermy mounts of green parrots (MNHN Zoothèque, 2018). (Photograph by Adrian Van Allen).

bird is laid flat on its back, wings tucked underneath, legs crossed and instead of glass eyes the sockets are filled with balls of cotton, and most importantly a label is tied around the bird's foot noting among other pieces of data its species name, location and date where it was collected and who collected it. The process of transforming a bird into a specimen is bound to this label, as a preserved bird skin without data becomes, in the words of one museum staff member, "very expensive compost"[10] – without it the bird isn't viable for scientific purposes.

The bird skins I stuffed at the MNHN were within this long tradition of making bird bodies into tools, a process of standardizing the imperfect animal body into a symmetrical, regulated object – one that could reveal relevant data, or as one ornithologist said, "to see variation looking down a drawer."[11] In other words, the form of a study skin as tool for scientific study allowed unusual characteristics to stand out when looking across a row of similarly stuffed birds. These details are important, as the material properties of making specimens is the foundation for understanding how

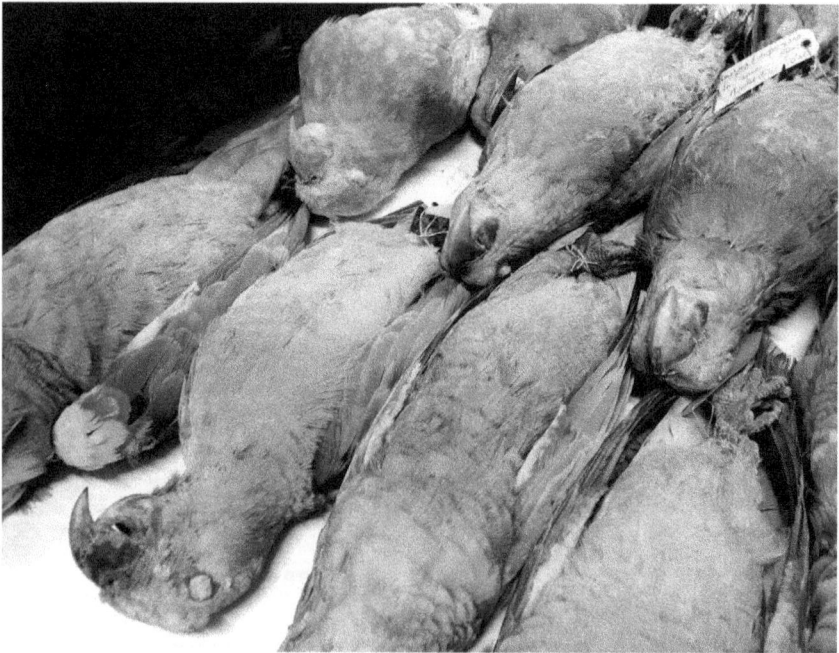

Figure 4.7 — Blue and yellow macaw specimens made from birds who once inhabited the Menagerie at the Jardin des Plantes, Paris (MNHN Department of Birds, 2018). (Photograph by Adrian Van Allen).

they were crafted to persist through time and produce specific kinds of knowledge.

Preserving feathers and skin, which by their nature are prone to degradation in humidity, sunlight and being eaten by insects, offers a considerable challenge. The same concerns for collections and their susceptibility to damage were voiced in the workrooms of the Paris MNHN during my fieldwork in 2017–2018 as well as in notes in some of the earliest taxidermy manuals from previous centuries (Belon 1555; Buffon 1800; Dufresne 1803; Turgot 1758). Crafting a bird body to persist through time presents a challenge, and it seems, always has. Yet the emergence of ornithology as a discipline is bound to the material production of these specimens – in their embodiment as representations of theoretical views of nature and forms of life. And they continue to be vital to the re-imagining of the natural world, as they are transformed from exotic curiosities to scientific tools, from objects marking the historical development of ornithology to resources for biotechnological mining.

Turning now to the Smithsonian National Museum of Natural History (NMNH) and a survey of the specimen prep kits assembled by technicians, I examine how types of time – narratives of reconstructed pasts and imagined futures – are shaped by individual makers through the materials themselves, thinking through the narratives the materials can tell and the stories to which they bear witness.

Histories of collecting and folding time: A bird skin as a time capsule

Washington DC, June 2015. I am in the Division of Birds at the Smithsonian National Museum of Natural History (NMNH). Long corridors lined with white metal cases stretch out into a labyrinth, stacked to the ceiling and filled with bird skins, nests, eggs and wings neatly arranged in rows. More than 640,000 bird specimens are housed here – the third largest bird collection in the world (Smithsonian NMNH, Division of Birds 2016).

In the Vertebrate Zoology Preparation Lab, I sit with a group of specimen preparators at a long table, surrounded by half-frozen birds taken from the freezer, our tool kits and in the centre of the table a communal bag of cryotubes and biorepository labels with individual barcodes.

Preparing a traditional bird study skin, according to several curators in the Smithsonian's Division of Birds, is like "keeping the wrapping paper and throwing away the gift."[12] That is, the most valuable 'body' of information is the actual bird's body, which is removed from the skin and discarded (Figure 4.8). The hollowed-out skin is then stuffed with cotton wool and sewn shut, the spine replaced with a wooden dowel. Some bones remain, such as the skull, some partial wing and leg bones. However, what is considered most valuable at this particular moment – taking samples of their heart, liver and muscle tissue to mark their existence in a specific point in time – were traditionally discarded. In current practices, tissue samples are taken and carefully preserved after which the majority of the body is disposed of as biological waste. However various curators, collection managers and technicians in Vertebrate Zoology are going, as one curator called it, "the way of the fishes."[13] The Division of Fish and Reptiles have traditionally 'pickled' their specimens, that is, 'fixed' them in formalin and then preserved them in sealed jars of either 70% or 90% ethanol. This has the advantage of preserving the entire organism, including its digestive tract and the organism's last meal. The preserved specimen then becomes a tiny microcosm of its environment, a moment stopped in time that preserves epigenetic (microorganisms both in and on the bird) as well as genetic data for future research.

"Time capsules, that's what collections are," one curator told me. "A window back in time, if you know how to get what you need out of them. And, of course, if you can get the permission to get it out in the first place."[14] The pickled specimen can be X-rayed, CT scanned or genetically sampled later – though the DNA may be fragmented by the formalin and requires special techniques to stitch the sequences back together and produce 'meaningful data'.[15] While these practices in different zoological disciplines have long histories, which have shaped their ways of making and ways of knowing, they each concentrate global biodiversity into a museum setting where the meaning and value of each specimen is negotiated.

In the prep lab, we pulled out piles of cigar boxes and Tupperware from the cabinets – individual kits of tools used for specimen preparation. Some were passed down, I was told, as people retired, but most people arrived with their own kit that they had assembled over time

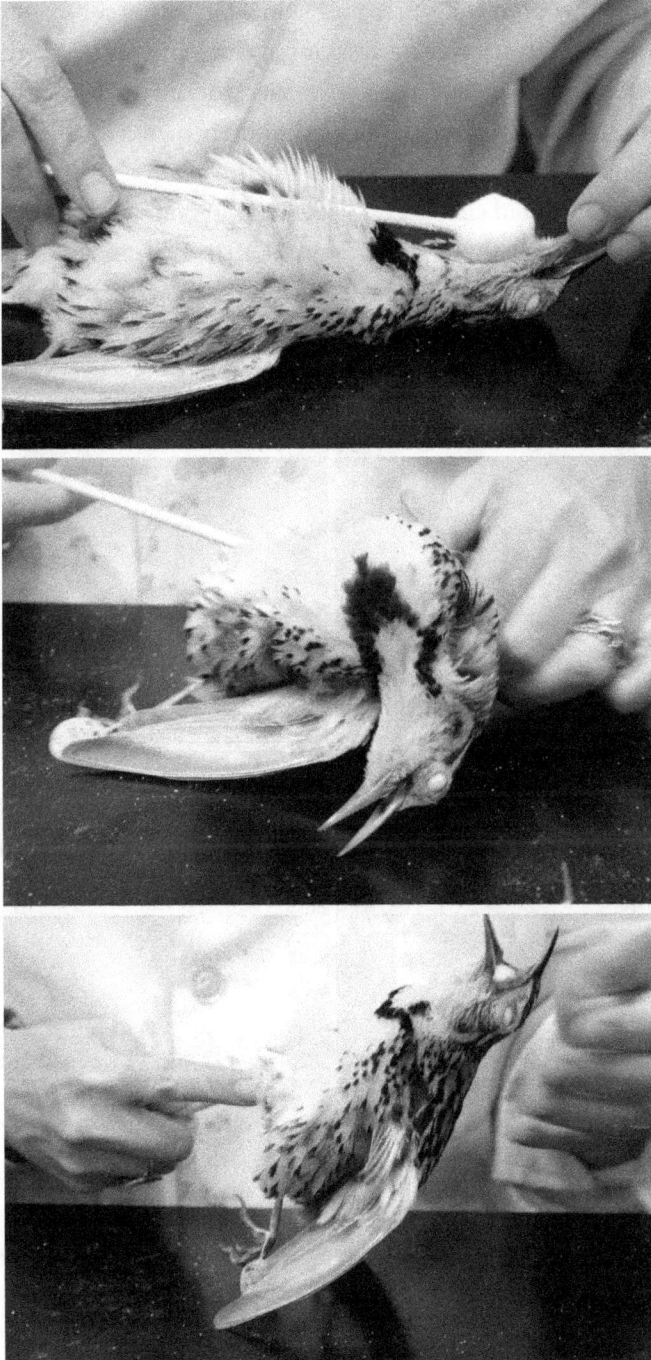

Figure 4.8 — Making a meadowlark (Vertebrate Zoology Prep Lab, Smithsonian NMNH, January 2015). (Photographs by Adrian Van Allen).

from learning to prep their first bird skins. Instead of a box of communal tools pulled out for weekly sessions at the Paris MNHN, here at the
Smithsonian everyone had their own kit, stocked with their own sets
of selected tools and materials, shaped to their own habits and views of
what was needed to prepare bird skins (Figure 4.9).

Opening up an old cigar box, the owner and I unpack the collection of
tools inside and lay them out (Figure 4.10). They include traditional
tools such as an array of scissors, scalpels and tweezers, linen thread and
sewing needles, an awl and paper tags ready to be tied around a bird's leg
to label it with an identifying collection number. However, we pull out

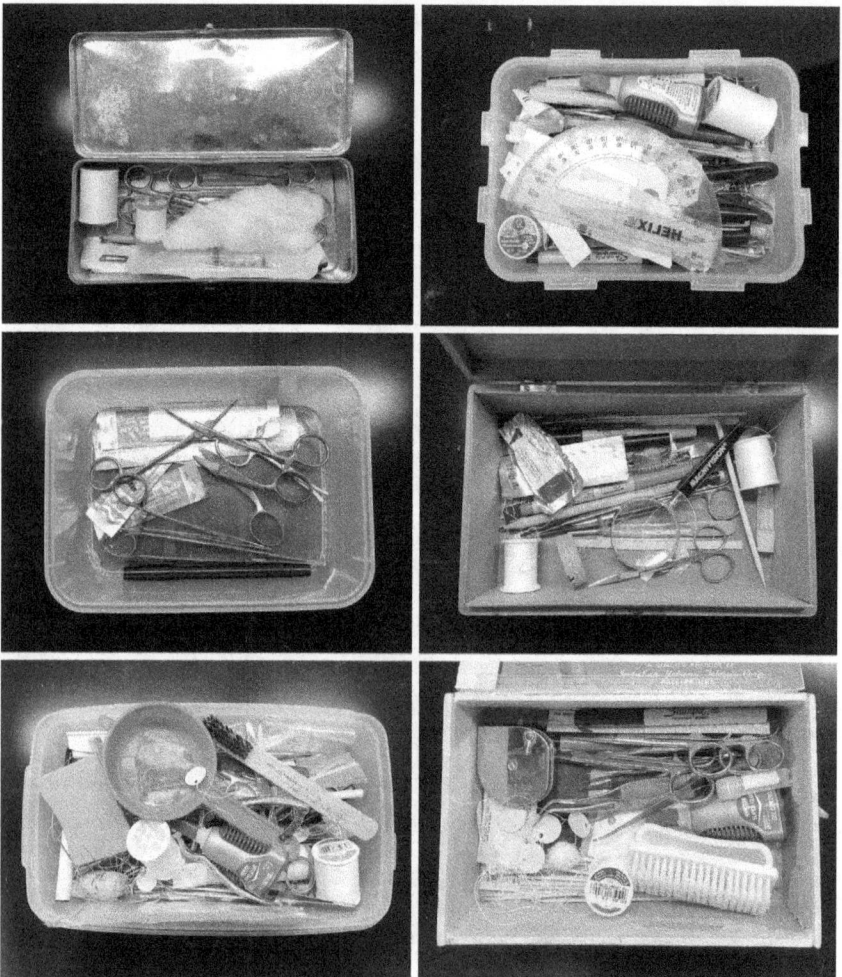

Figure 4.9 — Specimen preparation kits. (Photographs by Adrian Van Allen).

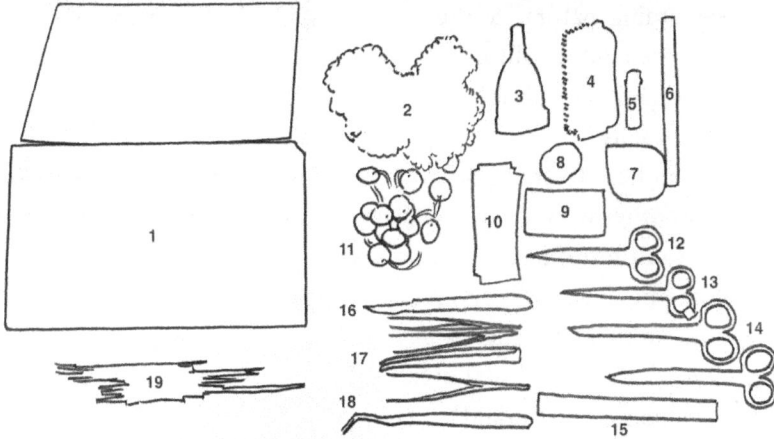

Figure 4.10 — Items in the specimen preparation kit: [1] cigar box; [2] cotton wool; [3] superglue, bottle with precision applicator tip; [4] brush for removing corncob 'dust' from feathers; [5] tissue tube; [6] Sharpie for marking tissue tube with collection number; [7] measuring tape; [8] cotton thread; [9] sewing needles; [10] scalpel blades; [11] identification tags, pre-strung with thread; [12] pointed scissors, medium; [13] pointed scissors, small; [14] round-tip scissors, two pairs; [15] plastic ruler, marked in mm; [16] scalpel; [17] tweezers (one featherweight), four pairs; [18] angled tweezers; [19] wooden dowels and bamboo skewers, to use in wings and as 'backbones' in smaller birds. (Vertebrate Zoology Prep Lab, Smithsonian NMNH, January 2015).

new additions as well: the cryotubes for tissue samples and the superglue. I ask about the superglue, and I'm told it is used to glue beaks shut. The muscles holding the beak together are removed during the cleaning process, leaving the loose upper and lower beaks prone to damage. "Not everyone uses it [the superglue]," the preparator tells me. "Some think it might add chemicals that could interfere with future genetic sampling... So some [preparators] use the old way and tie the beak closed with thread, but it takes more time... and I have a freezer full of birds to prep[are]."[16]

This collapsing of time – patina of old against new, juxtaposed but also in parallel use, different parts of a process changing ever-so-slowly – was a recurring theme in the preparation lab, and indeed across many spaces of the Smithsonian. At times, it became disorienting, this play of time stretching and folding back on itself. It was also in these moments of contrast – of twenty-first century superglue and tissue tubes alongside nineteenth-century cotton and wooden dowels – that caught my attention, signalling a shift in practices. However, these new practices were quickly becoming 'natural' to the museum staff. "I don't even notice anymore," one preparator told me. "Making study skins just looks normal to me now... Taking [tissue] samples will probably be that way too," he pauses, "eventually."[17]

It is useful here to think briefly about the material aspects of things in play, or in other words, how they are constantly generative of new circumstances and resulting consequences – such as collecting tissue samples, creating new data types to render the proliferating parts and pieces meaningful. Thinking through this 'liveliness', or at least the potential for liveliness as I held the internal organs of a bird in my hand, I extend this into the liveliness of material things-in-process. The heart of the matter, in this case, might be an actual heart. Or a very small section of a heart, accompanied by slices of muscle tissue and liver, carefully pushed into the bottom of a small plastic vial so they didn't squeeze out when the cap was screwed on as life was, once again, at risk of spilling over its frame.

In re-conceptualizing the boundaries between humans, things and technologies in the making and remaking of bird specimens, I want to critically engage the implicit control enacted over animate and inanimate life – what Clarke and Fujimura call a new "philosophy of becoming" (1992: 30). This opens up an understanding of social theory by emphasizing not only the play of words, but the interplay of the (bio)materials through which knowledge is produced and negotiated. The generative potential of a tissue sample, then, is balanced uneasily at the border of animate and inanimate. It was, on the one hand, just little bits of meat (Figure 4.11). On the other hand, these little bloody fragments held the potential for multiple technological and temporal 'unfoldings' into the future – the extracted DNA assessed and amplified, mapped and assembled, and then circulated to databases the world over, a form of 'proxy' for the individual animal, the species as a whole, or even as an indicator for the health of an ecosystem. Through skill and careful technique, new materials and new technologies become 'natural': just as making study skins becomes something that "just looks normal," as the preparator said, so too will making the tissue tubes become a standard and unremarkable object in the museum. Yet the ideas around what iteration of life is being preserved in the making of these objects – that is, the representation of a species made through the body of a bird, through a genome-quality tissue sample, or through the genomic data itself – is inextricably bound up with the making of those objects, freighted along with the feathers, tissue or data as they are hand-crafted.

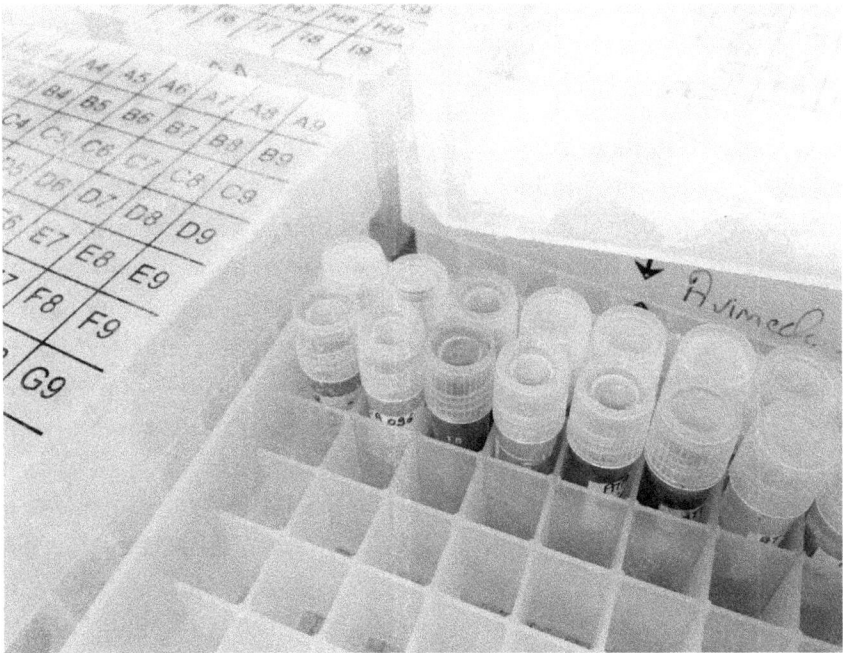

Figure 4.11 — A tray of frozen bird tissues. (MNHN Department of Birds, 2018). (Photograph by Adrian Van Allen).

Conclusion

Bird preservation in museums takes on multiple forms. Once, taxidermy mounts made from hollowed-out feathered skins were used to represent their species, yet now the bodies of birds are taken apart and put back together into a wide array of forms both morphological and molecular. These can include many recognizable and abstracted bird-derived objects: a stuffed bird skin, parasites combed from feathers, heavy metals extracted from liver, pollen carved from beaks, isotope samples from bone marrow, frozen tissue samples, DNA extracts, genetic and genomic data from DNA barcodes to whole assembled genome sequences, to the layered labels tied to a specimen's foot or stuck on a cryotube, collection data handwritten in log books to entries in public databases. This array of bird-derived parts and pieces form a networked tangle at varying stages of transformation from the morphological to the molecular, from the analogue of a handmade object with the hand of the maker visible, to the digital world of databases and protein sequences that are equally handmade, but where the handwork is obscured.

The 'rediscovery' of natural history collections by conservation biologists as sites for gaining new types of data – data types that were unthinkable when the collections were originally made 150, or even 50, years ago – has rapidly shifted the value of these animal-objects in the face of new demands by new audiences, in ways and directions beyond the valuations given by museum biologists. Valued now as sources of potential insight into historic climate change, population bottlenecks and extinction events, these natural history collections become 'windows into the past' that will ostensibly provide for our own species' future needs, according to the many scientists I interviewed and worked alongside at both museums. The collections are not just a way of marking time in the 'Anthropocene', of measuring human impact and configuring pieces to fill gaps that humans have made in the fabric of biodiversity. Natural history collections are also cultural artefacts of our species' multiple and ongoing redefinitions of what constitutes the 'natural world' – as defined in the Global North. As such they serve as a conduit for voicing what place in that iteration of 'nature' human beings could, or should, occupy.

A deeply motivating factor for conservation of biodiversity stems from the destruction of 'natural' habitats (Lowe 2006; Lowe and Münster

2016; Tsing 2015a). Though a deeper analysis of the 'naturalness' of many of these environments is clearly warranted, it is perhaps also productive to examine this move as articulating a perspective of the 'natural' world as merely a resource – one that has flowed from the Global South into the collections and laboratories of the Global North. Within the context of 'salvage' operations to biobank biodiversity before it disappears, it is also important to attend to the materials through which these claims are made and negotiated. As a curator at Paris MNHN articulated this concern, "We must get as much as we can from each specimen, because we do not know when we might be able to collect another."[18]

I follow the argument made by Donna Haraway (2015) that issues of naming the era of the Anthropocene have to do with "scale, rate/ speed, synchronicity and complexity" (2015: 159) more than the simple acknowledgment that human beings have radically reshaped the natural world over differently defined epochs of time. The recurring question in considering such systemic phenomena must be an attention to "when *changes in degree become changes in kind*, and what are the effects of bio-culturally, biotechnically, biopolitically, historically situated people (not Man) relative to, and combined with, the effects of other species assemblages and other biotic/abiotic forces?" (2015: 159 emphasis mine). In Haraway's interwoven multispecies world no one species acts alone; instead "assemblages of organic species and of abiotic actors make history... the evolutionary kind and the other kinds" (2015: 159). This brings to the forefront not simply ecological devastation brought about by human forces – a dominating version of the Anthropocene cast as the 'Age of Humans' – but instead shifts the focus down to specific assemblages of historically situated and materially grounded *people, places* and *things* and the effects of their interactions. This resonates with my own focus here on the details of practice, material interactions and the types of time involved in making specimens at the Smithsonian NMNH and Paris MNHN.

In 'Feral Biologies' (2015b), Anna Tsing suggests that the inflection point between the Holocene and the Anthropocene might be the wiping out of most of the spaces for refuge (what she calls 'refugia') where diverse "species assemblages" (ibid) can be reconstituted after major cataclysmic events such as massive loss of habitat, epidemics or an influx of invasive species that shift local ecologies. From one perspective the natural history museum projects under consideration here to collect and preserve 'all species of life' can be seen as a replica of a 'refugia', a constructed site of refuge from which to reconstitute the 'natural' world after a potential cataclysmic event in the future.

Within changing concepts of time such as the Anthropocene, humans have been re-situated into a larger temporal framework of 'deep time' (Ginn et al. 2018) and are now conceived as a geologic force, with our species' impact on environments archived through biosocial (or perhaps 'cryosocial', see: Hoeyer 2017; Kowal and Radin 2015; Radin and Kowal 2017) objects frozen in time such as ice cores (Antonello and Carey 2017), seed banks (Harrison 2017; Parry 2004), human (Radin 2017) and nonhuman genetic biobanks (Breithoff and Harrison 2020; Van Allen 2018, 2019b). New moral and ethical imperatives have emerged in response to these conceptions of time, shaping concepts of care in times of crisis such as natural history museum projects to 'preserve and understand the genomic biodiversity of life on Earth' (GGI 2019) in the face of increasing extinction rates.

The condition of possibility for such projects is the introduction of biotechnology into traditional museum collecting methods, where genomics becomes a salvational tool for making an archive of 'life' for the future. However, as the material world of Anthropocenic *'nature'* becomes a site of contesting interests and values, it is also the material *culture* of nature that is called into question as embodied in the practices for collecting and preserving natural history specimens. As Joanna Radin's (2013) account of frozen blood and tissue samples demonstrates, the value of genetic collections were archived as a form of latent values, to be thawed and used at some future moment in time.

Through analyzing two sets of objects – historic specimen preparation tools and inventories of contemporary prep kits – I have examined a different kind of 'time series', one that suggests that specimen preparation practices are archives of the past, not just in the body of the bird itself but of the techniques, materials and epistemic frameworks used to create the specimen. Ways of knowing the world are archived in the materials, as well as in the ways they are used. These are layered into the specimen and its network of abstracted parts and pieces through the process of transformation from a living thing into a museum specimen. As specimen preparators skin, sew, cut and freeze pieces of birds, they create specimens shaped by their labour as much as they are shaped by their ideas of care for uncertain environmental futures. It is within these practices that time is folded, as disciplinary histories are woven together with visions for various futures in the transformed bodies of birds. As specimens, these birds' bodies are being intricately crafted to offer up future potential, be it data to expand scientific knowledge, to protect species or to repair ecologies.

Notes

1. This chapter draws on ongoing research with staff at the Paris MNHN and the Smithsonian NMNH. I am indebted to the staff at both museums for their generosity and collaboration. Funding for this research has been generously provided by a Smithsonian Institution Peter Buck Fellowship (2014-2016), the Wenner-Gren Foundation (Grant No. 8977) and a Museé du quai Branly Postdoctoral Fellowship (2017-2018).

2. Genetic collecting projects in museums range from small collections made by curators for their own phylogenomic research, to institutional or global collecting programmes. These include projects such as the Global Genome Initiative at the Smithsonian (GGI 2019; see also Van Allen 2016, 2017), the Barcode of Life project to capture DNA snippets to identify all species (BOL 2016; see Ellis, Waterton and Wynne 2009; Waterton 2010; Waterton, Ellis and Wynne 2013), global public databases such as GenBank (GenBank 2018), as well as many family-specific projects such as the All Birds Barcoding Initiative (ABBI 2018) and the Bird 10,000 Genome Project (B10K Database 2018), set on drafting genomes of all existent bird species by 2020.

3. From interview notes with a curator, Smithsonian NMNH, February 5, 2015.

4. From interview notes with a curator, Smithsonian NMNH, February 5, 2015.

5. Interviews were conducted by author in a mix of French and English, interspersed with Latin names for birds. All translations are my own.

6. From interview notes with a specimen preparator, Paris MNHN, February 21, 2018.

7. From interview notes with a specimen preparator, Paris MNHN, February 21, 2018.

8. From interview notes with a specimen preparator, Paris MNHN, March 7, 2018.

9.   From interview notes with a curator, Paris MNHN, November 18, 2017.

10.   From interview notes with a collection manager, Smithsonian NMNH, June 20, 2015.

11.   From interview notes with a curator, Smithsonian NMNH, April 10, 2015.

12.   From interview notes with a curator, Smithsonian NMNH, April 10, 2015.

13.   From interview notes with a genetics project manager, Smithsonian NMNH, July 10, 2015.

14.   From interview notes with a collection manager, Smithsonian NMNH, February 1, 2016.

15.   From interview notes with an ancient DNA specialist, Smithsonian NMNH, August 2016.

16.   From interview notes with a specimen preparator, Smithsonian NMNH, February 2, 2015.

17.   From interview notes with a specimen preparator, Smithsonian NMNH, January 12, 2016.

18.   From interview notes with a curator, Paris MNHN, November 18, 2017.

References

ABBI. 2018. All Birds Barcoding Initiative. http://www.barcodingbirds.org/.

Antonello, A. and M. Carey. 2017. Ice Cores and the Temporalities of the Global Environment. *Environmental Humanities* 9(2): 181–203.

B10K Database. 2018. Bird 10,000 Genome Database Project. https://b10k. genomics.cn/index.html.

Barad, K. 2003. Posthumanist Performativity: Toward an Understanding of How Matter Comes to Matter. *Signs: Journal of Women in Culture and Society* 28(3): 801–31.

Belon, P. 1555. *L'Histoire de La Nature Des Oyseaux*. Paris: B. Prevost for G. Corrozet. https://www.biodiversitylibrary.org/bibliography/78886#/summary.

Berny, P., L. Vilagines, J.-M. Cugnasse, O. Mastain, J.-Y. Chollet, G. Joncour and M. Razin. 2015. VIGILANCE POISON: Illegal Poisoning and Lead Intoxication Are the Main Factors Affecting Avian Scavenger Survival in the Pyrenees (France). *Ecotoxicology and Environmental Safety* 118 (August): 71–82.

Bi, K., T. Linderoth, D. Vanderpool, J. M. Good R. Nielsen and C. Moritz. 2013. Unlocking the Vault: Next-Generation Museum Population Genomics. *Molecular Ecology* 22(24): 6018–32.

BOL. 2016. What Is DNA Barcoding? Barcode of Life (BOL). http://www. barcodeoflife.org/content/about/what-dna-barcoding.

Bowker, G. 2000. Biodiversity Datadiversity. *Social Studies of Science* 30(5): 643–83.

Breithoff, E. and R. Harrison. 2020. From Ark to Bank: Extinction, Proxies and Biocapitals in Ex-Situ Biodiversity Conservation Practices. *International Journal of Heritage Studies* 26(1): 37-55.

Brown, T. 1853. *The Taxidermist's Manual, or, The Art of Collecting, Preparing, and Preserving Objects of Natural History: Designed for the Use of Travellers, Conservators of Museums, and Private Collections / by Captain Thomas Brown.* 11th ed. Dublin: A. Fullarton and Co. https://www.biodiversitylibrary.org/ bibliography/42325#/summary.

Buffon, G.-L. Leclerc Comte de. 1800. *Histoire naturelle generale et particuliere.* Paris: de l'Imprimerie de F. Dufart.

Clarke, A. and J. Fujimura. 1992. *The Right Tools for the Job: At Work in Twentieth-Century Life Sciences.* Princeton: Princeton University Press.

Crutzen, P. J. and E. F. Stoermer. 2000. The Anthropocene. *Global Change Newsletter* 41: 17–18.

Daston, L. 2004. Type Specimens and Scientific Memory. *Critical Inquiry* 31(1): 153–82.

Dobres, M.A. 1999. Technology's Links and Chaînes: The Processual Unfolding of Technique and Technician. In *The Social Dynamics of Technology*, edited by C.R. Hoffman and M.A. Dobres, 124–146. Washington D.C.: Smithsonian Institution Press.

Droege, G., J. Coddington, K. Barker, J. J. Astrin, P. Bartels, C. Butler, D. Cantrill and F. Forest. 2013. The Global Genome Biodiversity Network (GGBN) Data Portal. *Nucleic Acids Research*, 1–6.

Dufresne, L. 1803. *Nouveau Dictionnaire d'histoire Naturelle.* Paris: Déterville.

Ellis, R. 2008. Rethinking the Value of Biological Specimens: Laboratories, Museums and the Barcoding of Life Initiative. *Museum and Society* 6 (2): 172–91.

Ellis, R., C. Waterton and B. Wynne. 2009. Taxonomy, Biodiversity and Their Publics in 21st Century Barcoding. *Public Understanding of Science* 19(4): 497–512.

Fabian, J. 1983. *Time and the Other: How Anthropology Makes Its Object.* New York: Columbia University Press.

Farber, P. 1977. The Development of Taxidermy and the History of Ornithology. *Isis* 68(4): 550–66.

Farber, P. 1982. *The Emergence of Ornithology as a Scientific Discipline: 1760-1850.* Dordrecht, Holland: Springer Netherlands.

Fujimura, J. 1996. *Crafting Science: A Sociohistory of the Quest for the Genetics of Cancer*. Cambridge; London: Harvard University Press.

Fujimura, J. 2003. Future Imaginaries: Genome Scientists as Socio-Cultural Entrepreneurs. In *Genetic Nature/Culture: Anthropology and Science Beyond the Two Culture Divide*, 176–99. Berkeley: University of California Press.

GenBank. 2018. GenBank - National Institute of Health Genetic Sequence Database. 2018. http://www.ncbi.nlm.nih.gov/genbank/.

GGBN. 2015. Global Genome Biodiversity Network. http://www.ggbn.org/ggbn_portal/.

GGI. 2019. Smithsonian Global Genome Initiative. https://ggi.si.edu/.

Ginn, F., M. Bastian, D. Farrier and J. Kidwell. 2018. Introduction: Unexpected Encounters with Deep Time. *Environmental Humanities* 10(1): 213–25.

Grealy, A., N. Rawlence, M. Bunce, A. Grealy, N. J. Rawlence and M. Bunce. 2017. Time to Spread Your Wings: A Review of the Avian Ancient DNA Field. *Genes* 8(7): 184.

Hanner, R., A. Corthals and R. DeSalle. 2009. Biodiversity, Conservation, and Genetic Resources in Modern Museum and Herbarium Collections. In *Conservation Genetics in the Age of Genomics*, edited by G. Amato, R. DeSalle, O. A. Ryder and H. C. Rosenbaum, 115–23. New York: Columbia University Press.

Haraway, D. 2015. Anthropocene, Capitalocene, Plantationocene, Chthulucene: Making Kin. *Environmental Humanities* 6: 159–65.

Haraway, D. 2016. *Staying with the Trouble: Making Kin in the Chthulucene*. Durham: Duke University Press.

Harding, S. 2008. *Sciences from Below: Feminisms, Postcolonialities, and Modernities*. Durham and London: Duke University Press.

Harrison, R. 2015. Beyond 'Natural' and 'Cultural' Heritage: Toward an Ontological Politics of Heritage in the Age of Anthropocene. *Heritage & Society* 8(1): 24–42.

Harrison, R. 2017. Freezing Seeds and Making Futures: Endangerment, Hope, Security, and Time in Agrobiodiversity Conservation Practices. *Culture, Agriculture, Food & Environment* 39(2): 80–89.

Hoeyer, K. 2017. Suspense: Reflections on the Cryopolitics of the Body. In *Cryopolitics: Frozen Life in a Melting World*, edited by J. Radin and E. Kowal, 205–14. Cambridge, MA: MIT Press.

Hykin, S. M., K. Bi and J. A. McGuire. 2015. Fixing Formalin: A Method to Recover Genomic-Scale DNA Sequence Data from Formalin-Fixed Museum Specimens Using High-Throughput Sequencing. *PLOS ONE* 10(10): e0141579.

Kalshoven, P. T. 2018. Piecing Together the Extinct Great Auk: Techniques and Charms of Contiguity. *Environmental Humanities* 10(1): 150–70.

Knorr-Cetina, K. 1999. *Epistemic Cultures: How the Sciences Make Knowledge.* New Haven and London: Harvard University Press.

Kohlstedt, S. G. 2005. Thoughts in Things. *Isis* 96(4): 586–601.

Kowal, E. and J. Radin. 2015. Indigenous Biospecimen Collections and the Cryopolitics of Frozen Life. *Journal of Sociology* 5(1): 63–80.

Kress, J. 2014. Valuing Collections. *Science* 346(6215): 1310.

Latour, B. 1988. *The Pasteurization of France.* Translated by Alan Sheridan and John Law. Cambridge, MA: Harvard University Press.

Law, J. 2010. The Materials of STS. In *The Oxford Handbook of Material Culture Studies*, edited by D. Hicks and M. Beaudry, 171–86. Oxford: Oxford University Press.

Lemonnier, P. 1986. The Study of Material Culture Today: Toward an Anthropology of Technical Systems. *Journal of Anthropological Archaeology* 5(2): 147–86.

Lorimer, J. 2015. *Wildlife in the Anthropocene: Conservation after Nature.* Minneapolis: University of Minnesota Press.

Lowe, C. 2006. *Wild Profusion: Biodiversity Conservation in an Indonesian Archipelago.* Princeton: Princeton University Press.

Lowe, C. and U. Münster. 2016. The Viral Creep: Elephants and Herpes in Times of Extinction. *Environmental Humanities* 8(1): 118–42.

Malkogeorgou, T. 2011. Folding, Stitching, Turning: Putting Conservation into Perspective. *Journal of Material Culture* 16(4): 441–55.

Miller, W., D. I. Drautz, J. E. Janecka, A. M. Lesk, A. Ratan, L. P. Tomsho, M. Packard, et al. 2008. The Mitochondrial Genome Sequence of the Tasmanian Tiger (Thylacinus Cynocephalus). *Genome Research* 19(2): 213–20.

Mol, A. 2003. *The Body Multiple.* Durham and London: Duke University Press.

Murray, K., L. Skerratt, G. Marantelli, L. Berger, D. Hunter, M. Mahoney and H. Hines. 2011. Cryopreservation and Reconstitution Technologies: A Proposal to Establish A Genome Resource Bank For Threatened Australian Amphibians. Appendix 2 to Murray, K., L. Skerratt, G. Marantelli, L. Berger, D. Hunter, M. Mahony and H. Hines 2011. *Guidelines for Minimising Disease Risks Associated with Captive Breeding, Raising and Restocking Programs for Australian Frogs.* A report for the Australian Government Department of Sustainability. http://olr.npi.gov.au.

Nachman, M. 2013. Genomics and Museum Specimens. *Molecular Ecology* 22(24): 5966–5968.

Neimanis, A., C. Åsberg and J. Hedrén. 2015. Four Problems, Four Directions for Environmental Humanities: Toward Critical Posthumanities for the Anthropocene. *Ethics & the Environment* 20(1): 67–97.

Parry, B. 2004. *Trading the Genome: Investigating the Commodification of Bio-Information*. New York: Columbia University Press.

Péquignot, A. 2002. Histoire de La Taxidermie En France de 1729–1928, Etude Des Facteurs de Ses Évolutions Techniques et Conceptuelles, et Ses Relations à La Mise En Exposition Du Spécimen Naturalisé. Ph.D. Dissertation, Paris: Muséum National d'Histoire Naturelle.

Pomian, K. 1990. *Collectors and Curiosities: Paris and Venice, 1500–1800*. Cambridge: Polity Press.

Pyke, G. H. and P. R. Ehrlich. 2010. Biological Collections and Ecological/ Environmental Research: A Review, Some Observations and a Look to the Future. *Biological Reviews* 85(2): 247–66.

Radin, J. 2013. Latent Life: Concepts and Practices of Human Tissue Preservation in the International Biological Program. *Social Studies of Science* 43(4): 484–508.

Radin, J. 2017. *Life on Ice*. Chicago: Chicago University Press.

Radin, J. and E. Kowal (eds). 2017. *Cryopolitics: Frozen Life in a Melting World*. Cambridge, MA: MIT University Press.

Rheinberger, H.-J. 1997. *Toward a History of Epistemic Things: Synthesising Proteins in the Test Tube*. Stanford CA: Stanford University Press.

Rocha, L.A., A. Aleixo, G. Allen, F. Almeda, C. C. Baldwin, M. V. L. Barclay, J. M. Bates, et al. 2014. Specimen Collection: An Essential Tool. *Science* 344 (6186): 814–15.

Rose, D. B. 2012. Multispecies Knots of Ethical Time. *Environmental Philosophy* 1: 127–140.

Rose, D. B. and T. van Dooren. 2011. Unloved Others: Death of the Disregarded in the Time of Extinctions. *Australian Humanities Review* 50 (May): 1–4.

Schipper, J., J. S. Chanson, F. Chiozza, N. A. Cox, M. Hoffmann, V. Katariya, J. Lamoreux, A. S. L. Rodrigues, S. N. Stuart and H. J. Temple. 2008. The Status of the World's Land and Marine Mammals: Diversity, Threat, and Knowledge. *Science* 322 (5899): 225–8075.

Schulze-Hagen, K., F. Steinheimer, R. Kinzelbach and C. Gasser. 2003. Avian Taxidermy in Europe from the Middle Ages to the Renaissance. *Journal Für Ornithologie* 144(4): 459–78.

Sholts, S. B., J. A. Bell and T. C. Rick. 2016. Ecce Homo: Science and Society Need Anthropological Collections. *Trends in Ecology & Evolution* 31(8): 580–83.

Smithsonian Institution, Living in the Anthropocene Consortia. 2015. Living in the Anthropocene | The Smithsonian Consortia. https://interdisciplinary.si.edu/collaboration-highlights/anthropocene/

Smithsonian NMNH, Division of Birds. 2016. Division of Birds: Department of Vertebrate Zoology: National Museum of Natural History: Smithsonian Institution. http://vertebrates.si.edu/birds/.

Strasser, B. 2010. Collecting, Comparing, and Computing Sequences: The Making of Margaret O. Dayhoff's Atlas of Protein Sequence and Structure, 1954–1965. *Journal of the History of Biology* 43(4): 623–60.

Suarez, A. V. and N. D. Tsutsui. 2004. The Value of Museum Collections for Research and Society. *BioScience* 54(1): 66–68.

Sunder Rajan, K. and S. Leonelli. 2013. Introduction: Biomedical Trans-Actions, Postgenomics, and Knowledge/Value. *Public Culture* 25(3): 463–75.

Tsing, A. 2015a. *The Mushroom at the End of the World: On the Possibility of Life in Capitalist Ruins*. Princeton: Princeton University Press.

Tsing. A. 2015b. Feral Biologies. presented at the inaugural conference of Anthropological visions of sustainable futures. Organized by M. Brightman and J. Lewis., London: Centre for the Anthropology of Sustainability (CAOS), University College London, February 13, 2015.

Tsing, A., N. Bubandt, E. Gan and H. A. Swanson (eds). 2017. *Arts of Living on a Damaged Planet: Ghosts and Monsters of the Anthropocene*. Minneapolis: University of Minnesota Press.

Turgot, E. F. 1758. *Mémoire Instructif Sûr La Manière Da Rassembler, de Preparer, de Conserver, et d'envoyer Les Diverses Curiosités d'histoire Naturelle*. Paris and Lyon: chez Jean Marie Bruyset.

Van Allen, A. 2016. Crafting Nature: An Ethnography of Natural History Collecting in an Age of Genomics. Ph.D. Dissertation, Berkeley: University of California, Berkeley.

Van Allen, A. 2017. Bird Skin to Biorepository: Making Materials Matter in the Afterlives of Natural History Collections. *Knowledge Organization* 44(7): 529–44.

Van Allen, A. 2018. Pinning Beetles, Biobanking Futures: Practices of Archiving Life in a Time of Extinction. *New Genetics and Society* 37(4): 387–410.

Van Allen, A. 2019a. Oiseaux, Plumes, Spécimens, Données: 'Embodied Archives' et Valeur Changeante Des Collections de Musée Dans l'anthropocène. In *Valeurs et Matérialité: Approches Anthropologiques*, Paris: Presses de l'Ecole Normale Supérieure-Musée du quai Branly. 127–47.

Van Allen, A. 2019b. Resurrecting Ferrets and Remaking Ecosystems. *Anthropology News*, Special Issue: *Animalia*, 60 (3): 20–23.

Waterton, C. 2010. Barcoding Nature: Strategic Naturalization as Innovatory Practice in the Genomic Ordering of Things. *The Sociological Review* 58: 152–71.

Waterton, C., R. Ellis and B. Wynne. 2013. *Barcoding Nature: Shifting Cultures of Taxonomy in an Age of Biodiversity Loss.* London: Routledge.

Winker, K. 2004. Natural History Museums in a Postbiodiversity Era. *BioScience* 54(5): 455–459.

Wonders, K. 1993. *Habitat Dioramas: Illusions of Wilderness in Museums of Natural History.* Uppsala: Almqvist and Wiksell International.

Chapter 5

Making Futures in End Times:
Nature Conservation in the Anthropocene

ESTHER BREITHOFF & RODNEY HARRISON

Introduction: Making nature in the Anthropocene

Responses to climate change and global warming could be said to "saturate our sense of the now" (Chakrabarty 2009: 197). They do so no more than in the current designation of 'our' time as the Anthropocene, a geological epoch in which anthropogenic activity is understood to have constituted the dominant influence on the Earth's climate, ecosystems and geology. But what does it mean to live in the 'Age of Humans'? On 20 June 2012, the short film *Welcome to the Anthropocene* opened the UN's Rio+20 summit on sustainable development, held in Rio de Janeiro, Brazil. Attended by over 45,000 participants, it was reported to have been the largest and most well-publicized such meeting to date. Over an image of a planet rotating slowly with a superimposed rising line graph which we later learn represents 'The Great Acceleration', the film is narrated with a woman's voice which intones calmly, "This is the story of how one species changed a planet." Citing the industrial revolution and the rise of globalization and population growth, it continues, "We have shaped our past, we are shaping our present and we can shape our future. You and I are part of this story. We are the first generation to realize this new responsibility. As the population grows to 9 billion we must find a safe operating space for humanity, for the sake of future generations. Welcome to the Anthropocene." In addition to the summit's establishment of the Sustainable Development Goals (SDGs) which have come to dominate the post 2015 development agenda, the outcome document of the summit, titled 'The Future We Want' (United Nations 2012) emphasized the emancipatory potential of human action to intervene within the series of ecological crises which the film introduced to its participants. Humans were thus elevated to the dual positions of gods and monsters, simultaneously the architects of their own destiny and the Earth's demise.

The notion that the Earth is currently in the midst of the 'Sixth Mass Extinction', resulting in what has been described as a 'biodiversity crisis' (Singh 2002) – a global loss of diversity in flora and fauna – is a fundamental part of current discussions relating to the impacts of anthropogenic climate change in the Anthropocene. Although it has been observed that species have regularly gone extinct throughout the millennia through non-anthropogenic (or at least, primarily non-anthropogenic) processes – processes which Darwin famously saw as slow and gradual but inevitable and progressive, driven by natural selection and the 'survival of the fittest' (e.g. see Sepkoski 2016) – human influence in the form of over-hunting and habitat depletion are today understood to be playing a central role in the current 'chronic disaster' (Westley 1997) of animal extinction rates (Fletcher 2008). Anthropogenic climate change is seen to constitute a key threat multiplier to these factors (Cahill et al. 2012). Paleobiologists warn that no matter on what scale these anthropogenic extinctions take place, biodiversity will need millions of years to recover (Kirchner and Weil 2000: 177).

The question we ask here is what does it mean to conserve 'nature' in the Anthropocene, or what Marris (2013) has termed a 'post-wild world'? The aim of this chapter is to explore some of the distinctive ways in which scientists and conservationists are responding to these challenges and how we might critically investigate the socio-cultural work which surrounds such efforts before returning to some of the ways in which the recognition of the Anthropocene as the age of humans both troubles and is troubled by such efforts. Working across natural and cultural heritage, our work is informed by observations of the ways in which research in what we might call 'the climate change era' forces a dissolution of the distinction between natural and cultural history (Chakrabarty 2009). Here we intersect with a new critical engagement with nature conservation (e.g. Benson 2010; Lorimer 2015) and extinction studies (e.g. Bird Rose 2013; Heise 2016; van Dooren 2016; Bird Rose, van Dooren and Chrulew 2017) in exploring the distinct social and cultural frameworks which produce 'natural heritage', and the ways in which 'cultural heritage' is not outside of, but integrally a part of them (e.g. Harrison 2015; DeSilvey 2017). Our work also connects both conceptually and empirically with recent anthropological engagements with 'futures' (e.g. Appadurai 2013; Salazar et al. 2017; Harrison et al. 2020), and with current creative academic engagements with global climatological and environmental change (e.g. Haraway 2016; Tsing 2015; Tsing et al. 2017) and the multiple worldings (cf. Barad 2007; de la Cadena and Blaser 2018; Omura et al. 2018; for heritage see Breithoff 2020) of their entangled conservation practices.

Frozen futures

Empirically, we focus on the future-making practices inherent in the work of global agrobiodiversity conservation and non-human animal endangered DNA cryopreservation, drawing on research with the Nordic Genetic Resource Centre (NordGen), the Svalbard Global Seed Vault (SGSV) and the Frozen Ark Project (FA). In doing so, we observe a contemporary shift in the meaning of the practice of collecting, archiving and safeguarding such plant and animal biomaterials. From an initial 'heroic' narrative that cast such biobanks in a static, dormant role – isolated arks to carry endangered DNA into an uncertain future (Doyle 1997; Watson and Holt 2001; Bowkett 2009; Chrulew 2017) – we detect a recent shift to a more active function which acknowledges their potential for reanimation of genetic material in future biosocial and biopolitical programmes, including the well-publicized 'restoration' of the ICARDA seed bank in Syria, and in so-called 'de-extinction' initiatives. We suggest that the role of such institutions has transformed from repository to speculative reinvestment: the 'arks' that stored and safeguarded genetic samples for survival within an endangerment narrative (cf. Turner 2007) have altered to become 'investment banks' where genetic materials can be actively reworked and revived to build new futures (see also Bowker 2005b; Heatherington 2012; Chrulew 2017).

Where the new forms of biocapital generated with such repositories seem to reflect reformulations of late capitalist values (e.g. Doyle 1997; Shukin 2009; Thacker 2010), in this chapter we consider the ways in which a critical perspective on the operations of these enterprises might help us to bring new insights to bear on the latent possibilities contained within these reservoirs of cold stored and frozen seeds and DNA. We suggest that unravelling the details of the temporal orientations of conservation practices and their underpinning sociotechnical and biopolitical processes helps us to understand how conservation practices of different kinds are not normative, but vary across time and space, actively shaping different kinds of future worlds. In doing so, we draw on approaches to the study of archives and collections, which emphasize the ways in which their collecting and ordering practices not only reflect but actively intervene within and shape the worlds they order (see further discussion in Bennett et al. 2017). Our examination of global agrobiodiversity conservation and endangered DNA cryopreservation programmes reveals the complexity of temporal aspects of biodiversity conservation, as well as the complicated ways in which conservation practices both 'archive' diversity and generate and accumulate latent forms of biocapital (Helmreich 2008; Rajan 2006) in their aim to secure genetic resources for the future.

The predicted loss of two-thirds of the world's vertebrae population by 2020 (WWF 2016), and a similarly bleak outlook for invertebrate species, has intensified biodiversity conservation efforts globally. These take the form of both in-situ conservation programmes (e.g. through the designation of protected areas) and ex-situ captive breeding programmes (e.g. in zoos and aquaria). More recently, these ex-situ conservation efforts have accelerated, as a result of the DNA 'revolution', through the development of organized archives of non-human animal and plant biomaterials which aim to document and preserve genetic information on the biology, ecology and evolutionary history of threatened plants, mammals, birds and reptiles in the form of viable cells and DNA preparations, before it is irretrievably lost (Corley-Smith and Brandhorst 1999; Watson and Holt 2001; Friese 2013; Costa and Bruford 2018). Genetic resource banking – the freezing of plant and animal genetic material for ex-situ storage and its use in research within a present-day and potential future context – has emerged as a response to what has been understood to be a contemporary extinction crisis, and in many cases cryobanks have come to be seen as the only and last resort for recording and storing biological material from endangered species for potential future retrieval.

The practice of freezing and storing biological material (including blood, germplasm, embryos, tissues and somatic cells of non-human animals) in genetic resource banks for the advancement of human medicine and the development of agro-industries is not a new development in scientific research (see further discussion in Radin 2015, 2017; Radin and Kowal 2017). Nonetheless, it is only recently that ex-situ cold and cryogenic practices have become a leading and driving force in biodiversity research within the context of endangered species conservation (see Gemeinholzer et al. 2011; Wildt et al. 1997; see also Howard et al. 2016; Wisely et al. 2015, on the ferret biobank) – with biobank facilities such as the Smithsonian, the San Diego Frozen Zoo® and the genetic repository at the Natural History Museum in London collecting blood, tissues, cell cultures, eggs, spermatozoa and embryos specifically for conservation purposes, and the global expansion of regional and national seedbanks for agrobiodiversity conservation. According to its website, the San Diego Zoo Institute for Conservation Research now stores "the largest and most diverse collection of its kind in the world" with over "10,000 living cell cultures, oocytes, sperm and embryos representing nearly 1,000 taxa, including one extinct species, the po'ouli". Since its foundation in 1975, the San Diego Frozen Zoo® has become an irreplaceable and continuously expanding source of biological information for significant scientific advancements in fields such as conservation, medicine, assisted reproduction, evolutionary biology, physiology and wildlife medicine (Chemnick, Houck and Ryder 2009).

In a paper on the Frozen Zoo® concept published in 1984, Benirschke advocates that "biologists at zoological gardens have a unique opportunity – if not an *obligation* – to preserve materials for scientific study. At a time when biomedical capabilities are expanding rapidly, we find ourselves in the position that biological resources are dwindling rapidly. Many forms of life are at the point where *extinction is imminent*, yet the animal or plant has not become understood in any of its major biological ways" (1984: 325, our emphases). Benirschke's words convey an urgency to not only save dwindling genetic material for scientific study in the present but to safeguard it for an undetermined future in which humans will be in a better position to extract from it as-yet uncovered information. (They also provide the key to understanding the role of such facilities in contributing to the growth of new forms of biocapital, as we will discuss later in the chapter.) Here, cryobanks become the harbourers of 'time-travelling resource[s]' (Radin 2017), which are both enactors of, and produce templates for, 'futures in the making' (Adam and Groves 2007; Turner 2007). Genetic resources of endangered animals, for example, have enabled developments in reproductive technologies to maintain genetic diversity which have already produced promising 'real-life' results in a number of conservation programmes (e.g. Howard et al. 1992; Wildt et al. 1997). As such, frozen zoos and other non-human biobanks are driving ongoing research into cloning, de-extinction and re-introduction of endangered and once extinct species (see further discussion in O'Connor 2015; Shapiro 2015; Pilcher 2016). Cryobanks thus facilitate human intervention in the categorization and manipulation of biological diversity in standardized data management systems, turning the 'wild' into 'managed natures' (Buller 2014) and thus opening up seemingly endless possibilities for what Donna Haraway calls the 'reinvention of nature' (Haraway 1991). These developments are likely to have significant impacts on what we might now, in the light of the recognition of the Anthropocene epoch, term the 'Human Planet' (Lewis and Maslin 2018).

The Svalbard Global Seed Vault (SGSV)

The SGSV is currently the world's largest secure seed storage facility, established in 2008 by the Royal Norwegian Ministry of Agriculture and Food; the Global Crop Diversity Trust (now known as the 'Crop Trust'), an independent international organization based in Germany (established as a partnership between the United Nations Food and Agriculture Organization (FAO) and the Consultative Group on International Agricultural Research (CGIAR)); and the Nordic Genetic Resource Centre (NordGen). At a cost of US$9 million to the Norwegian government, the construction of the SGSV began in 2005 as a result of the recommendations of the 2004 International Treaty on Plant Genetic Resources for Food and Agriculture, which created a global ex-situ system for the conservation of agricultural plant genetic resource diversity. Situated on the island of Spitsbergen in the Svalbard archipelago, it received its first deposits of seeds in 2008.

NordGen is responsible for the day-to-day operations of the facility and maintains a publicly accessible database documenting its samples. NordGen's website (NordGen 2016) provides the details of its operations, as follows. The site reports that the SGSV holds in its frozen repository approximately 850,000 accessions and 54.7 million seeds, provided by 233 countries and 69 depositor institutions. Each accession represents a sample taken of a specific living crop population from a specific geographic location at a specific point in time, and is usually made up of approximately 500 individual seeds. Depositing institutions first dry the seed accessions to limit their moisture content to 5–6%, and then seal them inside an individual airtight aluminium bag. These bags are packed into standard-sized crates and stacked on shelving racks within one of the three separate, identical storage vaults, each measuring approximately 9.5 × 27 metres, which are refrigerated to maintain a constant temperature of −18°C. These vaults have been excavated approximately 120 metres into the side of a sandstone mountain at a height of 130 metres above sea level; entry to the vaults is via a 100-metre entrance tunnel. Equal parts bunker and frozen 'ark', the dramatic façade includes a commissioned artwork, *Perpetual Repercussion* by Dyveke Sanne, which "renders the building visible from far off both day and night, using highly reflective stainless steel triangles of various sizes" (Government of Norway 2015; see Figure 5.1). The cold climate and permafrost ensure that even if power is lost, the storage vaults would remain frozen for a significant period of time, even taking into account the possible effects of climate and sea level changes. "Designed for [a] virtually infinite lifetime", it is perceived to be "robustly secured against external hazards and climate change effects" (Government of Norway 2015).

The SGSV is not a conventional seed bank but was conceived of as part of a global system to facilitate the secure storage of a duplicate 'backup' of seed accessions held in national and regional repositories.

> Worldwide, more than 1,700 genebanks hold collections of food crops for safekeeping, yet many of these are vulnerable, exposed not only to natural catastrophes and war, but also to avoidable disasters, such as lack of funding or poor management. Something as mundane as a poorly functioning freezer can ruin an entire collection. And the loss of a crop variety is

Figure 5.1 — Entrance to the Svalbard Global Seed Vault. (Photograph by Rodney Harrison).

as irreversible as the extinction of a dinosaur, animal or any form of life (Crop Trust 2016a).

These backup sets of seeds are stored free of charge and are held as part of an international agreement in which the seeds remain the property of the depositing institution, and are available for withdrawal only by that institution, at any time. It is thus not an active genebank, but a literal 'vault' containing a secure stock of duplicate accessions, which can be used if seed stocks from the depositing institution become depleted or lost. The need for such a facility seemed clearly demonstrated when, in September 2015, scientists from the International Centre for Agricultural Research in Dry Areas (ICARDA) who had lost access to their genebank facility in Aleppo, Syria, requested the return of seeds deposited in the SGSV, to reconstruct their collection in a new facility in Lebanon. This first withdrawal of seed samples from the SGSV as a result of the ongoing conflict in Syria was reported widely in the media, and seemed to indicate that the SGSV was already fulfilling a purpose that had previously been assumed would arise in a more distant future (most often framed within the temporal horizon of medium- to long-term global climate change; see Fowler 2008), thus justifying the significant investment in this global 'insurance policy'. The manager of the new ICARDA gene-bank facility in Terbol, Bekaa, was reported to have said of the withdrawal of seed samples, "It [SGSV] was not expected to be opened for 150 or 200 years... It would only open in the case of major crises but then we soon discovered that, with this crisis at a country level, we needed to open it" (Alabaster 2015).

Banking diversity, making futures, securing hope

In articulating the need for such a repository, the SGSV's mission is framed within what we might see as a fairly conventional articulation of the endangerment sensibility and its accompanying entropic view of the relationship between diversity and time. The Crop Trust, as the charitable organization responsible for funding the ongoing operations of the SGSV and the preparation and shipment of seed from developing countries, perhaps articulates this most clearly in its explanation of the SGSV's purpose: "The purpose of the Svalbard Global Seed Vault is to provide insurance against both incremental and catastrophic loss of crop biodiversity held in traditional seed banks around the world. The Seed Vault offers 'fail-safe' protection for one of the most important natural resources on Earth". It continues: "Crop diversity is the resource to which plant breeders must turn to develop varieties that can withstand pests, diseases and remain productive in the face of changing climates. It will therefore underpin the world food supply... the Seed Vault will ensure that unique diversity held in genebanks in developing countries is not lost forever if an accident occurs" (Crop Trust 2016b). In these statements we see all of the conventional articulations of an entropic view of diversity, including the potential loss of diversity through catastrophic incidents and the need to build resilience in the face of such changes.

However, the situation becomes somewhat more complicated when we consider the operation of the SGSV in relation to the global system of agrobiodiversity conservation, and in particular, the relationship of the materials stored in the SGSV to the specific conservation targets of agrobiodiversity conservation practices. As Sara Peres (2016) shows, seed banks were originally developed as part of a strategy to ensure the maintenance of crop genetic diversity in the face of widespread adoption of a small number of high-yielding crop varieties during the agricultural industrialization and modernization of the twentieth century. The freezing of seeds would enable the maintenance of agrobiodiversity without the need for ongoing cultivation of old crop varieties, resulting in an 'archive' of the evolutionary histories of crop varieties that might be of use to future generations of agricultural scientists and farmers.

The notion of 'genetic erosion' fundamentally underpins this global system. First coined at the 1967 FAO/International Biological Program Technical Conference on the Exploration, Utilization and Conservation of Plant Genetic Resources (Pistorius 1997: 2), the concept gained strength from its resonance with the, by then, well-known concept of soil erosion, suggesting that the full range of both wild and domesticated genetic diversity, threatened with "erosion" by agricultural modernization programmes, was fundamental to future food security (see Fenzi and Bonneuil 2016: 74-6). 'Landraces', localized genetic variants of crop species resulting from both cultural and natural selection processes, were seen to represent a bank of genetic diversity that held potential for future crop improvement to both mediate the effects of future climate change and develop resilience to future diseases (e.g. see further discussion in Hummer 2015).

Peres (2016), drawing on the work of Parry (2004) and van Dooren (2009), goes on to show that the present system of genebanks is the outcome of debates in the 1960s and 1970s surrounding the most appropriate methods of agrobiodiversity conservation – in situ or ex situ – in which the frozen seeds held in seed banks across the world came to act as 'proxies' for crops. These debates were closely related to, and indeed stimulated, the development of broader technologies of ex-situ cryogenic, as well as other cold and frozen preservation practices, across a large number of different fields of conservation (see Radin 2016, 2017; Radin and Kowal 2017). Elaborating on the temporal aspect of seeds as proxies, Peres argues that frozen seeds could become records or 'archives' of a crop's evolutionary history, because they were preserved statically and latently, and as such they might be 'recalled' in the future (see also Bowker 2005a):

Seed banks can therefore be imagined as repositories that enabled the 'recall' of genetic diversity, both by committing it to memory and by allowing it to be recovered from cold storage for use. By evoking both these meanings, the concept of recall conveys how the conservation of old landraces is entangled with concerns regarding their future use. Seed banks thus function as archives that make records of the past of crops accessible in the future (Peres 2016: 102).

It is worth thinking through in more detail the concepts of the archive and of the relationship between the seed, its genetic material and the bio-social record of a crop's evolutionary history. Peres (2016) suggests that seeds are individual records of a crop's evolutionary history; from this framing, we extrapolate that the seed functions as the 'document' within the accession 'folder', which is a component of the genebank as 'archive'. However, we want to suggest a more complicated, nested relationship in which we might consider each seed to also function as a form of biosocial archive in its own right. We suggest this is the case in the sense that each seed holds within its genetic material records of localized crop experimentation and natural and cultural selection which, although partial and iterative, describe histories of agricultural activity which may extend back in time to the earliest prehistoric experimentation with domestication of crop species. These seeds could thus be characterized, as van Dooren (2007: 83) does, as archives of "inter-generational, inter-species, human/ plant kinship relations". In relation to the ICARDA accession withdrawal, the genebank manager was also quoted as saying, "When you trace back the history of these seeds, [you think of] the tradition and the heritage that they captured... They were maintained by local farmers from generation to generation, from father to son and then all the way to ICARDA's genebank and from there to the Global Seed Vault in Svalbard" (Alabaster 2015, n.p.). Whilst each individual seed may only record the outcomes of particular processes of natural and cultural selection, in the sense that these are 'inscribed' in the genetic material of the seed itself, holding these seeds at low temperatures would potentially halt the genetic erosion that might occur in situ through a combination of natural and cultural processes. Thus, the cumulative (meta-) archive of the SGSV conserves not only genetic agrobiodiversity, but also individual archives (seeds) which contain a series of specific biological-historical accounts (genes) of multispecies biosocial relations.

If the nature of the SGSV is complicated by this articulation of a more intricate, nested relationship of document to folder to archive, it is even further complicated by its relationship with time, and with the forms of diversity it holds in its repository. In freezing crop seeds as archives that map global genetic diversity from different points in time, each of which contains echoes or fragments of the diversity of past multispecies biosocial processes, the SGSV intervenes in the normative, entropic decay of diversity, 'banking' a record of past and present genetic diversity in frozen, arrested time. As in Radin's (2017) account of frozen blood and tissue samples discussed earlier, the values of these collections are banked as latent values, which are only to be realized at some future moment in time. In conjunction with ongoing processes of in-situ agrobiodiversity maintenance, themselves subject to continuing processes of natural and cultural selection that alter contemporary global agrobiodiversity, the vault's collection reverses the entropic process of diversity decay by increasing global crop genetic diversity. It does this because in-situ conservation (working through time) goes on producing other, new forms of agrobiodiversity while ex-situ conservation (working through frozen time) maintains older diversity into the future, thus increasing global diversity overall.

The Crop Trust suggests that "the Vault is the ultimate insurance policy for the world's food supply, offering options for future generations to overcome the challenges of climate change and population growth. It will secure, for centuries, millions of seeds representing every important crop variety available in the world today. It is the final back up" (Crop Trust 2016a). But the notion of a 'backup' here, which implies that duplicate accessions remain (biologically and socially) functionally equivalent, belies the complicated biosociotechnical and discursive shifts that occur within the repository which, along with the possibility of further genetic changes within cold storage (e.g. Soleri and Smith 1999), mean that which is deposited is fundamentally transformed by the process, creating something significantly different in ex-situ conservation when compared to that which is conserved in situ. In this sense, the operations of the SGSV seem to hold much in common with other archives,

Figure 5.2 — Shelves storing boxed samples of the world's seeds
inside the SGSV. (Photograph by Rodney Harrison).

where the materials contained are reconfigured and acquire new forms
of significance through their archival deposition (e.g. Stoler 2009). They
also have in common the idea of the archive as a place in which different
forms of relations are ordered and shaped, and which in turn shape and

order the worlds to which these archives refer (e.g. Joyce 1999; Bowker 2005a; Bennett et al. 2017). As such, the SGSV as meta-archive also constitutes its own biosocial record of specific, historically embedded, neoliberal practices of multispecies relationships, i.e., the attempts to mediate modernized agriculture through ex-situ conservation that emerged in the latter part of the twentieth century. The vault contains many 'times' and their associated biosocial relations – from the Neolithic through to our own 'end' times in the Anthropocene. Each of these times is folded together, neatly filed and frozen alongside the others in aluminium packaging (see Figure 5.2).

These complicated temporal operations within the repository, in turn, contribute to the accumulation of forms of biocapital by SGSV that are different to those values that accrue within the national and regional genebanks providing their 'duplicate' samples to the SGSV. These biocultural values draw not only on the added prestige derived from belonging to the 'global' seed vault – as part of the 'final' backup – and from the specific stories (e.g. the Syrian withdrawal) associated with objects contained within it, but also, through processes of genetic shift, to the addition of novel forms of biodiversity to the frozen, latent life contained within its archive. If the metaphor of a 'backup' is only partially accurate then, its designation as a *bank* in this process of the creation and accumulation of new forms of biocapital seems far more apposite (see also Bowker 2000, 2005b). This is a notion we will now explore in more detail in relation to another biobanking initiative, The Frozen Ark.

The Frozen Ark

The UK registered charity The Frozen Ark is at the forefront of 'saving cells and DNA of endangered species' before they become extinct. The goal is for the Frozen Ark to become both a physical and an open-access virtual biobank that stores, manages and safeguards biological material from the world's threatened species, and to connect researchers on a global level. Founding partners include the London Natural History Museum and the Zoological Society of London, as well as the University of Nottingham, which provides laboratory and office space and serves as the seat of the Frozen Ark, while research is now mainly being carried out at Cardiff University. At the time of writing its consortium of zoos, aquaria and other conservation bodies counted twenty-seven national and international partners from all over the world (Costa and Bruford 2018). The apocalyptic message conveyed by the project's logo, a "stylised ark on stormy seas" (Chrulew 2017), is both clear and urgent: in the face of anthropogenic ecological loss, the collecting, storing and managing of biological material from endangered species might be the only chance for humanity and the species with which we co-habit the planet. Yet, unlike Noah's Ark which, according to the Genesis flood narrative carried a male and a female of all the world's animals to save them from extinction by drowning, the Frozen Ark is a 'cryogenic' or 'technoscientific ark' (Parry 2004) that adheres to its website's motto of "saving cells and DNA of endangered species" – materials which act as ex-situ proxies of the living species they were taken from.

The University of Nottingham currently provides two −80°C freezers storing just over 700 blood and tissue samples obtained from endangered non-human animals from UK-based zoos and aquaria. The charity's collection in Nottingham consists of over 700 samples from a number of different animals including the scimitar horned oryx (extinct in the wild), the Colombian spider monkey, pileated gibbon, siamang gibbon, lar gibbon, snow leopard and Malayan tapir (all endangered). These are held as cultured cells, tissue and gametes stored in liquid nitrogen (see Figures 5.3 and 5.4). When our researchers visited the Nottingham laboratory we were shown how information on all the samples stored there is organized in physical file folders and includes, amongst other details, an internal identification number, a universal zoo number, the species, type and location of sample, what it is preserved in, sample quality and, where applicable, a Whatman FTA card. The Frozen Ark's interim Director, based in Cardiff, indicated that the ultimate objective is to form a confederated model that functions as both a physical and virtual infrastructure, storing and managing the genetic material from endangered species, sampled in the wild and in zoos and aquaria, from all over the world. At the time of writing, CryoArks, a Cardiff-based and BBSRC-funded (UK Biotechnology and Biosciences Research Council) initiative resulting from a collaboration between The Frozen Ark and some of its partner institutions as well as the UK node of the EAZA (European Association

Figure 5.3 — Inside one of the Frozen Ark's −80°C freezers, University
of Nottingham. (Photograph by Esther Breithoff).

of Zoos and Aquaria) biobank, is in the process of being established. Due
to limitations imposed by the Nagoya Protocol, which, according to the
Convention on Biological Diversity website, ensures 'the fair and equitable sharing of benefits arising from the utilization of genetic resources,

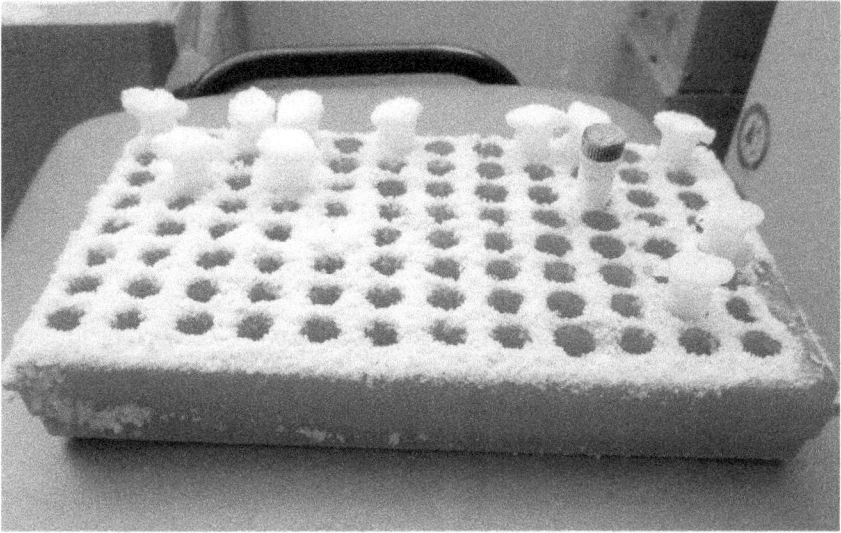

Figure 5.4 — Cryopreserved DNA samples stored in a −80°C freezer in the Frozen
Ark laboratory, University of Nottingham. (Photograph by Esther Breithoff).

thereby contributing to the conservation and sustainable use of biodiversity', CryoArks will be mainly focused on the UK and Ireland, whereas The Frozen Ark has a global remit – to this end it has already started cataloguing samples of extinct, endangered and threatened[1] species held by consortium members, and is aiming to increase the number of, and coordination between, consortium members. Unlike other biobanks around the world (e.g. San Diego Frozen Zoo®, Smithsonian Biobank, Svalbard Global Seed Vault), which intend to form a single point on Earth where genetic material from all over the world is being stored inside a central biobank, the Frozen Ark aims to be a Nagoya-compliant backup storage facility for institutions that, due to various reasons, cannot store their own samples or would like to have duplicates of existing collections and to hold centralized records relating to a distributed network of physical biobanks which store biosamples of endangered non-human animals.

Based on interviews with the charity's staff undertaken during Breithoff's lab placement in Cardiff and subsequently with staff based in Nottingham, The Frozen Ark's concern for preservation of genetic material for future generations initially outweighed active conservation efforts. With species going extinct all over the world and the dramatic anticipated loss of genetic information, The Frozen Ark eventually decided to change from acting purely as a repository to become an active collection. This decision seems to have been influenced partially by the emergence of new experimental genetic work, but also reflected a change in philosophy about the Ark's role. "The focus was always for the future," reflects Jude Smith, who has been the charity's administrator from the beginning, "but as we've gone on, it has become really obvious that the future is here now, you know, it's now" (interview with Esther Breithoff, August 31, 2017).

This new approach, described as more "pragmatic" by current Interim Director, Professor of Biodiversity and Conservation geneticist Mike Bruford (interviews with Esther Breithoff, July 18 and 21, 2017), recognizes the need to boost the profile of the charity in order to deliver on its promises for the future: the collection, safeguarding and management of biological and genetic material from endangered species for both anticipated and unanticipated future uses. The vision is for the Frozen Ark to become an active and ethical facility for genomic resource management that helps identify and prioritize which animal species are at risk of extinction and thus in need of sampling, and develop the most effective techniques of collecting, storing and managing biological material. In its educational role, the Frozen Ark supports institutions both in the UK and abroad with setting up their own biobank facilities and/or successfully managing already existing repositories. According to the charity, its main goals are:

> (i) coordinating global efforts in animal biobanking; (ii) sharing expertise; (iii) offering help to organizations and governments that wish to set up biobanks in their own countries; (iv) providing the physical and informatics infrastructure that will allow conservationists and researchers to search for, locate and use this material wherever possible without having to resample from wild populations (Costa and Bruford 2018).

In the current absence of coordination and lack of shared protocols and databases between different biobanks nationally and internationally, The Frozen Ark plans on setting up a virtual stand-alone open-access database connecting existing biobanks on a global level. This would facilitate increased access to research material for researchers and conservationists internationally. The Frozen Ark sees its role in safeguarding and managing genetic diversity as part of a joint effort between ex-situ and in-situ conservation practices. Cryostoring biomaterial of endangered species in freezers and liquid nitrogen tanks – although space effective – does however come with a high carbon footprint, which one could suggest ultimately increases the threat of extinction to the animals it was designed to protect. The Frozen Ark website emphasizes that establishing and maintaining a global biobank at present is also a costly undertaking that has been suffering from a lack of funding since its inception.

> Time is running out for many species. Conservation efforts will undoubtedly save some but we must preserve the genetic record of all endangered species for our future. Time is also running out for the Frozen Ark, which has been running with volunteers on a shoestring budget for several years. Help us save Nature's genetic heritage so that future generations can enjoy the natural world as we have all done (The Frozen Ark 2018).

Like the endangered species whose biological material it aims to secure in the race against irretrievable loss of biodiversity, The Frozen Ark itself also senses a risk of its own endangerment in articulating these difficulties of establishing long-term funding to secure its future operations. These issues of uncertainty relating to the securing of ongoing financial resources for the organization's research and collections were a regular topic of discussion – in the laboratory, in conferences and in more formal interview contexts – and form another of the various ways in which the urgency of the work of the organization, and biodiversity conservation more generally, is expressed.

Playing God in the Anthropocene:
Biodiversity, cryopreservation and future-making

In 1993, the Stephen Spielberg film *Jurassic Park* seemed to offer an improbable view of an alternative future in which long extinct species could be regenerated from ancient DNA. We have shown that initiatives to collect and store the raw materials for such a process in the form of frozen blood, tissue and other human and nonhuman animal organic materials have a much longer genealogy. However, recent developments in genetic rescue programmes which aim to revive extinct and threatened animal species suggest such genomic engineering is scientifically possible. Several projects that sound equally implausible – including work currently being undertaken by Revive and Restore (https://reviverestore.org/) to clone extinct passenger pigeons and woolly mammoths – are likely to realize results within the next decade (e.g. see Jørgensen 2013; Shapiro 2015). Sherkow and Greely (2013) explain that the three approaches which appear most likely to yield results are back-breeding, in which selective breeding is used to produce the phenotypes of extinct species; cloning using cryopreserved tissue; and genetic engineering using whole genome sequencing and the editing of DNA in cells from genetically similar extant animals. In many ways these projects constitute a realization of the latent futures which are resourced by frozen zoos and cryopreservation technologies. The move within The Frozen Ark away from perceiving its role primarily as a passive collecting institution for the future, to one of active experimental conservation in the present, exemplifies the ways in which such collections resource the development of new realities in which the possibilities of reviving extinct species through hybridization with extant ones is increasingly becoming fulfilled. But in their enabling of certain forms of what Vidal and Dias helpfully term "restitution fantasies" (2016: 1) they also reinforce dominant (although not uncomplicated – see Dibley 2012, 2015) forms of anthropocentrism which remain barely hidden within the Anthropocene chronotope (cf. Pratt 2017) in the fulfilment of humanity's ultimate mastery over nature: the ability to resurrect the species that we have ourselves rendered extinct. The quest for such a reality is embodied in The Frozen Ark's own creation narrative in which the founders' attempts to save the Partula land snail through more conventional methods of captive breeding are unsuccessful and force them to turn to cryopreservation for future hybridization and de-extinction programmes as the last hope for this totemic species.

From ark to bank: Biodiversity and biocapital

It is in the transformation of these latent possibilities into new economic (as well as ecological) realities that we are able to determine shifts in the nature of biobanking facilities and the forms of value they both generate and are caught up within (Shukin 2009). A significant literature in science studies, which develops and expands upon Foucault's 'late' work on biopower/biopolitics, has traced the development of what Cooper (2008) terms the 'bioeconomy' since the 1970s in the specific relations of biotechnology, neoliberal politics and economic policy (e.g. see Doyle 1997; Thacker 2006; Rose 2006; Waldby and Mitchell 2006; Shukin 2009; Franklin 2013; Cooper and Waldby 2014). Central to the bioeconomy has been the emergence and evolution of a range of new forms of 'biocapital'. We draw on Helmreich's (2008; see also Rajan 2006) definition of biocapital as the surplus values generated by the commodification and circulation of forms of biological life within economic systems. Helmreich points out, however, that biocapital is understood and deployed in a number of different ways by scholars across science studies and itself may manifest in a range of different forms, as parts of different sociomaterial assemblages. It is the ways in which biocapital emerges flexibly and replicates itself across these different sociomaterial assemblages which concerns us here. Given the significance of the study of concepts of value to critical heritage studies, we might ask how cryobanks such as The Frozen Ark have contributed to the development of new forms of value? And in what ways are those new values accumulated and distributed within the bioeconomy?

In his influential paper which originally developed the concept of the Frozen Zoo in 1984, Benirschke observes the relationship between the growth of cryopreservational technologies and the dwindling biological resources these are produced to conserve. As biodiversity (bearing in mind that this concept is itself plastic and subject to shifts in meaning) diminishes, the value of these banked biomaterials increases both individually and collectively. As we have argued in relation to the work of seed biobanks, these processes are forms of *speculative biocapital accumulation*, banking on, yet simultaneously imaginatively resourcing, the development of the biotechnologies which will realize these future values. Thus extinction, biobanking, biocapital and biodiversity

176 Esther Breithoff & Rodney Harrison

come to be linked in a complicated network of values within the emerging bioeconomy. In its speculation on, and investment in the anticipation of loss, the work of The Frozen Ark (and the field of biodiversity cryopreservation more generally) can also be understood to represent a response to neoliberal economics in the ways in which it constitutes an optimization of the use of space and resources. Cryobanking "represents a technically viable method for helping to conserve species biodiversity, without having to maintain large captive populations of each organism" (Hosey, Malfi and Pankhurst 2009: 319 as quoted by Chrulew 2017), nor, indeed, the designated landscapes in which these organisms might conventionally be preserved (as national parks, for example). As Chrulew goes on to surmise from these comments in his own discussion of The Frozen Ark, "the forms of preservation and exchange made possible by the frozen zoo transform the relationships between humans, animals, and technologies, reorganizing space and time beyond familiar constraints in the interests of optimal efficiency and diversity" (Chrulew 2017: 297).

The ability of biodiversity conservation to designate conservation proxies which are immutable, combinable mobiles (in the Latourian sense) is thus central to the ways in which biobanks function within a bioeconomy to accumulate biocapital. As Harrison (2017) has observed of the seeds in ex-situ seed banks, while these are conceptualized as copies of biomaterials held in other collections, or as we qualify here, not so much as copies as fragments of the original sample which remain authentic at the level of the DNA – indeed, as Chrulew (2017) notes, doubles of doubles held in captivity which are themselves doubles of wild animals – they are not, in fact, duplicates, as their presence within these particular biosocial archives allows them to accumulate new forms of value, and indeed, possible new genetic characteristics which do not directly replicate those from which they were originally copied. This is again reflected in the change of perception of the function of the Frozen Ark from repository – where frozen biomaterials would be collected untouched for the future, to speculative reinvestment – where such biomaterials would be part of active and ongoing genetic experimentation with saving threatened species and potentially reversing extinction, in particular through the generation of hybrids which combine genetic materials from both living and extinct species. Finally, cryobanking reconfigures relationships between life and death. Talking in the context of frozen genetic material from humans, Lemke observes that:

> 'human material' transcends the living person. The person who dies today is not really dead. He or she lives on, at least potentially. Or more precisely, parts of a human being – his or her cells or organs, blood, bone marrow and so on – can continue to exist in the bodies of other people, whose quality of life they improve or who are spared death through their incorporation. The organic materials of life are not subordinate to the same biological rhythms as the body is. These materials can be stored as information in biobanks or cultivated in stem cell lines. Death can be part of a productive circuit and used to improve and extend life. The death of one person may guarantee the life and survival of another. Death has also become flexible and compartmentalized (Lemke and Flett 2012: 95).

Similarly, biotechnologies employed by The Frozen Ark allow for the breaking down of species into a range of components at the biomolecular level which allow for almost endless recombination (Doyle 1997; Chrulew 2017), further complicating the question of the relevant units by which biodiversity might be measured, and the relative values of such units and their proxies. The importance of the late-capitalist context of these developments cannot be overstated. This extension of life and expansion of what constitutes biological reproduction is a function of what Cooper refers to as the bioeconomy's transformation of biological life into surplus value (Cooper 2008; see also Shukin 2009; Thacker 2010). As in the case of the Svalbard Global Seed Vault (see Harrison 2017), the operations of The Frozen Ark can be understood to accumulate and generate surplus value through reversing what are perceived to be 'natural' as well as humanly produced entropic processes of biodiversity decay (Sepkoski 2016); but importantly, the new forms of value which it produces are not simply inherent to its proxies themselves, but also derive from the latent (cf. Radin 2017) potential for new and experimental forms of life they may be used to produce. In this sense,

The Frozen Ark contributes to what Radin terms a form of 'planned hindsight' (Radin 2015) – it realizes its own technofutures through its collecting policies in the present. Its latent generation of future value in the form of biocapital requires direct speculation upon the extinction and biodiversity loss it is created to secure the present against. The Frozen Ark counterintuitively depends upon the future biodiversity loss which it works against, but simultaneously anticipates, in its present operations.

Discussion and conclusion

Biobanks such as SGSV and The Frozen Ark have come to play a significant role in nature conservation efforts globally and are a driving force behind biodiversity research (see Arbeláez-Cortés et al. 2015; Gemeinholzer et al. 2011; Fletcher 2008; Segreto et al. 2010; Lermen et al. 2009). Efforts to conserve endangered plant and animal species range from cold storage and cryo-preservation to digital preservation (Pizzi et al. 2013) and experiments in hybridization and de-extinction. Fertility preservation strategies based on cryopreservation are engaged to promote the sustainability and protection efforts of rare and endangered animal species, particularly "to assist managing or 'rescuing' the genomes of genetically valuable individuals" (Comizzoli and Holt 2014; see also Comizzoli 2015a and Comizzoli 2015b). The lack of public knowledge about such ex-situ biobanks and their purpose constitutes a significant problem (Gaskell and Gottweis 2011). Etchegary et al. (2013, 2015) argue for the active involvement of the public in discussions on genetic research in order for their views and expectations to be acknowledged by the research community. Gaskell and Gottweis propose that "science communication must go beyond the simple dissemination of basic information" (Gaskell and Gottweis 2011: 160). Elsewhere, Gottweis, Gaskell and Starkbaum contend the "development of a communications strategy that reaches and informs potential study participants and the public and gives them the basics of what biobanks are good for" (2011: 739) is urgently required. Within this context, the principle of reciprocity should be at the core of biobanking: scientists should engage with the public, be transparent about their research and give feedback to research participants to allow people "to feel like they are part of something larger and that their donation feeds into a mutual, respectful relation-ship" (ibid). In order to both strengthen communication between bio-bank builders and participants, and to give the latter more general input, researchers are proposing various forms of public engagement, such as "a wiki-governance model for biobanks that harnesses Web 2.0, and which gives citizens the ability to collaborate in biobank governance and policy-making" (Dove, Joly and Knoppers 2012: 158). But we argue that con-servationists must go further than this too, to acknowledge the ways in which these and aligned practices not only preserve but actively *remake* nature in the Anthropocene, building new, divergent worlds and signifi-cantly influencing their futures.

If biodiversity conservation is viewed through the lens of a criti-cal exploration of the forms of value it generates and the interactions of those forms of value on and with one another, ex-situ biobanks (like the FA and SGSV) can no longer be thought of as dormant genetic 'arks' but rather as 'investment banks' which accumulate and produce values through speculation upon the forms of extinction which they themselves seek to resist through their reconfiguration of post-genomic life. One might argue that the newly emergent bioeconomy discussed here con-stitutes the logical product of a recognition of the Anthropocene as the period in which humans have become the primary force of global geo-logical and climatological change (e.g. Lewis and Maslin 2018). Rather than imagining that these biopreservational technologies and institutions are passively 'preserving' biodiversity, we might suggest instead that they are engaged in a complicated process of *making new forms of value* and contributing directly to the development of a new late-capitalist bio-economy. This has significant implications for how we understand the future of 'nature'. Such developments point to the significance of critical approaches to contemporary ex-situ nature conservation practices, and suggest that rather than *conserving* nature in the light of human interven-tion, that such initiatives are themselves part of the accelerated human intervention in the management of planetary species diversity which might be seen to both define and be defined by new post-Anthropocene worlds, in which biodiversity conservation practices constitute active processes of *making* (human) natures.

Acknowledgements

The research presented in this chapter draws on field visits, interviews and other collaborations with The Frozen Ark and SGSV undertaken by the authors and Sefryn Penrose as part of a broader comparative study of natural and cultural diversity conservation practices, one of four major areas of thematic foci for the 'Heritage Futures' research programme. Heritage Futures was funded by an Arts and Humanities Research Council (AHRC) 'Care for the Future: Thinking Forward through the Past' Theme Large Grant (AH/M004376/1), awarded to Rodney Harrison (as principal investigator), Caitlin DeSilvey, Cornelius Holtorf, Sharon Macdonald (as co-investigators), Antony Lyons (as senior creative fellow), Martha Fleming (as senior postdoctoral researcher), Nadia Bartolini, Sarah May, Jennie Morgan and Sefryn Penrose (as named postdoctoral researchers). Three PhD students were additionally funded as in-kind support for the research programme by their respective host universities: Kyle Lee-Crossett (University College London), Bryony Prestidge (University of York) and Robyn Raxworthy (University of Exeter). Martha Fleming left to focus on other responsibilities during the research programme's first year; the team of researchers was subsequently joined by Esther Breithoff as postdoctoral researcher and by Hannah Williams as administrative assistant and events coordinator, a role which was in turn later filled by Kyle Lee-Crossett, and in its final year, by Harald Fredheim (as postdoctoral researcher). Heritage Futures also received generous additional support from its host universities and partner organizations. See www.heritage-futures.org for further information. Rodney Harrison's work on this chapter was also supported by his AHRC Heritage Priority Area Leadership Fellowship Grant (AH/P009719/1). See www.heritage-research.org for further information.

Notes

1.   These terms have specific technical definitions which relate to the categories established by the IUCN Red List of Endangered Species.

References

Adam, B. and C. Groves. 2007. *Future Matters: Action, Knowledge, Ethics.* Leiden: Brill.

Alabaster, O. 2015. Syrian Civil War: Svalbard 'Doomsday' Seeds Transferred to Lebanon to Preserve Syria's Crop Heritage. *The Independent Newspaper,* 10 October 2015. http://www.independent.co.uk/news/world/middle-east/syrian-civil-war-svalbard-doomsday-seeds-transferred-to-lebanon-to-preserve-syrias-crop-heritage-a6689421.html.

Appadurai, A. 2013. *The Future as Cultural Fact: Essays on the Global Condition.* London: Verso.

Arbeláez-Cortés, E., M. F. Torres, D. López-Álvarez, J. D. Palacio-Mejía, Á. M. Mendoza and C. A. Medina. 2015. Colombian Frozen Biodiversity: 16 Years of the Tissue Collection of the Humboldt Institute. *Acta Biológica Colombiana* 20(2): 163–73.

Barad, K. M. 2007. *Meeting the Universe Halfway: Quantum Physics and the Entanglement of Matter and Meaning.* Durham, NC; London: Duke University Press.

Benirschke, K. 1984. The Frozen Zoo Concept. *Zoo Biology* 3: 325–28.

Bennett, T., F. Cameron, N. Dias, B. Dibley, I. Jacknis, R. Harrison and C. McCarthy 2017. *Collecting, Ordering, Governing: Anthropology, Museums and Liberal Government.* Durham, NC: Duke University Press.

Benson, E. 2010. *Wired Wildness: Technologies of Tracking and the Making of Modern Wildlife.* Baltimore, MD: Johns Hopkins University Press.

Bird Rose, D. 2013. *Wild Dog Dreaming: Love and Extinction.* Charlottesville, VA: University of Virginia Press.

Bird Rose, D., T. van Dooren and M. Chrulew (eds). 2017. *Extinction Studies: Stories of Time, Death, and Generations.* New York: Columbia University Press.

Bowker, G. C. 2000. Biodiversity Datadiversity. *Social Studies of Science* 30: 643–84.

Bowker, G. C. 2005a. *Memory Practices in the Sciences.* Cambridge, MA: MIT Press.

Bowker, G. C. 2005b. Time, Money and Biodiversity. In *Global Assemblages: Technology, Politics and Ethics as Anthropological Problems,* edited by A. Ong and S. J. Collier, 107–23. Cambridge, MA: Blackwell.

Bowkett, A. E. 2009. Recent Captive-Breeding Proposals and the Return of the Ark Concept to Global Species Conservation. *Conservation Biology* 23: 773–76.

Breithoff, E. 2020. *Conflict, Heritage and World-Making in the Chaco: War at the end of the Worlds?* London: UCL Press.

Buller, H. 2014. Animal geographies I. *Progress in Human Geography* 38(2): 308–318.

de la Cadena, M. and M. Blaser (eds). 2018. *A World of Many Worlds*. Durham, NC: Duke University Press.

Cahill, A. E., M. E. Aiello-Lammens, M. C. Fisher-Reid, X. Hua, C. J. Karanewsky, H. Y. Ryu, G. C. Sbeglia, F. Spagnolo, J. B. Waldron, O. Warsi and J. J. Wiens 2012. How Does Climate Change Cause Extinction? *Proceedings of the Royal Society. B. Biological Sciences* 280: 20121890.

Chakrabarty, D. 2009. The Climate of History: Four Theses. *Critical Inquiry* 35(2): 197–222.

Chemnick, L. G., M. L. Houck and O. A. Ryder. 2009. Banking of Genetic Resources. In *Conservation Genetics in the Age of Genomics*, edited by G. Amato, O.A. Ryder, H. Rosenbaum, and R. DeSalle, 124–30. New York: Columbia University Press.

Chrulew, M. 2017. Freezing the Ark: The Cryopolitics of Endangered Species Preservation. In *Cryopolitics: Frozen Life in a Melting World*, edited by J. Radin and E. Kowal, 283–306. Cambridge, MA: MIT Press.

Comizzoli, P. 2015a. Biotechnologies for Wildlife Fertility Preservation. *Animal Frontiers* 5(1): 73–78.

Comizzoli, P. 2015b. Biobanking Efforts and New Advances in Male Fertility Preservation for Rare and Endangered Species. *Asian Journal of Andrology* 17(4): 640–45.

Comizzoli, P. and W. V. Holt. 2014. Recent Advances and Prospects in Germplasm Preservation of Rare and Endangered Species. In *Reproductive Sciences in Animal Conservation*, edited by W. Holt, J. Brown and P. Comizzoli, 331–56. New York: Springer.

Cooper, M. E. 2008. *Life as Surplus: Biotechnology and Capitalism in the Neoliberal Era*. Seattle, WA: University of Washington Press.

Cooper, M. and C. Waldby. 2014. *Clinical Labor: Tissue Donors and Research Subjects in the Global Bioeconomy*. Durham, NC: Duke University Press.

Corley-Smith, G. E. and B. P. Brandhorst. 1999. Preservation of Endangered Species and Populations: A Role for Genome Banking, Somatic Cell Cloning, and Androgenesis? *Molecular Reproduction and Development* 53: 363–67.

Costa, M. and M. Bruford. 2018. The Frozen Ark Project: Biobanking and Endangered Animal Samples for Conservation and Research. *Inside Ecology*, 12 January 2018. https://insideecology.com/2018/01/12/the-frozen-ark-project-biobanking-endangered-animal-samples-for-conservation-and-research/

Crop Trust. 2016a. Svalbard Global Seed Vault. https://www.croptrust.org/what-we-do/svalbard-global-seed-vault/

Crop Trust. 2016b. FAQ about the Seed Vault. https://www.croptrust.org/what-we-do/svalbard-global-seed-vault/faq-about-the-vault/

DeSilvey, C. 2017. *Curated Decay: Heritage beyond Saving*. Minneapolis, MN; London: University of Minnesota Press.

Dibley, B. 2012. 'The Shape of Things to Come': Seven Theses on the Anthropocene and Attachment. *Australian Humanities Review* 52: 139–53.

Dibley, B. 2015. Anthropocene: The Enigma of 'The Geomorphic Fold'. In *Animals in the Anthropocene: Critical Perspectives on Non-Human Futures*, edited by HAR, 19–32. Sydney: Sydney University Press.

Dove, E. S., Y. Joly and B. M. Knoppers. 2012. Power to the People: A Wiki-Governance Model for Biobanks. *Genome Biology* 13(5): 158.

Doyle, R. 1997. *On Beyond Living: Rhetorical Transformations of the Life Sciences*. Stanford, CA: Stanford University Press.

Etchegary, H., J. Green, E. Dicks, D. Pullman, C. Street and P. Parfrey. 2013. Consulting the Community: Public Expectations and Attitudes about Genetics Research. *European Journal of Human Genetics* 21(12): 1338–43.

Etchegary, H., J. Green, P. Parfrey, C. Street and D. Pullman. 2015. Community Engagement with Genetics: Public Perceptions and Expectations about Genetics Research. *Health Expectations* 18(5): 1413–25.

Fenzi, M. and C. Bonneuil. 2016. From 'Genetic Resources' to 'Ecosystems Services': A Century of Science and Global Policies for Agrobiodiversity Conservation. *Culture, Agriculture, Food and Environment* 38(2): 72–83.

Fletcher, A. L. 2008. Mendel's Ark: Conservation Genetics and the Future of Extinction. *Review of Policy Research* 25: 598–607.

Fowler, C. 2008. *The Svalbard Global Seed Vault: Securing the Future of Agriculture*. The Global Crop Trust. https://blogs.worldbank.org/sites/default/files/climatechange/The%20Svalbard%20Seed%20Vault_Global%20Crop%20Diversity%20Trust%202008.pdf

Franklin, S. 2013. *Biological Relatives: IVF, Stem Cells, and the Future of Kinship*. Durham, NC: Duke University Press.

Friese, C. 2013. *Cloning Wild Life: Zoos, Captivity, and the Future of Endangered Animals*. New York: New York University Press.

The Frozen Ark. 2018. Home. https://www.frozenark.org/.

Gaskell, G. and H, Gottweis. 2011. Biobanks Need Publicity. *Nature* 471 (7337): 159–60.

Gemeinholzer, B., G. Dröge, H. Zetzsche, G. Haszprunar, H.-P. Klenk, A. Güntsch, W. G. Berendsohn and J.-W. Wägele. 2011. The DNA Bank Network: The Start from a German Initiative. *Biopreservation and Biobanking* 9: 51–55.

Gottweis, H., G. Gaskell and J. Starkbaum. 2011. Connecting the Public with Biobank Research: Reciprocity Matters. *Nature Reviews Genetics* 12(11): 738–39.

Government of Norway. 2015. Svalbard Global Seed Vault: More about the Physical Plant. Last updated February 23, 2015. https://www.regjeringen.no/en/topics/food-fisheries-and-agriculture/landbruk/svalbard-global-seed-vault/mer-om-det-fysiske-anlegget/id2365142/.

Haraway, D. J. 1991. *Simians, Cyborgs, and Women: The Reinvention of Nature.* London: Free Association.

Haraway, D. J. 2016. *Staying with the Trouble: Making Kin in the Chthulucene.* Durham, NC; London: Duke University Press.

Harrison, R. 2015. Beyond 'Natural' and 'Cultural' Heritage: Toward an Ontological Politics of Heritage in the Age of Anthropocene. *Heritage and Society* 8(1): 24–42.

Harrison, R. 2017. Freezing Seeds and Making Futures: Endangerment, Hope, Security, and Time in Agrobiodiversity Conservation Practices. *Culture, Agriculture, Food and Environment* 39(2): 80–89.

Harrison, R., C. DeSilvey, C. Holtorf, S. Macdonald, N. Bartolini, E. Breithoff, L.H. Fredheim, A. Lyons, S. May, J. Morgan and S. Penrose. 2020. *Heritage Futures: Comparative approaches to natural and cultural heritage practices.* London: UCL Press.

Heatherington, T. 2012. From Ecocide to Genetic Rescue: Can Technoscience Save the Wild? In *The Anthropology of Extinction: Essays on Culture and Species Death*, edited by G. M. Sodikoff, 39–66. Bloomington, IN: Indiana University Press.

Heise, U. K. 2016. *Imagining Extinction: The Cultural Meanings of Endangered Species.* Chicago, IL: University of Chicago Press.

Helmreich, S. 2008. Species of Biocapital. *Science as Culture* 17: 463–78.

Hosey, G., V. Malfi and S. Pankhurst. 2009. *Zoo Animals: Behaviour, Management, and Welfare.* Oxford: Oxford University Press.

Howard, J.G., A. M. Donoghue, M. A. Barone, et al. 1992. Successful Induction of Ovarian Activity and Laparoscopic Intrauterine Artificial Insemination in the Cheetah (Acinonyx Jubatus). *Journal of Zoo and Wildlife Medicine* 23(3): 288–300.

Howard, J.G., C. Lynch, R. M. Santymire, P. E. Marinari and D. E. Wildt. 2016. Recovery of Gene Diversity Using Long-Term Cryopreserved Spermatozoa and Artificial Insemination in the Endangered Black-Footed Ferret: Black-Footed Ferret Gene Restoration. *Animal Conservation* 19: 102–11.

Hummer, K. E. 2015. In the Footsteps of Vavilov: Plant Diversity Then and Now. *Horticultural Science* 50(6): 784–8.

Jørgensen, D. 2013. Reintroduction and De-Extinction. *BioScience* 63: 719–20.

Joyce, P. 1999. The Politics of the Liberal Archive. *History of the Human Sciences* 12(2): 35–49.

Kirchner, J. W. and A. Weil. 2000. Delayed Biological Recovery from Extinctions throughout the Fossil Record. *Nature* 404: 177–80.

Lemke, T. and I. Flett. 2012. Second Nature: In the Age of Biobanks. *The Yearbook of Comparative Literature* 58: 188–92.

Lermen, D., B. Blömeke, R. Browne, et al. 2009. Cryobanking of Viable Biomaterials: Implementation of New Strategies for Conservation Purposes. *Molecular Ecology* 18(6): 1030–33.

Lewis, S. L. and M. A. Maslin. 2018. *The Human Planet: How We Created the Anthropocene*. London: Pelican Books.

Lorimer, J. 2015. *Wildlife in the Anthropocene: Conservation after Nature*. Minneapolis, MN: University of Minnesota Press.

Marris, E. 2013. *Rambunctious Garden: Saving Nature in a Post-Wild World*. New York: Bloomsbury.

NordGen. 2016. Seed Portal of the Svalbard Global Seed Vault. http://www.NordGen.org/sgsv/index.php?page=sgsv_information_sharing

O'Connor, M. R. 2015. *Resurrection Science: Conservation, De-Extinction and the Precarious Future of Wild Things*. New York, NY: St. Martin's Press.

Omura, K., G. J. Otsuki, S. Satsuka and A. Morita (eds). 2018. *The World Multiple: The Quotidian Politics of Knowing and Generating Entangled Worlds*. Abingdon and New York: Routledge.

Parry, B. 2004. *Trading the Genome: Investigating the Commodification of Bio-Information*. New York: Columbia University Press.

Peres, S. 2016. Saving the Gene Pool for the Future: Seed Banks as Archives. *Studies in History and Philosophy of Biological and Biomedical Sciences* 55: 96–104.

Pilcher, H. 2016. *Bring Back the King: The New Science of De-Extinction*. London: Bloomsbury Press.

Pistorius, R. 1997. *Scientists, Plants and Politics: A History of the Plant Genetic Resources Movement*. Rome: International Plant Genetic Resources Institute.

Pizzi, F., A. M. Caroli, M. Landini, N. Galluccio, A. Mezzelani and L. Milanesi. 2013. Conservation of Endangered Animals: From Biotechnologies to Digital Preservation. *Natural Science* 5: 903–13.

Pratt, M. L. 2017. Coda: Concept and Chronotope. *Arts of Living on a Damaged Planet*, edited by A. L. Tsing, N. Bubandt, E. Gan and H. A. Swanson, G169-G174. Minneapolis, MN: University of Minnesota Press.

Radin, J. 2015. Planned Hindsight: The Vital Valuations of Frozen Tissue at the Zoo and the Natural History Museum. *Journal of Cultural Economy* 8: 361–78.

Radin, J. 2016. Planning for the Past: Cryopreservation at the Farm, Zoo, and Museum. In *Endangerment, Biodiversity and Culture*, edited by F. Vidal and N. Dias, 218–38. Abingdon and New York: Routledge.

Radin, J. 2017. *Life on Ice: A History of New Uses for Cold Blood*. Chicago, IL: University of Chicago Press.

Radin, J. and E. Kowal (eds). 2017. *Cryopolitics: Frozen Life in a Melting World*. Cambridge, MA: MIT Press.

Rajan, K. S. 2006. *Biocapital: The Constitution of Postgenomic Life*. Durham, NC; London: Duke University Press.

Rose, N. 2006. *The Politics of Life Itself: Biomedicine, Power, and Subjectivity in the Twenty-first Century*. Princeton, NJ: Princeton University Press.

Salazar, J. F., S. Pink, A. Irving and J. Sjöberg (eds). 2017. *Anthropologies and Futures: Researching Emerging and Uncertain Worlds*. London; New York: Bloomsbury Academic.

Segreto, R., K. Hassel, R. Bardal and H. K. Stenøien. 2010. Desiccation Tolerance and Natural Cold Acclimation Allow Cryopreservation of Bryophytes without Pretreatment or Use of Cryoprotectants. *Bryologist* 113(4): 760–70.

Sepkoski, D. 2016. Extinction, Diversity, and Endangerment. In *Endangerment, Biodiversity and Culture*, edited by F. Vidal and N. Dias, 62–86. Abingdon: Routledge.

Shapiro, B. 2015. *How to Clone a Mammoth: The Science of De-Extinction*. Princeton, NJ: Princeton University Press.

Sherkow, J. S. and H. T. Greely. 2013. What If Extinction Is Not Forever? *Science* 340: 32–33.

Shukin, N. 2009. *Animal Capital: Rendering Life in Biopolitical Times*. Minneapolis, MN: University of Minnesota Press.

Singh, J. S. 2002. The Biodiversity Crisis: A Multifaceted Review. *Current Science* 82: 627–38.

Soleri, D. and S. E. Smith. 1999. Conserving Folk Crop Varieties: Different Agricultures, Different Goals. In *Ethnoecology: Situated Knowledge/Located Lives*, edited by V. D. Nazarea, 135–54. Tucson: University of Arizona Press.

Stoler, A. L. 2009. *Along the Archival Grain: Epistemic Anxieties and Colonial Common Sense*. Princeton, NJ; Oxford: Princeton University Press.

Thacker, E. 2006. *The Global Genome: Biotechnology, Politics, and Culture*. Cambridge, MA: MIT Press.

Thacker, E. 2010. *After Life*. Chicago, IL: University of Chicago Press.

Tsing, A. L. 2015. *The Mushroom at the End of the World: On the Possibility of Life in Capitalist Ruins*. Princeton, NJ: Princeton University Press.

Tsing, A. L., H. Swanson, E. Gan and N. Bubandt (eds). 2017. *Arts of Living on a Damaged Planet*. Minneapolis, MN; London: University of Minnesota Press.

Turner, S. S. 2007. Open-Ended Stories: Extinction Narratives in Genome Time. *Literature and Medicine* 26: 55–82.

United Nations. 2012. *The Future We Want. Resolution adopted by the General Assembly on 27 July 2012, A/RES/66/288*. http://www.un.org/ga/search/view_doc.asp?symbol=A/RES/66/288&Lang=E

van Dooren, T. 2007. Terminated Seed: Death, Proprietary Kinship and the Production of (Bio)Wealth. *Science as Culture* 16(1): 71–94.

van Dooren, T. 2009. Banking Seed: Use and Value in the Conservation of Agricultural Diversity. *Science as Culture* 18(4): 373–95.

van Dooren, T. 2016. Authentic Crows: Identity, Captivity and Emergent Forms of Life. *Theory, Culture & Society* 33(2): 29–52.

Vidal, F. and N. Dias. 2016. Introduction: The Endangerment Sensibility. In *Endangerment, Biodiversity and Culture*, edited by F. Vidal and N. Dias, 1–38. Abingdon: Routledge.

Waldby, C. and R. Mitchell. 2006. *Tissue Economies: Blood, Organs, and Cell Lines in Late Capitalism*. Durham, NC: Duke University Press.

Watson, P. and W. V. Holt. 2001. *Cryobanking the Genetic Resource: Wildlife Conservation for the Future?* London: Taylor and Francis.

Westley, F. 1997. 'Not on Our Watch': The Biodiversity Crisis and Global Collaboration Response. *Organization & Environment* 10: 342–60.

Wildt, D. E., W. F. Rall, J. K. Critser, S. L. Monfort and U. S. Seal. 1997. Genome Resource Banks. *BioScience* 47: 689–98.

Wisely, S. M., O. A. Ryder, R. M. Santymire, J. F. Engelhardt and B. J. Novak. 2015. A Road Map for 21st Century Genetic Restoration: Gene Pool Enrichment of the Black-Footed Ferret. *Journal of Heredity* 106: 581–92.

WWF (World Wildlife Fund). 2016. Living Planet Report: Risk and Resilience in a New Era. https://wwf.panda.org/wwf_news/?282370/Living-Planet-Report-2016

Chapter 6

Heritage as Critical Anthropocene Method

Colin Sterling

I worry that we've already swallowed the idea of the Anthropocene
and stopped considering the importance of it; the profound shock that
it should cause has already been diffused into just one more idea game
that we play. (Robinson 2017a: 146)

The 'profound shock' of the Anthropocene, as science fiction writer Kim Stanley Robinson describes it here, has reverberated in multiple directions at once over the past two decades. From the almost quasi-mythical first utterance of the concept by atmospheric chemist Paul Crutzen at a conference in Mexico City in 1999, Anthropocene ripples have been felt in politics, philosophy, art, history, pedagogy, political theory and popular culture, not to mention the various fields of science concerned with coding, measuring, analyzing and interpreting the Earth System. While Robinson locates the shock of the concept in the sci-fi like premise that humanity now constitutes a geological force on the same level as volcanoes, earthquakes and asteroid strikes, others have registered a more profound disquiet with the term itself, which they see as alarming in its hubris and underlying occlusions. Alternative labels such as Plantationocene (Tsing 2015), Pyrocene (Pyne 2015), Necrocene (McBrien 2016), Chthulucene (Haraway 2015) and – the most prevalent – Capitalocene (Moore 2017; Haraway 2016; Demos 2017) document an obligation to constantly disentangle and qualify the 'Anthropos' of the Anthropocene (strangely the 'cene' – or *kainos*, ancient Greek for 'new' – is deemed less toxic in such thinking, as if the constant demand for novelty was not also intrinsic to the capitalist system). As Joanna Zylinska has recently argued, "even though we are nowhere near solving the Anthropocene's climate issues, in some areas of critical theory we already find ourselves post-Anthropocene, it seems" (2018: 5).

These morphological transformations contain important lessons about the need for alternative histories and vocabularies to confront the Anthropocene as a material-discursive force in the world, but they also have something of the 'idea game' about them, positioning the Anthropocene as a *chronotope* (Bakhtin 1981): "a particular configuration of time and space that generates stories through which a society can examine itself" (Pratt 2017: G170). The idea of the Anthropocene seems always to invite this spatio-temporal reflexivity. Perhaps unsurprisingly, then, the concept has given rise to a host of 'thought experiments' across the arts and sciences since the turn of the century. We might mention here Jan Zalasiewicz's popular science book *The Earth After Us* (2008), in which extra-terrestrial geologists study the planet 100 million years from now to find traces of 'our human empire', or Daisy Hildyard's lyrical account of the dissolving boundaries between all life on Earth in *The Second Body* (2017). The best work in this vein tests the very foundations of the Anthropocene hypothesis without losing sight of its wide-ranging ethical and ontological implications. More than simply 'idea games', the emergence of humanities and arts-based research focused on the Anthropocene is part of a wider move to open up scientific discourse to sustained critique, the aim here being to "disrupt specialist divisions, democratize debate and pose critical questions of political significance to discussions on environmental developments" (Demos 2017: 12). My aim in this chapter is to show how discursive and material formations common to heritage practice – including the museum, the monument and the ruin – are being leveraged to pose such critical questions, drawing together science fiction imaginaries and historical methodologies in exhibitions, artworks and wider critical-creative research. A small set of micro-examples is investigated here to sketch out the main attitudes and principles common to such projects, which range from philosophical experiments to spatial interventions. This approach repositions heritage as a critical *method* that seeks to challenge and potentially redirect the emerging Anthropocene chronotope.

Why focus on heritage in this way? Implicitly concerned with issues of preservation, memory, salvage and storytelling, heritage thinking and practice has chiefly confronted the Anthropocene as an existential threat: a new condition of Earthly survival that radically undermines civilizational processes of inheritance and renewal. It is in this context that Roy Scranton, writing in *Learning to Die in the Anthropocene*, implores "humanity" to build biological and cultural "arks" to carry forward "endangered genetic data" and "endangered wisdom" alike (2015: 109). This biblical task has been taken up in various quarters in recent years, from the Austria-based 'Memory of Mankind' project (see Harrison et al. 2020) to the Arch Mission, which aims to disseminate human knowledge throughout the galaxy. As Claire Colebrook reminds us, however, the "unprecedented fragility" that many have expressed in the shadow of the Anthropocene is testament to nothing so much as the sudden precarious outlook of a small section of the planet that "draws its resources from elsewhere, transfers its waste and violence, and *then* declares that its mode of existence is humanity as such" (2017, original emphasis). The various biobanks and cultural arks, digital archives and messages to the future that constitute the 'heritage' of the Anthropos can only ever be seen as partial and highly contingent (see further discussion in Breithoff and Harrison, this volume). It would not be too much to say that this is heritage in capitulation with the Anthropocene – there's nothing to be done, so let's just make sure evidence of our existence persists in some form, ready to be 'taken on' and 'passed down' in some unspecified future scenario.

While these experiments in existential survival continue apace, a different mode of thinking *with heritage* has emerged in relation to the Anthropocene/Capitalocene/Chthulucene (Haraway 2015). Whether implicitly or explicitly, the tools and aesthetics of heritage – from museums and archives to ruins and memorials – have been deployed by various artists and writers to help (re)conceptualize the historical formation and future implications of the Anthropocene across disciplinary boundaries. We find this mode of critique in texts by Anna Tsing (2015) and Bronislaw Szerszynski (2017), for example, where the ruin and the monument respectively are reimagined for the Anthropocene epoch. The creative work of Tomás Saraceno, FICTILIS and Gustafsson & Haapoja meanwhile brings new historical-curatorial configurations to bear on issues of climate change, territorial boundaries and nature-culture relationships: all central questions for Anthropocene research beyond the narrow confines of geology and stratigraphy. Working in diverse media and without a specific manifesto or agenda, such critical Anthropocene projects play with the discursive and affective potency of heritage to imagine alternative ways of living and acting that are inherently oppositional to the apocalyptic motif of the ark. Rather than emphasizing stasis, preservation and continuity, much of the work I take up here resonates with Rodney Harrison's characterization of heritage as "collaborative, dialogical and interactive, a material-discursive process in which past and future arise out of dialogue and encounter between multiple embodied subjects in (and with) the present" (2015: 27). Moreover, these thought experiments build from a widespread recognition that the material geographies of the Anthropocene – including "waste sites, mining shafts and extraction zones" – may in themselves constitute "the new museums of humanity" (Yusoff 2017). Just as the Anthropocene historicizes the present by imagining humanity's descendants (or some alien equivalent) interpreting Earth's strata to locate the exact point at which 'we' began to transform the planet, so critical thinking has increasingly framed the Anthropocene concept itself as a museological problem, with all the questions of history, narrative, collecting, display and power this categorization implies.

As Bruno Latour has argued, the critic "is not the one who debunks, but the one who assembles... not the one who lifts the rugs from under the feet of the naïve believers, but the one who offers the participants arenas in which to gather" (2004: 246). It is in this spirit that the uptake of heritage as a methodology for peeling apart and reconfiguring the *work* of the Anthropocene interests me. To understand ruins, memorials, museums and other *lieux de mémoire* (Nora 1989) as 'gathering spaces' for critical Anthropocene thinking is to recognize the renewed vitality of institutional frameworks and aesthetics which are always somehow backwards looking *and* forwards facing. In this sense the projects and proposals I focus on in the present chapter – some realized, some hypothetical – are *more-than* thought experiments, both in the way they design and develop fully embodied moments of encounter, and in the creative worlds they call into being: spaces for imagining alternative subjectivities, ecologies and social formations (Guattari 1989). Heritage in this context cannot be reduced to a set of processes or specific agendas for engaging with the past in the present. It is rather a fluid and emergent phenomenon gesturing towards an unstable future – a future, that is, in which the things, stories and places currently categorized as 'heritage' might well be discarded, vilified or fundamentally reimagined. The projects I am interested in here thus pose vital questions about what heritage is or might be; a task that goes far beyond policy and preservation to touch on issues of time, identity and the inevitable historicization of the present (Jameson 2013).

What is a museum?

In the heady world of cultural masterplanning and urban regeneration, the idea of placing a 'museum' at the heart of any new development has been viewed with suspicion since at least the start of the new millennium, when a spate of projects sought to reimagine civic architecture and design. This is especially true of the 'old world', which gave birth to the museum as a form of control and a display of power (Bennett 1995). Keen to demonstrate their forward-facing, post-industrial, post-modern credentials, city planners, developers and architects have pursued various alternatives to the museum model, from 'cultural hubs' and 'urban forums' to 'history centres' and 'heritage laboratories'. This is not to say that no new museums have been built over the past two decades (a quick glance at China and the Arabian Gulf would soon undermine this argument), only that the confidence and certainty with which museums were imagined, designed and constructed in the nineteenth century has, in many places, given way to a more indeterminate cultural-historical landscape.

At the same time, within the back offices and display rooms of established museums, a minor revolution has taken place. The idea of exhibiting an anthropological or social-historical artefact in a glass case with a singular and uncontested narrative would now be unthinkable to most museum professionals. There must be layers of interpretation, multivocal perspectives, questions not answers (Vergo 1989). Processes of collecting, curating, conserving and exhibition making have emerged as key testing grounds in post-colonial, feminist and anti-hegemonic critique, with museums often positioned as catalysts and lightning rods for wider discussions around identity, history and collective memory (Macdonald 2013). As Fiona Cameron has argued with specific reference to the problem of climate change, unpicking the ontologies and assemblages of the museum offers a valuable technique for addressing issues of social relevance in a more immediate and engaging fashion (Cameron 2015).

While the institutional foundations of the museum have thus been called into question over the past two decades, the very concept of the museum as a critical-creative framework has simultaneously gained considerable traction (James Putnam's 2009 book *Art and Artifact: The Museum as Medium* provides an excellent overview of this trend in the art world). Rather than see these two developments as paradoxical, I think it is better to understand them as mutually reinforcing: questioning what a museum *is* means opening up the ontology of the concept to experimental configurations. There are many examples to draw on here, from the Museum of Broken Relationships – an online collection of objects associated with doomed love stories, now with a permanent physical presence in Zagreb and Los Angeles – to Orham Pamuk's remarkable Museum of Innocence – a collection of ordinary objects amassed by the author as he developed a novel of the same name; the two narrative forms feeding into one another through a dialectic of prose and artefact, curated 'thing' and imagined story. This category of the non-museum might also contain the itinerant outsider art installation the Museum of Everything, or indeed the Museum of Failure, 'a collection of failed products and services from around the world'. A complete genealogy of such institutions is beyond the scope of this chapter, but we might locate an early ancestor in the Museum of Jurassic Technology, founded in California in 1988 by husband and wife David Hildebrand Wilson and Diana Drake Wilson. Described as 'a museum about museums', the Wilsons' esoteric collection uses familiar tactics of lighting, labels and scholarly references "to inspire wonder not just at the objects (real or invented) but at the nature of museums themselves, the way they select items from the world and allow us to recognize them as strange and wonderful" (Rothstein 2012). Outside of a formal philosophical programme, Pamuk's 'Modest Manifesto for Museums' offers some indication of the wider project towards which these and other similar institutions contribute. Here the writer suggests that "the measure of a museum's success should not be its ability to represent a state, a nation or company, or a particular history. It should be its capacity to reveal the humanity of individuals" (Pamuk 2012: 56). Although marginal to the mainstream work of museums, it is worth noting that both the Museum of Innocence and the Museum of Broken Relationships have been recognized for the novelty of their approach, with the former named European Museum of the Year in 2014 and the latter receiving the Kenneth Hudson award for 'Europe's Most Innovative Museum' in 2011.

The critical force of such counter-institutions lies in their subversion of a familiar cultural apparatus, which is quite different from rejecting such practices outright. Where the laboratories, hubs and forums of the early twenty-first century seemed to accept Adorno's characterization of museums as "family sepulchres" (1981: 175) – and thus sought to invent new models of 'heritage engagement' to replace these dusty relics – the *museum that is not a museum* acknowledges the complex history of such institutions as a first step towards marking out a space of critique *inside* this tradition. The very nomenclature of 'the museum' is vital to this work, immediately invoking a set of spatial, discursive and aesthetic conventions against which a counter-proposal may be registered. As several artist and activist groups have recognized in recent years, the reflexive nature of the Anthropocene concept shares many points of reference with this museological re-framing – an overlap that offers fertile ground for rethinking the spatial-discursive form of this strange cultural artefact.

The Museum of Capitalism is a case in point here. Responding to the well-worn assertion that it has become easier to imagine the end of the world than the end of capitalism, the artist group FICTILIS sought to imagine a prospective museum that might educate current and future generations on the ideology, history and legacy of this particular world system (Figure 6.1). As their 'mission statement' affirms, the Museum of Capitalism (which held its inaugural exhibition in Oakland, California in 2017), strives to

> broaden public understanding of capitalism through multi-faceted programs: exhibitions; research and publication; collecting and preserving material evidence, art and artefacts related to capitalism; commemorations, reenactments and

other events; distribution of education materials and teacher resources; and a variety of public programming designed to enhance understanding of capitalism and related issues, including those of contemporary significance.

The vocabulary of 'public understanding' and artefact accumulation here situates the museum within a familiar tradition of hegemonic cultural institutions designed to educate the masses about a culture, place or historical moment fundamentally different from their own. In this context, however, such a viewpoint is initiated from *within* the system or society to be displayed – a form of ironic detachment that is necessary for the critical work of self-analysis to unfold. As the curators ask in the accompanying catalogue: "What better form than a museum to call progress into question, and how better to reorient ourselves in the present than with an institution we already use to orient ourselves toward the past?" (FICTILIS 2017: 14).

This deconstruction of the museum model is played out in numerous ways. The catalogue for example opens with a satirical exegesis on the museum's discursive foundations, moving from a selection of mission statements from comparator institutions such as the Museum of

Figure 6.1 — Museum of Capitalism: Oakland, 2017. (Photograph by Brea Mcanally).

Communism in Prague and the Museum of Apartheid in Johannesburg, to a lengthy inner monologue on the thorny issue of how to even define a 'Museum of Capitalism'. The interpretation meanwhile enacts another mode of defamiliarization, viewing common objects (a baseball cap, a mug, a coin) as if they are the remnants of a now defunct culture, or recontextualizing everyday things in strange settings, as in Evan Desmond Yee's *Core Samples* series, which shows modern artefacts encased in geological strata (Figure 6.2). Photographs, artworks, toys, archival documents and personal effects all have a role to play here in mapping out and redirecting visitors towards an embryonic post-capitalist future.

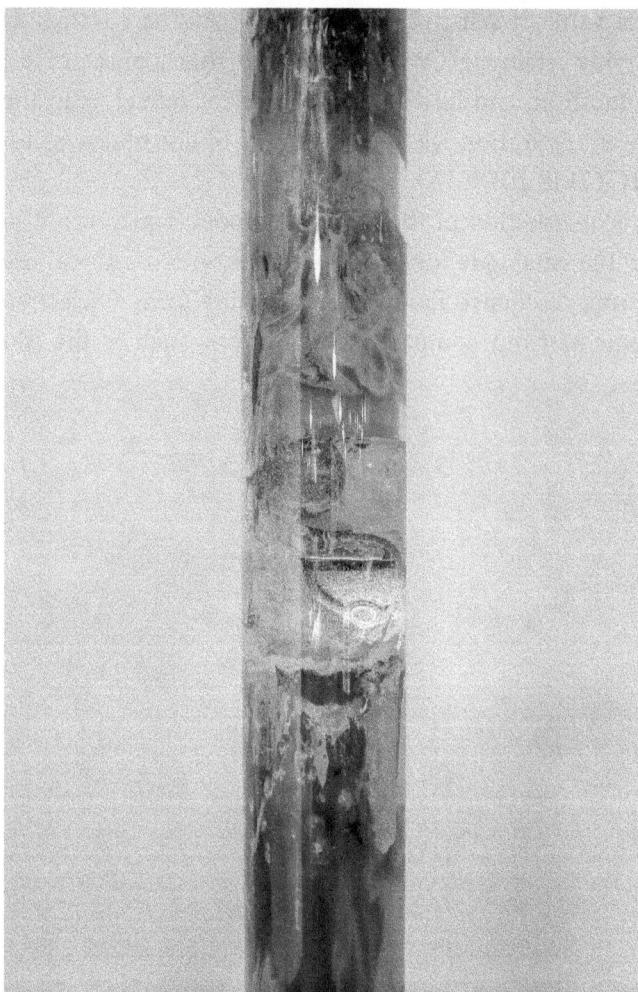

Figure 6.2 — Evan Desmond Yee, Core Sample #1, 2017. (Photograph by Museum of Capitalism).

This sense of reorientation also surfaces in the Museum of Nonhumanity, a multi-channel video installation created by writer Laura Gustaffson and visual artist Terike Haapoja, which presents the history of 'animalization' over the past 2,500 years. Using text, archival images and sounds, this immersive work – first shown in Helsinki in 2016 – explores the construction of humanity and animality as 'binary moral categories': a boundary distinction which, the museum argues, has provided the foundation for exploitation and abuse across human and non-human worlds (Figures 6.3 and 6.4). Against this backdrop, the museum becomes a "utopian institution and an initiative for imagining future narratives," performing a backwards glance from a reality in which society has "moved forward from the oppressive and destructive human-animal divide" (Gustaffson and Haapoja 2016). The museum in this context again serves as a kind of memorial to a system and form of living the curators wish to move beyond. This historicizing impulse questions the logics and ontological formations of the present from an *anticipated* future. Here it is worth noting the close relationship the Museum of Nonhumanity establishes between animalization and capitalism, which Gustaffson and Haapoja describe as "the mechanism through

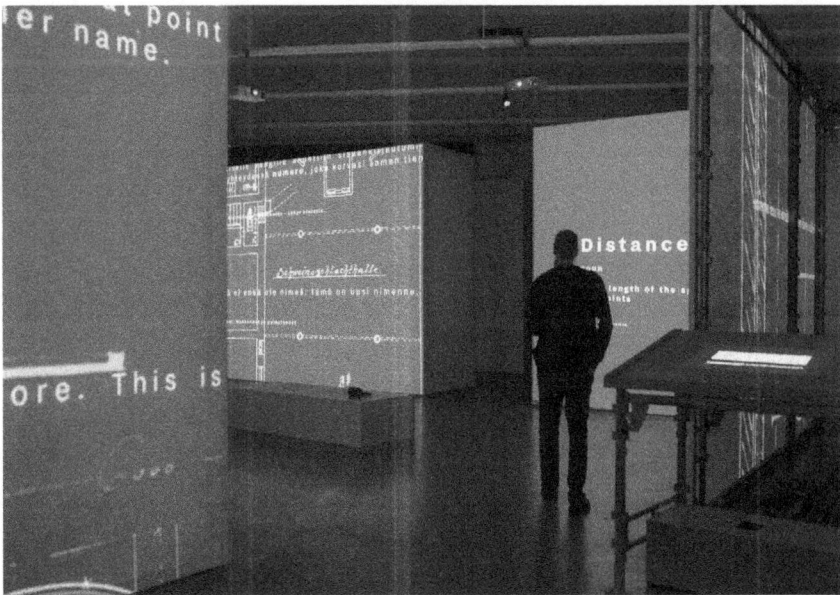

Figure 6.3 — Museum of Nonhumanity, Installation view.
(Photograph by Terike Haapoja, MONH).

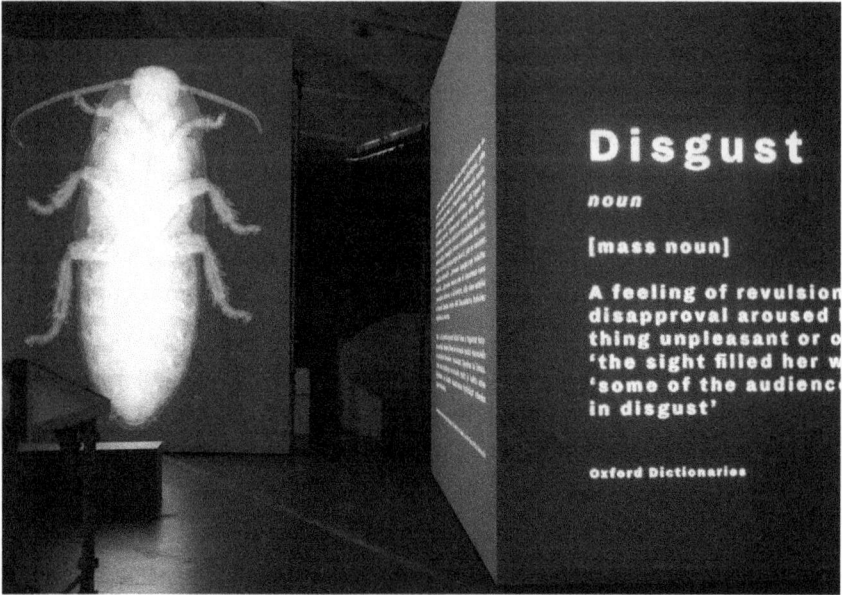

Figure 6.4 — Museum of Nonhumanity, Installation view.
(Photograph by Terike Haapoja, MONH).

which bodies of all species become available to exploitation as resources, as disposable, or something to control" (2016). "It is not possible to fight for a systemic change without fighting against the notion of animality," they maintain, "because the logic of animalization forms the foundation of racist, imperialist, patriarchal capitalism" (ibid).

Alongside their visual-spatial form – the one object-oriented, the other predominantly digital and immersive – both the Museum of Capitalism and the Museum of Nonhumanity place considerable emphasis on programming as part of a measured attempt to rethink and repurpose the common tropes of modern museum practice. The Museum of Capitalism for example hosts regular 'artefact donation events' and – more broadly – aims to bring artists, educators and activists together through a range of public events focused on different manifestations of the capitalist system, from gentrification to the links between oil and tourism. As the museum website states, such occasions invite audiences 'to inhabit an indeterminate, imaginary future in order to better recognize the historical specificity, idiosyncrasy and contingency of the present'. The Museum of Nonhumanity takes a similar approach with seminars that seek to 're-imagine the future through the past' and others that aim to 'decentre history' via artistic interventions. To paraphrase Latour, these gathering spaces provide active lessons in the critical power of the assembly, drawing on a multiplicity of voices to generate new ways of being and acting in the world that constantly question and reconfigure the inheritances 'we' have been left with. There are clear resonances here with Ursula Biemann's posthumanist vision for the museum, which acknowledges the difficulty but also the necessity of imagining a more-than-human 'common world' heritage in the age of the Anthropocene:

> For a future where human-nonhuman relations are less violent, less destructive, the past will have to be reassembled. This sort of rewriting of history resembles somewhat the rewriting of post-colonial history. Only this time, it is not a matter of admitting formerly excluded groups of human populations to the theatre of significance, it means to radically decentre the human figure altogether. It is difficult to imagine such a place and yet this is what is at stake now. What we can already say is that a common future that we share with everything else would be equally rooted in cultural and natural narratives; the collections of this common world, our heritage, would necessarily include at once cultural and natural histories. Perhaps from there, we can envision a less divided future that can harbour a post-human way of being in the world (2016: 60–61).

While both the Museum of Capitalism and the Museum of Nonhumanity are built around an embodied sense of encounter and assembly, the performative nature of the counter-museum finds its radical apotheosis in the Museo Aero Solar, an open source international community initiated by Tomás Saraceno in 2007 which is dedicated to the transformation of airborne travel. The museum itself is many things in this instance: an evolving balloon-like sculpture made from thousands of reused plastic bags, which – when heated by the sun – will float in the air free of any fossil-driven propulsion; a gathering of individuals and ideas focused on a single goal yet distributed across time and space, connected via social media and digital collectives; a mode of recycling as collection that transcends territorial boundaries, both 'in the air' and 'on the ground', with branded plastic assembled from multi-national chains and hyper-local companies alike. At the core of this work lies a commitment to reconfigure the imprint of humans on the Earth: a form of deterritorialization that questions the seductive nature of twenty-first century 'hypermobility' and the socio-material realities any form of fossil-fuel powered travel relies upon. Museo Aero Solar transforms one of the most emblematic traces of the Anthropocene – the plastic bag – from an object of disdain to a transversal symbol of hope and possibility. In the words of architect Pierre Chabard, the floating "museums" created as part of this project are "ambiguous, dynamic, less subversive than transgressive...sublime parasites or radical enterprises of diverting our inherited world" (2015). Here we find an echo of the Museum of Jurassic Technology and of the museum enterprise as a whole, which has always depended on isolating objects from the world so that they might be seen and understood in a new light.

Although tackled separately, capitalism, nonhumanity and the politics of air collide in these new museological imaginaries. The Anthropocene is both a backdrop and a rallying point for this work, framed differently through the logics of the Capitalocene, the Chthulucene and the 'Aerocene' – Saraceno's wider artistic project which aims to generate a new ethics with and for the Earth's atmosphere. This demarcation avoids what Richard Pell, curator of the Center for Postnatural History, has called the "fuzzy rhetoric" of the Anthropocene (2015: 314). Better to assemble "core samples", Pell argues (ibid), building collections and archives one object at a time. Rather than see such projects as museological silos separating nature from culture, and history from the ongoing present, I think it is more useful to understand their emergence as part of a transversal reckoning with the Anthropocene as a totalizing concept and inescapable reality. Just as Comte de Buffon implored early geologists to "excavate the archives of the world" and "assemble in one body of proofs all the evidence of physical changes that enable us to reach back to the different ages of Nature" (quoted in Szerszynski 2017: 116), so the material-discursive *idea* of the museum is now increasingly deployed to disentangle and recombine social and natural histories across different registers. The separations necessary to grasp and ultimately challenge the Anthropocene must always be offset by this ongoing interconnectivity (Braidotti 2013).

Figure 6.5 — Museo Aero Solar. Installation at Anthropocene Monument.
(Photograph © LesAbattoirs by Sylvie Leonard for Tomás Saraceno)

By linking the German word *museal* ('museumlike') to the mauso-
leum, Adorno sought to emphasize the deathly atmosphere of the traditional museum, a place where culture could be 'neutralized'. The museal
in this reading "describes objects to which the observer no longer has a
vital relationship and which are in the process of dying. They owe their
preservation more to historical respect than to the needs of the present" (1981: 175). The projects I am interested in here seem to reverse
this polarity: their museality is founded on the ambiguous and emergent
status of things and narratives, rather than on familiar notions of stasis
and obsolescence. This is because the stories of capitalism, atmospheric
pollution and animalization are not dead or dying but rather constantly
unfolding (an argument could be made that many of the supposedly 'complete' stories we find in contemporary museums are far from resolved,
including colonialism, racism, conflict etc, but this would take another
essay to work through in detail). The temporality thus evoked is one of
the future anterior: looking forward to look back, the present becomes
history in the new museums of the Anthropocene, which operate as progenitors of *what might be* even as they adopt a viewpoint that is at once
historical and memorializing. This temporal and political complexity is

not a convenient by-product of their self-designation as museums, but rather an integral property of their status as anticipatory mechanisms. They drag the capitalist, anthropocentric, territorialized present kicking and screaming into a near-utopian future which looks back with no small degree of incredulity at the madness of 'our time'. As Robinson writes, "we are the primitives of an unknown civilization" (2017b: 150).

In his recent book *The Great Derangement*, Amitav Ghosh asks how climate change and the sixth mass extinction will be perceived by future audiences:

> When readers and museum-goers turn to the art and literature of our time, will they not look, first and most urgently, for traces and portents of the altered world of their inheritance? And when they fail to find them, what should they – what can they – do other than to conclude that ours was a time when most forms of art and literature were drawn into the modes of concealment that prevented people from recognizing the realities of their plight? (2016: 11).

This line of inquiry brings the inherent reflexivity of the Anthropocene narrative closer to home. Rather than looking to the geological strata of a post-human Earth one hundred million years from now, Ghosh imagines a future scenario in which historical knowledge is still paramount. What the museums of capitalism, nonhumanity and the aerocene do is slightly different, however. Instead of waiting for the present to be past – and thus suitable as a topic for musealization – they critique this future inheritance from within, "experimenting with what life can or might be in both its virtual and future anterior modes" (Yusoff 2017). Such a "geoaesthetics... allows life to surpass itself" (ibid). Not predictive or speculative so much as hopeful, the museums of the Anthropocene bring difference into being, seeking worlds beyond "the farthest points our own thought can reach" (Jameson 2013: 308). What needs underlining here, however, is that this utopian gesture does not leave us in a banal future gazing back at the present as the past, but rather situates heritage as a vital component in the messy work of disentangling the Anthropocene at the moment of its very emergence. The museum in this context is a space for sensing, breathing, tasting, smelling and touching the Anthropocene (Zylinska

2018: 64): a familiar space made unfamiliar to better question the fragile uncertainties of an unfolding present.

Ruin redux: After the end of the world

The notion of *living with* rather than *gazing at* heritages in the making can be elaborated further with specific reference to the Anthropocene ruin. Following the work of Claire Colebrook and Anna Tsing, there are two main points to develop here. The first concerns the absence of 'readers' in the post-human future that the Anthropocene forces us to confront; the second relates to the possibility of 'livable collaborations' in the precarious spaces generated by late capitalism. Both of these threads take us away from a narcissistic, anthropocentric view of a world 'gone to ruin' to consider alternative perspectives on decay, abandonment, vulnerability and entanglement: all central to the emergence of ruination as a core theme in modernity and post-modernity (Hell and Schönle 2010; Dillon 2014). This tracks the growth and spread of heritage as a mode of relating to the past in the present – an historical connection which implies a certain level of reciprocity beyond mere temporal coincidence. As with the museological experiments outlined above, the Anthropocene ruin thus gestures towards the central role that heritage more broadly might play in reconceptualizing this new geological epoch, as well as turning a critical lens back on the work of heritage itself.

I would not be the first to note that an apocalyptic zeitgeist has gripped much of popular culture and the humanities in recent years (Berger 1999; Germanà and Mousoutzanis 2014). Within this context, 'end of the world' narratives have been given a "new sense of direction after becoming linked to the Anthropocene" (Zylinska 2018: 4). The prospect of mass extinctions, floods, scorched earths and civilizational collapse is played out with grim regularity in films, graphic novels, works of literature and mainstream television (streaming services driven by algorithms report that apocalyptic horror is among the most popular and therefore lucrative genres). Heritage is generally allotted either a hopeful or a satirical role in such works. The zombie-infested world of *The Walking Dead*, for example, shows survivors banding together to visit a museum holding pre-industrial technology in the hope of bringing certain objects back into use – items that may finally allow humanity to 'rebuild'. In the 2009 film *The Age of Stupid* meanwhile a future archivist views 'historic' news footage from the early twenty-first century to understand why humanity failed to address climate change when it had the chance. Both future worlds treat the residues of the past as vital resources for understanding and/or transforming the present, but there is little space in these imaginary settings for viewing the contemporary world – *our world* – as anything more than the prelude to catastrophe and collapse.

In this sense there is something almost refreshing about the postapocalypse, which emphatically denies the memorial veneration that so much of heritage strives towards. Indeed, I do not think it would be too much to claim that much of what we take to be 'heritage' is a form of narcissism, most commonly founded on a desire to see some aspect of one's self or one's culture taken forward into the future. Searching for ancestors and building museums are both inherently future-oriented processes: they actively produce a world in which the narratives and things of today (which include those aspects of the past that the present currently cares about) will matter to the people of tomorrow. Past and future generations are bound together in this unfolding assemblage, but it is the present that sets the terms of reference. To some degree this has always been the case. Why else would emperors, kings and tyrants of all persuasions build vast mausoleums, or inscribe their names in stone and bronze, other than to ensure that the future acknowledges their existence in some capacity? We need to be careful here, however, not to conflate this desire for immortality with the self-referential attitude of our current time. As Boris Groys puts it,

> Our contemporary age seems to be different from all other historically known ages in at least one respect: never before has humanity been so interested in its own contemporaneity. The Middle Ages were interested in eternity, the Renaissance was interested in the past, modernity was interested in the future. Our epoch is primarily interested in itself (2016: 137).

While questions of imminent climatic, civilizational breakdown seem to undermine the memorializing capacities of the future – and thus resituate the present as prologue – the longer geological temporalities of the Anthropocene turn attention back on to the here-and-now by pre-emptively monumentalizing humanity as a "future fossil" (Yusoff 2013). The problem here however is that the Anthropocene as strata "asks us to think and perceive *as if* our world would be readable in the absence of what we now take to be readers" (Colebrook 2014: 34, original empha-sis). As Colebrook continues, "all our traces (literary and otherwise) would remain but without human context or concept. The archive would be a dead letter... The people would be missing, leaving something like a maximal force of dissemination that would also be a maximal force of inertia" (2014: 37-8). Here we are in the realm of Eugene Thacker's *world-without-us*, a "spectral and speculative" planet that we can never experience and that seems to act "as a limit that defines what we are as human beings" (2011: 5). Where the Romantic cult of ruins liked to imagine future travellers exploring the cities of London and Paris in an advanced state of decay, the Anthropocene ruin is denied even this gaze, separated as it must be from any such readers or visitors that we might recognize as human. Here it is worth noting that Zalasiewicz's *Earth After Us* makes an explicit link between these two memorializing outlooks: looking far into the future, Zalasiewicz declares, "the remains of our human empire should soon crumble away and decay, leaving scarcely a footprint on the sands of geological time. Our legacy would be as pitiful as that of Ozymandias' mighty kingdom in Shelley's poem, reduced to a shattered statue amid the boundless desert wastes" (2008: 2).

As the reference to Shelley indicates, picturing the contemporary world in ruins is a persistent trope in modernist and post-modernist lit-erature and aesthetics – one that draws together heritage and science fic-tion imaginaries. The constraint of such future ruins, however, is that they tend to rehearse a familiar set of assumptions about the value of material-semiotic formations as they decay and collapse. Most commonly, proph-ecies of London, Paris, New York or any other emblematic modern space in an advanced state of decay simply project Alois Riegl's early twentieth-century categorization of monument types into some unknown future. Such ruins may be assigned "historical value" – that is, treated as docu-ments that can reveal something about the time in which they were cre-ated – or more often "age value," which involves "an affective pleasure in signs of natural processes of disintegration and decay" (Szerszynski 2017:

118). Thinking with heritage after the end of the world, in the framework of what I have been calling a critical Anthropocene method, means looking beyond such tropes. A new theory and imaginary of the future ruin is needed, one founded not on art history or 'cultural' memory, but on a transversal and posthumanist conception of more-than-human unravelling and recomposition. Colebrook underlines the problem with discovering such an imaginary in the Anthropocene strata:

> At first the capacity to view ourselves as if from a post-human future, seems to diminish the self, creating a sublime distance whereby we annihilate ourselves for the thought of a life and readability to come. The reading of a past that is not ours (or our capacity to touch and reach out to what is not ourselves) seems to open the self to the not-self, to a radically post-human future. But the same gesture of alterity is also auto-archiving and auto-affecting. We now, narcissistically, imagine the tragedy of the post-human future as one in which death and absence will be figured through the unreadability of our own fragments, as though our self-alienation through archive and monument yields some sentiment that we ought to remain as readers of ourselves (2014: 40).

Building on this 'sublime distance', I think there is cause to ask what alternative forms of *prospective* decay and unravelling might be leveraged to rethink heritage as a critical apparatus? This would need to acknowledge that processes of inheritance are to some extent inescapable, binding together past, present and future in an affective embrace, but also that the *project* of heritage is contingent and malleable. To think *with* heritage in the face of the world's ending should not mean aligning oneself with a self-absorbed quest for immortality, but instead asking what the inherent alienation offered by this scenario might offer to our evolving sense of history, of memory, of vulnerability and of inheritance.

For a historical culture built on narcissism, the prospect of the end of the world can be difficult to accept. Perhaps this is why, in many quarters, the spectre of systemic collapse and mass extinction has been greeted not with fierce opposition but with something approaching reflexive panic. What place does memorialization have in a world where human life is no longer tenable? Or – somehow even more difficult to stomach – how might a culture fundamentally different from our own engage with the legacies 'we' bestow? Such questions have come to dominate discussions of the Anthropocene amongst memory scholars, with Richard Crownshaw in particular developing the framework of speculative memory to understand dystopian literature that deals with the future anterior of climate change: "these fictions are useful not only for their memorative disposition but also for their melancholic orientations towards the fossil-fuelled worlds they imagine untenable or in ruins" (2018: 501). This notion of remembrance beyond or in excess of the human dovetails with a broader ecological turn in the discipline (Groes 2016; Rigney 2017), bringing Jameson's work on 'historical futures' into dialogue with Deleuze and Guattari's theory of the assemblage and the sympoietics of Donna Haraway for a broader picture of memory across human and more-than-human worlds.

The apocalyptic tenor of the Anthropocene debate clearly owes much to earlier visions of civilizational collapse, but there is an added environmental twist to these discussions in the face of the sixth mass extinction. While such debates are often criticized for denying or occluding previous and ongoing catastrophes of 'world-ending' scale – from the decimation of Indigenous populations as Western modernity spread around the globe, to previous extinction events in the Triassic, Pleistocene and Devonian eras – the current moment of existential crisis is marked by a realization that *both* these worlds now seem on the verge of cataclysm: the Earth System and Integrated World Capitalism are intimately bound in the historical formation and future unravelling of the Anthropocene epoch (Guattari 2009; Moore 2017). As a result, transdisciplinary dialogue has grown exponentially, with artists, political economists, historians, geologists, biologists, cultural critics and many others contributing to an increasingly public conversation about the past, present and future of the planet and its human and non-human inhabitants. Academic symposia, popular media, contemporary art and activist interventions collide in this new arena, which implicitly questions the separation of nature from society in the pursuit of a new relationship for and with the Earth. The most interesting work in this emerging space of critical utopian-dystopian-apocalyptic thinking engages with an expanded sense of precarity drawn from feminist and post-colonial literature while also questioning the dominant anthropocentricism of humanist philosophies (Braidotti 2013).

The work of Anna Tsing is exemplary in this regard. In her 2015 book *The Mushroom at the End of the World*, Tsing explores entangled landscapes of precarity and ruination – states of the world which are seen as symptomatic of late-capitalist modernity. Precarity, it is argued here, is now "an earthwide condition" (2015: 5): "in contrast to the mid-twentieth century, when poets and philosophers of the global north felt caged by too much stability, now many of us, north and south, confront the condition of trouble without end" (2015: 2). The failure of progress narratives is central to this picture. Where previously the promises of modernity seemed to align with a progressive politics of emancipation, freedom and security, progress in the age of climate breakdown and mass extinction seems less clear, less… progressive. Constant growth is no longer reasonable let alone sustainable. Without a sense of progress – even in its patchy and unequal form – the ruins of the world become more apparent; indeed, they assume a pivotal role in what Tsing calls "collaborative survival" (2015: 19). These ruins are inherently natural and cultural: they emerge through a combination of ecological collapse and capitalist exploitation. The question of collaboration meanwhile crosses social and natural domains to focus on issues of racism, sexism, imperialism and environmental justice alongside *and in conjunction with* concerns around biodiversity and conservation. Survival for Tsing can only occur if we acknowledge that "staying alive" requires "livable collaborations… working across difference" (2015: 28). Ruins in this context are spaces of precarity and of potential resurgence not simply because they lie outside the typical frameworks of capitalist control, but also because the very processes of ruination force us to imagine and negotiate "life in human-damaged environments" (2015: 131). As Tsing concludes,

> Without stories of progress, the world has become a terrifying place. The ruin glares at us with the horror of its abandonment. It's not easy to know how to make a life, much less avert planetary destruction. Luckily there is still company, human and not human. We can still explore the overgrown verges of our blasted landscapes (2015: 282).

Responding directly to Tsing's desire for liveability in ruins, Zylinska has recently put forward an "alternative microvision" for a "feminist counterapocalypse that might take seriously the geopolitical unfoldings on our planet while also thinking our relations *to* and *with* it precisely as *relations*" (2018: 53, original emphasis). As well as opposing the seductive fantasies of the post-apocalypse, this framework seeks liberation from the competitive and overreaching masculine subjectivities of the present (2018: 59). "If unbridled progress is no longer an option," Zylinska asks, "what kinds of coexistence and collaborations do we want to create in its aftermath?" (ibid).

This question leads us back to the inherent precarity of the ruin – always on the verge of collapse, always entangled with a multiplicity of forces and materialities. While the pre-Anthropocene future ruins imagined by artists and writers such as Hubert Robert and Alfred Franklin evoked the grandeur and mystery of the present as history (Figure 6.6; Franklin 1875), Tsing's revised conceptualization seems to lead us in a very different direction. Living with ruins means more than simply acknowledging that decay might provide an aesthetic backdrop to our

Figure 6.6 — Hubert Robert, 1796. Imaginary view of the Grande Galerie in the Louvre in ruins.

lives: it requires a fundamental reorientation towards decomposition and the complex multispecies worldings likely to emerge in spaces of neglect and despair (Haraway 2016). There are clear points of overlap here with Caitlin DeSilvey's theorization of 'curated decay' (2017), which pursues a form of more-than-human, anti-egotistical heritage built around entropy rather than preservation. This alternative imaginary occupies a critical juncture between the dystopian satire and the forgetful future, suggesting a mode of speculative ruination better suited to the work of heritage in the Anthropocene epoch.

Antinomies of the Anthropocene

In the summer of 2010, as part of the dOCUMENTA(13) arts festival held in Kassel, Germany, artist Amy Balkin launched an initiative to inscribe Earth's atmosphere on the UNESCO World Heritage List. This would be done on "an emergency basis, consistent with the aims and goals of the World Heritage Convention" (Balkin 2015: 341). Balkin's goal here was straightforward enough: to highlight the "outstanding universal value of Earth's Atmosphere" in the hope of finding a "common interest" for the international community in protecting and preserving the atmosphere for "present and future generations" (ibid). The World Heritage 'site' in this context would transcend territorial boundaries, stretching around the entire planet, from sea level to the Kármán Line – an altitude around 100km above sea level which commonly represents the border between the Earth's atmosphere and outer space. As Balkin soon discovered, however, the process of inscribing a new site on the World Heritage List requires backing from a specific nation or coalition of State Parties. As host of dOCUMENTA(13), Germany was first invited to lead such a coalition – this was rejected by the Federal Minister for the Environment, Nature, Conservation and Nuclear Safety. A different approach was then taken, with 186 invitation letters sent to all UNESCO State Parties. The only positive response came from Dr Ana Maui Taufe'ulungaki, Minister of Education, Women's Affairs and Culture for the Kingdom of Tonga, who said the country unfortunately lacked the resources to initiate and lead a nomination process (2015: 345). Two years later, 90,000 signed postcards calling for the Earth's atmosphere to be listed as a World

Heritage Site were shipped to Peter Altmaier, Germany's new Federal Environment Minister. A letter of reply re-confirmed that Germany would not lead a coalition for inscription.

Balkin's experiment vividly demonstrates one of the key tensions in the fight against climate change: namely, that the political structure of the nation state and the international systems that have grown out of this, including the UN, are incapable of addressing the crisis on their own. Ghosh identifies this as the main obstacle to reversing the "Great Derangement" we currently find ourselves in, with climate change representing, "in its very nature, an unresolvable problem for modern nations in terms of their biopolitical mission and the practices of governance that are associated with it" (2016: 160). Protecting and preserving heritage sites within the UNESCO system is beset by the same problems (Meskell 2018), exacerbated by questions of identity, ownership, memory and strong connections to certain territorial spaces. As Ghosh makes clear, however, the contagion of climate breakdown and the Anthropocene "has already occurred, everywhere":

> the ongoing changes in the climate, and the perturbations that they will cause *within* nations, cannot be held at bay by reinforcing man-made boundaries. We are in an era when the body of the nation can no longer be conceived of as consisting only of a territorialized human population: its very sinews are now revealed to be intertwined with forces that cannot be confined by boundaries (2016: 144, original emphasis).

While the (proposed) act of listing a planetary wide heritage 'site' opens up this messy tension to renewed critical scrutiny, it cannot account for the uneven historical responsibilities and distributed consequences of an emergent Anthropocene epoch. Indeed, as the lack of responses to Balkin's request illustrates, it is precisely an unwillingness to take responsibility for a global phenomenon that stifles action at the level of the nation state. I think part of the problem here lies in the supposed finality that listing, protecting and – crucially – *stewarding* a site of heritage classically implies within UNESCO's bureaucratic framework (a framework that is similar though not entirely comparable to the UN's approach on climate change). Over 1,000 sites are now included on the

World Heritage List, and while the reasons for their inscription will differ from country to country, and from site to site, once added to the list they are bound by an international set of rules around care and conservation. Discord, debate, collaboration and uncertainty are problematic within a system that demands 'outstanding universal value' to be demonstrated and consistently upheld.

In the shadow of the Anthropocene, then, we are confronted by the antinomies of a certain conception of territory, akin to Fredric Jameson's antinomies of realism: a "historical and even evolutionary process in which the negative and the positive are inextricably combined, and whose emergence and development at one and the same time constitute its own inevitable undoing, its own decay and dissolution" (2013: 6). The stronger such a concept becomes – think walls, Brexit and the so-called migration 'crisis' – "the weaker it gets; winner loses; its success is its failure" (ibid). Such contradictions are intrinsic to the 'wicked problem' of climate change. Seen through the lens of the Anthropocene, this territorial paradox is matched by various other antinomies, including those of remembrance and of the subject. Where I depart from Jameson, however, is the claim that these contradictions represent "the farthest points our thought can reach… the opposition beyond which we cannot think" (2013: 308). The microexamples I have put forward in this chapter demonstrate precisely the continued importance of thought experiments that attempt to overcome such antinomies. One final case study may help to underline this point.

In 2014 Bronislaw Szerszynski, Bruno Latour and Olivier Michelon launched the *Anthropocène Monument* project at Les Abattoirs Museum of Contemporary Art in Toulouse, comprising an exhibition and colloquium bringing together twenty artists from around the world to imagine what form a monument to the Anthropocene might take (Figures 6.7 and 6.8; see Szerszynski 2017). Many of the ideas put forward for the exhibition incorporated emblematic markers of the Anthropocene as a geological series, including "minerals derived from plastics, or contemporary artefacts that might be disinterred and interpreted by future archaeologists" (2017: 126). Others sought to follow the logic of the Anthropocene and "blur the distinctions between natural and cultural entities" by "playing on the monumentalizing effects of decay and

Figure 6.7 — Anthropocene Monument, les Abattoirs. Installation view. (Photograph © LesAbattoirs by Sylvie Leonard).

Figure 6.8 — Anthropocene Monument, les Abattoirs. Installation view showing *Terra-Forming: Engineering the Sublime* by Adam Lowe and Jerry Brotton. (Photograph © LesAbattoirs by Sylvie Leonard for Factum Arte).

ruination" (ibid) – a nod to the future anterior temporality of this inherently reflexive concept. What is most useful, however, for understanding the extent to which a new monumental system might undermine
the antinomies of the Anthropocene is the resistance to certainty found
across many of the proposals. As Szerszynski records, most of the artists
seemed implicitly to navigate away from the traditional realm of the static
monument and opt instead for "*Gegendenkmäler*, counter-monuments,
which were variously mobile, dispersed, transient or demanded interaction, and that thus served not to consolidate cultural memory but to provoke communicative memory, debate and action" (ibid). I am reminded
here of Chantal Mouffe's call for an "agonistic museum," one that might
"facilitate the expression of dissent, helping people to better understand the contradictions of the world" (2017: 79), and also of Michael
Landzelius' "politics of dis(re)membering", which – following Deleuze
and Guattari –aims to supplant the lineage mentality of heritage with the
"rhizome of disinheritance" (2003: 210). While I do not believe that we
can 'disinherit' the Anthropocene in the sense that we ignore its material
legacies or hope its consequences dissipate, we might imagine an alternative framework through which the concept is taken on and passed down.
An Anthropocene monumental system here would have to,

> challenge the viewer to wrestle with the paradoxes and
> responsibilities involved in being a member of a species that,
> albeit unevenly, is achieving geological consequentiality...
> any monumental system for the Anthropocene would need to
> signify that this epoch-in-the-making will be actively woven
> from multiple stories and diverse imagined futures distributed
> around the globe (Szerszynski 2017: 128).

I want to end on this image of an epoch-in-the-making to stress the
openness that still clings to the Anthropocene concept. Whether looked
at in terms of museology, ruination or memorialization, heritage is valuable here for the way it forces us to confront the sticky inheritances of the
past in the present – legacies that 'we' have a differential responsibility for,
but that also provide an opportunity to (re)shape the future. The reflexive mode of the Anthropocene epoch aligns with the cyclical temporalities of heritage: the past is somehow always *more present* than the present,

while the future is both a projection and an attractor, shaping how we think and behave in relation to the contemporary world. Building on this, heritage as critical Anthropocene method means imagining new modes of preservation, new forms of curatorial practice and new processes of monumental-territorial inscription. Each of these gestures will inevitably give rise to key questions around subjectivities, social formations and material ecologies (Guattari 1989), but they are also vital frameworks through which to transform the work of heritage itself. A familiar, narcissistic, human-centred view of heritage is slowly giving way to this new imaginary, the contours of which we might begin to discern in strange encounters with floating museums, and uncertain monuments to a time still to come.

References

Adorno, T. 1981. *Prisms*. Cambridge, MA: MIT Press.

Bakhtin, M. M. 1981. *The Dialogic Imagination: Four Essays*. Austin: University of Texas Press.

Balkin, A. 2015. Public Smog. In *Art in the Anthropocene: Encounters among Aesthetics, Politics, Environments and Epistemologies*, edited by H. Davis and E. Turpin, 341–46. London: Open Humanities Press.

Bennett, T. 1995. *The Birth of the Museum: History, Theory, Politics*. London; New York: Routledge.

Berger, J. 1999. *After the End: Representations of Post-Apocalypse*. Minneapolis: University of Minnesota Press.

Biemann, U. 2016. Late Subatlantic Poetry in Times of Global Warming. In *Ecologising Museums*, edited by L'Internationale, 54–62. Paris: L'Internationale.

Braidotti, R. 2013. *The Posthuman*. Cambridge: Polity.

Cameron, F. 2015. The Liquid Museum: New Institutional Ontologies for a Complex, Uncertain World. In *The International Handbooks of Museum Studies: Museum Theory*, edited by A. Witcomb and K. Message, 345–361. Chichester, West Sussex: Wiley-Blackwell.

Chabard, P. 2015. Air Crafted Architecture. *Aerocene Newspaper*. https://aerocene.org/newspaper-chabard/

Colebrook, C. 2014. Archiviolithic: The Anthropocene and the Hetero-Archive. *Derrida Today* 7(1): 21–43.

Colebrook, C. 2017. End-times for Humanity. *Aeon*, June 1, 2017. https://aeon.co/essays/the-human-world-is-not-more-fragile-now-it-always-has-been

Crownshaw, R. 2018. Speculative Remembrance in the Anthropocene. *Memory Studies* 11(4): 498–515.

Demos, T. J. 2017. *Against the Anthropocene: Visual Culture and Environment Today*. Berlin: Sternberg.

DeSilvey, C. 2017. *Curated Decay: Heritage Beyond Saving*. Minneapolis; London: University of Minnesota Press.

Dillon, B. 2014. *Ruin Lust: Artist's Fascination with Ruins, from Turner to the Present Day*. London: Tate.

FICTILIS (eds). 2017. *The Museum of Capitalism*. New York: Inventory Press.

Franklin, A. 1875. *Les Ruines de Paris en 4875*. Paris: P. Daffis.

Germanà, M. and A. Mousoutzanis (eds). 2014. *Apocalyptic Discourse in Contemporary Culture: Post-Millennial Perspectives on the End of the World*. New York: Routledge.

Ghosh, A. 2016. *The Great Derangement: Climate Change and the Unthinkable*. Chicago and London: University of Chicago Press.

Groes, S. (ed). 2016. *Memory in the Twenty-First Century: New Critical Perspectives from the Arts, Humanities, and Sciences*. London: Palgrave Macmillan.

Groys, B. 2016. *In the Flow*. London; New York: Verso.

Guattari, F. 1989 [2014]. *The Three Ecologies*. Translated by I. Pindar and P. Sutton. London and New York: Bloomsbury.

Guattari, F. 2009. *Soft Subversions: Texts and Interviews 1977–1985*. Los Angeles: Semiotext(e).

Gustaffson, L. and T. Haapoja. 2016. *Museum of Nonhumanity: Info*. http://www.museumofnonhumanity.org/info.

Haraway, D. J. 2015. Anthropocene, Capitalocene, Plantationocene, Chthulucene: Making Kin. *Environmental Humanities* 6: 159–65.

Haraway, D. J. 2016. *Manifestly Haraway*. Minneapolis; London: University of Minnesota Press.

Harrison, R. 2015. Beyond 'Natural' and 'Cultural' Heritage: Toward an Ontological Politics of Heritage in the Age of the Anthropocene. *Heritage & Society* 8(1): 24–42.

Harrison, R., C. DeSilvey, C. Holtorf, S. Macdonald, N. Bartolini, E. Breithoff, L.H. Fredheim, A. Lyons, S. May, J. Morgan and S. Penrose. 2020. *Heritage Futures: Comparative Approaches to Natural and Cultural Heritage Practices*. London: UCL Press.

Hell, J. and A. Schönle (eds). 2010. *Ruins of Modernity*. Durham, NC; London: Duke University Press.

Hildyard, D. 2017. *The Second Body*. London: Fitzcararaldo.

Jameson, F. 2013. *The Antinomies of Realism*. London; New York: Verso.

Landzelius, M. 2003. Commemorative Dis(Re)membering: Erasing Heritage, Spatializing Disinheritance. *Environment and Planning D: Society and Space* 21: 195–221.

Latour, B. 2004. Why Has Critique Run out of Steam? From Matters of Fact to Matters of Concern. *Critical Inquiry* 30(2): 225–48.

Macdonald, S. 2013. *Memorylands: Heritage and Identity in Europe Today*. London; New York: Routledge.

McBrien, J. 2016. Accumulating Extinction: Planetary Catastrophism in the Necrocene. In *Anthropocene or Capitalocene? Nature, History, and the Crisis of Capitalism*, edited by J. W. Moore, 116–37. Oakland, CA: PM Press.

Meskell, L. 2018. *A Future in Ruins: UNESCO, World Heritage, and the Dream of Peace*. Oxford: Oxford University Press.

Moore, J. W. 2017. The Capitalocene, Part I: On the Nature and Origins of our Ecological Crisis. *Journal of Peasant Studies* 44(3): 594–630.

Mouffe, C. 2017. An Antagonistic Conception of the Museum. In *The Museum of Capitalism*, edited by FICTILIS, 79. New York: Inventory Press.

Nora, P. 1989. Between Memory and History: Les Lieux de Mémoire. *Representations* 26 (Spring): 7–24.

Pamuk, O. 2012. *The Innocence of Objects*. New York: Abrams.

Pell, R. W. 2015. PostNatural Histories: Richard W. Pell in Conversation with Emily Kutil and Etienne Turpin. In *Art in the Anthropocene: Encounters among Aesthetics, Politics, Environments and Epistemologies*, edited by H. Davis and E. Turpin, 299–316. London: Open Humanities Press.

Pratt, M. L. 2017. Coda: Concept and Chronotope. In *Arts of Living on a Damaged Planet*, edited by A. L. Tsing, N. Bubandt, E. Gan and H. A. Swanson, G169–G74. Minneapolis, MN: University of Minnesota Press.

Putnam, J. 2009. *Art and Artifact: The Museum as Medium*. Revised Edition. London: Thames Hudson.

Pyne, S. J. 2015. The Fire Age. *Aeon Essays*. https://aeon.co/essays/how-humans-made-fire-and-fire-made-us-human.

Rigney, A. 2017. Materiality and Memory: Objects and Ecologies. A Response to Maria Zirra. *Parallax* 23(4): 474–78.

Robinson, K. S. 2017a. Angry Optimism in a Drowned World: Interview with Kim Stanley Robinson. In *Després de la fi Del Mon*, edited by CCCB, 145–49. Barcelona: Centre de Cultura Contemporania de Barcelona.

Robinson, K. S. 2017b. Think of yourself as a planet. In *Després de la fi Del Mon*, edited by CCCB, 149-50. Barcelona: Centre de Cultura Contemporania de Barcelona.

Rothstein, E. 2012. Where Outlandish Meets Landish. *New York Times*, January 9. https://www.nytimes.com/2012/01/10/arts/design/museum-of-jurrasic-technology-shows-its-wild-side-review.html.

Scranton, R. 2015. *Learning to Die in the Anthropocene*. San Francisco: City Light.

Szerszynski, B. 2017. The Anthropocene Monument: On Relating Geological and Human Time. *European Journal of Social Theory* 20(1): 111–31.

Thacker, E. 2011. *In the Dust of This Planet: Horror of Philosophy Vol. 1*. Winchester: Zero Books.

Tsing, A. L. 2015. *The Mushroom at the End of the World: On the Possibility of Life in Capitalist Ruins*. Princeton, NJ; Oxford: Princeton University Press.

Vergo, P. (ed). 1989. *The New Museology*. London: Reaktion.

Yusoff, K. 2013. Geologic Life: Prehistory, Climate, Futures in the Anthropocene. *Environment and Planning D: Society and Space* 31: 779–95.

Yusoff, K. 2017. Epochal Aesthetics: Affectual Infrastructures of the Anthropocene. *e-flux*. https://www.e-flux.com/architecture/accumulation/121847/epochal-aesthetics-affectual-infrastructures-of-the-anthropocene/.

Zalasiewicz, J. 2008. *The Earth After Us: What Legacy Will Humans Leave in the Rocks?* Oxford: Oxford University Press.

Zylinska, J. 2018. *The End of Man: A Feminist Counterapocalypse*. Minneapolis: University of Minnesota Press.

II

Territories

Chapter 7

WATERKINO and HYDROMEDIA:
How to Dissolve the Past to Build a More Viable Future

Joanna Zylinska

Figurations

The opening premise of my article is the seemingly obvious yet nonetheless vital fact that water is a key element of our planetary habitat and a condition of our earthly survival. Taking up this volume's call to revisit the human and nonhuman past with a view to outlining a more viable future, I want to examine water's fluid ontology and the forms of life it enables. Specifically, my argument positions water as shared human-nonhuman heritage and a site of geo-cultural memory, while recognizing that water always comes to us mediated. With this, I adopt the critical apparatus of media theory to think about geology, heritage, history and memory in terms of dynamic processes rather than solid objects. I also propose two figurations – HYDROMEDIA and WATERKINO – as conceptual tools that will allow us to view cultural practices as constitutively entangled with their environments. The figure of HYDROMEDIA highlights that water only ever becomes something in relation to its container, body or place. It is thus a quintessentially communicative medium, although its language and purview transcend the human systems of communication. The figure of WATERKINO, in turn, encapsulates a genre of films which are not just about water, but which also mobilize water as a medium of both communication and world-formation. The chapter traces this agential aspect of water by analyzing two artefacts: *The Pearl Button*, a 2015 film by Chilean director Patricio Guzmán in which water is seen as a carrier of life, death and memory, and *Even the Rain*, a 2010 film by Spanish director Icíar Bollaín focused on the 'water wars' in Bolivia. It is in this encounter with cinematic events unfolding in the Global South, outside the dominant nexus of visibility and power – while still being part of global media flows – that the possibility of developing a new mode of engaging with our geo-political vulnerabilities is sought. The ultimate aim of my chapter is to outline a more fluid, and less solidly Western, theory of planetary viability and post-Anthropocene ethics. I do nevertheless remain mindful of Nicole Starosielski's call to move beyond the perception of water in its oceanic and other arrangements only in terms of "fluidity" and to see it also as "a social space" (2012: 165), encapsulating a complex "matrix of power relations" (2012: 150).

What are hydromedia?

The concept of 'hydromedia' offers a mode of understanding water as a dynamic process that temporarily stabilizes into various forms: tears, clouds, rain drops, rivers, oceans, but also, less obviously perhaps, devices, machines, systems, networks and infrastructures – in other words, media. This argument inscribes itself in the framework of environmental media theory (see Cubitt 2005 and 2016; Hjorth et al. 2016; Maxwell and Miller 2012; Parikka 2015), which has developed out of the recent interest on the part of media scholars in the material aspects of the production, consumption and distribution of media. The question concerning the life and death of our media objects and infrastructures has provided ethico-political impetus for the study of media decomposition and waste in the context of the wider environmental destruction of our planet. At the same time, the analysis of media in environmental terms has expanded the very notion of 'media' beyond its conventional understanding based on broadcast practices, to embrace other forms of communication and linkage between a variety of human and non-human agents.

Research into the ecological aspects of media has typically foregrounded the more solid aspects of technological degradation, with water being seen as one of the casualties of the contamination process. In *Digital Rubbish* Jennifer Gabrys has looked at the consumption patterns of media users which are anchored in the logic of planned obsolescence, resulting in the regular overproduction of media waste which is then disposed outside of the Global North. She explains that "Just as the production of electronics involves the release of numerous hazardous materials into the environment, so recycling and dumping of electronics unleashes a tide of pollutants, from lead and cadmium to mercury, brominated flame retardants, arsenic and beryllium that *spread through the soil and enter the groundwater*. From manufacture to final decay, *electronics seep into the aquifer* and subsoil, settling into longer orders of time and more enduring chemical-material conditions" (2011: 142, emphasis added). Threats to the stability of the hydrologic cycle are also of concern to Larissa Hjorth and colleagues, who link our media consumption patterns to "the rise in global ocean temperatures," as a result of which "the ocean has more potential to generate powerful tropical winds and cyclones" (2016: 42). Important as these analyses are in highlighting the anthropogenic influence on environmental degradation and climate change, they also inadvertently install the subject of the Anthropocene – i.e. Anthropos, the supposedly genderless transhistorical 'human' – at the centre of action. Due to this ontological uncertainty, whereby it is never clear whether it occupies the role of a substrate, resource or indeed medium, water has taken on the more solid form of a bedrock, or Gabrys' aquifer, in many analyses of ecomedia. In this perspective, something seems to happen *to* water through *our* excessive use of media, but the mediatic agency of water itself recedes to the background.

Yet we need to be mindful of the fact that water constitutes around 60% of the human body, which means that not only are we *connected to* water but that, by and large, *we are water*. As well as functioning as a dominant component of our bodies, with 77–78% of the brain being made of water, we enter into many other watery relations through the atmosphere (rain, clouds, snow), nutrition and other forms of consumption. As Jamie Linton points out in *What Is Water? A History of a Modern Abstraction*, "That we live and think by virtue of engagement with and participation in the water process means that we cannot identify water as something apart from ourselves except through violence of abstraction" (2010: 224). In his attempt to liberate water from its objectified status premised on the reduction of its flowy ontology to a quantifiable economic resource, Linton reminds us that there are in fact *many waters*, and that 'water as such' only ever appears comingled with other substances and materials. We could therefore re-tell the history of media (or, indeed, of any other human cultural practice or artefact) as a story of water(y) relations. Water literally sustains the seemingly nebulous digital infra-structures, which enable the production, distribution and storage of our media content today. As Sean Cubitt explains in *Finite Media* (2016: 18),

> Typically 1,160 servers will fit into a shipping container, com-plete with batteries, power, cabling, water-cooling and fans. Each container draws as much as 250 kilowatts of power. The containers themselves, in one facility dating back to 2005, are stacked and networked in buildings holding 45 contain-ers, each drawing down 10 megawatts apiece (including additional cooling and water pumps), which now has three such buildings. The design was subject of a patent applied for early in 2008.

Cubitt (2016: 19) also informs us that "In 2008, Google registered patents for floating wave-powered server farms (the floating part is signifi-cant because of the quantities of water required to cool the servers down)."

An attempt to link water with media in a more agential sense, where water becomes a productive (and potentially destructive) agent in the ecomedia process, has been offered by media theorist Max Haiven, who has reported that "[t]he lowest estimates of the quantity of purified water consumed in the production of a personal computer is about 1,500 litres, about twice what an adult should drink in a year" (2013: 213). However, we could go even further to claim, as suggested earlier, that water is not just used *to produce media* but also that, alongside computers and other electronics, *water itself is a medium*. This idea builds on the infrastruc-tural understanding of media as communication networks such as rail-roads and trade routes proposed in the middle of the twentieth century by Harold Innis – a scholar who "thought the fact of media more impor-tant than what was relayed" (Peters 2015: 18) – and was subsequently picked up by a junior colleague of Innis' at the University of Toronto, Marshall McLuhan. As explained by John Durham Peters, whose book, *The Marvelous Clouds: Toward a Philosophy of Elemental Media*, borrows from Canadian media theory, as well as its modulation in the work of Friedrich Kittler, "To study media, you cannot just study media… To understand media we need to understand fire, aqueducts, power grids, seeds, sewage systems…" (Peters 2015: 29). Peters launches a powerful defence of the expansion of the concept of media beyond message-bearing institutions and proposes we see media instead as "vessels and environments" (2015: 2). This conceptual expansion is arguably just a return to a pre-media studies understanding of media, namely their nineteenth-century conceptualization in terms of natural elements such as water and earth, fire and air (see Peters 2015: 2). For Peters the key task of media theory today lies in reconnecting media to their infrastructures and environments. This task is made ever more urgent by the exigencies of the Anthropocene. Even though this approach partakes of the intellectual trajectory of posthumanism, whereby the human is dethroned from their central position as the source and destination of all action and all meaning in the world, Peters reminds us that "there are profound and urgent reasons not to forget the enormous pressures that human beings are exerting on sea, earth, sky, and all that dwells in them" (2015: 121).

Placing water at the centre of media study (as well as 'media studies') becomes a logical consequence of deciding to take the environmental imperative outlined above seriously, which means addressing our relation to water as both concept and matter – and thinking of better ways of living with water. It also means engaging with economies of water scarcity and water waste, while also raising questions for the reduction of water to a resource for the human. It bears reiterating that water will never *stay* at the centre of anything because its fluid ontology means it is never *just* an object or an infrastructure: it is first and foremost a process, a movement and a reaching out – even though, as the editors of the Canadian anthology *Thinking with Water* remind us, "All water is situated. Moreover, we are all situated in relation to water" (Chen, MacLeod and Neimanis 2013: 8). Interestingly, in proposing such a relational understanding of this most elementary of media which also demands we pay attention to the kinds of relations we enter into with water, and that we see water itself entering into, Cecilia Chen, Janine MacLeod and Astrida Neimanis introduce the concept of "mediation" (2013: 8).

Mediation becomes for them a device that can help us grasp just how "water animates our bodies and economies," but also how it "permeates the ways we think" (2013: 10). There are arguably similarities here between their notion and the way Sarah Kember and I have theo-rized mediation in *Life after New Media: Mediation as a Vital Process*. For us, mediation is not "a translational or transparent layer or intermediary between independently existing entities" but rather a complex, hybrid and all-encompassing process in which we humans partake, alongside other organisms and processes (2012: xv). Seen from this perspective, water is a dynamic medium that *makes humans* – and that goes into the making of our world: not just computers, as highlighted by Haiven, but also food chains, transportation and communication networks, cities, homes. At the same time, humans are engaged in the making and remak-ing of water into what Linton (2010) calls 'a modern abstraction' fixed by the H_2O formula, a commodity and a resource for our sustenance (source of irrigation and electricity; navigation channel; bottled bever-age) – although we are not the only water-making agents, of course. Water is therefore always part of hydromediations: multiple naturecultural pro-cesses through which it temporarily stabilizes into "media, agents, rela-tions and networks" (Kember and Zylinska 2012: xv), but from which it also always potentially overflows to form new connections – and new dis-solutions. Water, as Linton puts it, is "shockingly promiscuous": "it goes and bonds with practically everything once it escapes the lab" (2010: 4).

Watery filmmaking

The decision to examine two films as sites of hydromediation in this chapter may need justifying, given that, in disciplinary terms, some scholars still see the study of film (usually undertaken under the dis-crete umbrella of 'film studies') as more attuned to the methodologies of other hermeneutic-textual disciplines such as literature or history than to 'media studies'. Yet, following in the footsteps of scholars such as Friedrich Kittler (1999), Sean Cubitt (2005) and Giuliana Bruno (2014), I want to suggest that the study of film requires the multidisci-plinary apparatus of media studies because the latter allows us to see film precisely as a 'medium'. Going beyond the semiotic and the hermeneu-tic, a media-driven approach allows us to extend the analysis of film to its technological and material aspects. It also allows us to look at film infra-structures in terms of their production, distribution and consumption practices. A mediatic perception of film is premised on the recognition of film's historical interlocking and material kinship with other media: photography, literature, comic strips. Developments such as 'expanded cinema', 3D cinema, computer games and virtual reality (VR) foreground as well as strengthen this kinship. With the concept of hydromedia applied to the reading of film here, I want to grasp and articulate the liquid and transformative aspect of all the naturecultural relations in the world – of which we are part. But I also aim to locate us – be it specifically media scholars, media users or simply living-breathing organisms composed predominantly of water – in the media fold, while expanding the definition of media beyond standard communication devices and practices.

There are good reasons for beginning a project on hydromedia with the analysis of film – not only because of film's kinship with other media but also because, as Gilles Deleuze put it in *Cinema 1*, film carries "the promise or implication of another state of perception: a more-than-human perception, a perception not tailored to solids, which no longer had the solid as object, as condition, as milieu. A more delicate and vaster perception, a molecular perception, peculiar to a 'cine-eye'" (1986: 80). Deleuze focuses his analysis of water in film on the pre-war French school of directors such as Renoir, Grémillon, Vigo, Renoir and Epstein. He justifies those directors' predilection for water scenes and water themes by this elementary medium's ability to fulfil simultaneously the aesthetic, narrative and social documentary requirement. On the aesthetic level, writes Deleuze, "water is the most perfect environment in which movement can be extracted from the thing moved, or mobility from movement itself" (1986: 77). The abstraction of running water creates for Deleuze a *sine qua non* cinematic experience which transcends the cognitive reception of the story or even of the images, while also connecting us to the world outside film. Indeed, the image of water on screen can "give us the real as vibration in its deepest sense" (1986: 78). For Deleuze, water thus reveals itself as the original cinematic mediation, an opening to the sensation of materiality which both contains experience in a medium (a film, a scene, a frame) and enables emotions and affects which cannot be easily framed. Water-on-film thus literally moves the viewer; or even, to return to the Bergsonian argument that Deleuze builds on in his book, it makes the viewer feel alive.

Figure 7.1 — Still from *H2O*.

H2O (1929), a 12-minute experimental silent film by US director
and photographer Ralph Steiner (Figure 7.1), provides an illustration of
Deleuze's point, while also expanding his argument beyond the context of
French cinema – and beyond narrative. *H2O* offers a meditation on both
water and its capture on film, and thus also on the very process of filmmaking as an attempt to capture movement, to contain it in rectangular
frames and to stich those frames back together into an experience of life
(see Bergson 1944: 169-82). The ebbs and flows of watery movement,
presented in the form of light-and-shadow zigzags, arabesques, shimmers,
pulsations and rotations, reach out from the screen to the viewer, whose
own body is composed of, and being moved by, water flows. Yet there
is something pure, or rather purified, in Steiner's meditative piece, with
dehistoricized water flows reduced to their aesthetic aspect. We may want
to pick up here on a desire expressed by one of the most interesting writers on both water and the materiality of the social world, Ivan Illich, in his
book, H_2O *and the Waters of Forgetfulness*, "to question the beauty intrinsic
to H_2O" (1985: 3) and emphasize instead the "historicity of matter." Illich
reminds us that this seemingly nebulous water, elevated, for instance, in

fine art – his examples include paintings by Degas and Courbert – is "the stuff that circulates through indoor plumbing" (1985: 1). Significantly, this link between the nebulous and the social, between movement and matter, is already present in Deleuze, for whom water-on-film, as signalled earlier, offers not just an aesthetic experience but also a socio-political one. The liquid abstract, writes Deleuze, also stands for the concrete environment in which a different way of life, and a different way of sensing and understanding life, can be imagined and enacted. Drawing on films by Grémillon, he suggests that "the proletarian or the worker reconstitutes everywhere... the conditions of a floating population, of a sea people, capable of revealing and transforming the nature of the economic and commercial interests at play in a society" (Deleuze 1986: 78). We could therefore go so far as to say that what I propose to call 'watery filmmaking'. read in Deleuzian terms, leads to political cinema per se, because it captures, aesthetically and narratively, the fluid experience of workers caught up in the flows of capital. More importantly, it also reveals the possibility of the liquidation of the existing socio-political conditions by showing them as inherently unstable. Last but not least, watery filmmaking facilitates a shift beyond the familiar frame of reference, aka the Western colonial mind set, which arguably still permeates our philosophy, history – and art history. Grand as the claim may sound, watery filmmaking could thus ultimately become a device for decolonizing the Western mind and eye.

The two films I look at in this chapter can therefore perhaps be seen not just as geo-cultural locations but also as sites of thought from which a water-rich picture of the world, in all its entanglements, spillages and overflows, can spring. This search for a better picture of the world as outlined from the perspective of water is more than just an intellectual exercise: it partakes of what Mielle Chandler and Astrida Neimanis have termed water's "facilitative capacity" (2013: 62), which allows for the raising of ethical questions about our relationships with human and non-human others, and about the processes through which these relations are configured, maintained and redrawn. Given that the two zones of hydro-mediation I am looking at in this chapter are *about* water – Guzmán's film is about the Pacific ocean off the borders of Chile and its entanglement with the region's natural and political history, while *Even the Rain* deals with the water supply in Bolivia and an attempt by a global corporation to privatize it – my engagement with them here could be seen as an ostensible return to the more dominant, hermeneutic tradition of film *and* media studies, where media are primarily analyzed in terms of their content. But I have chosen these two sites because, even though they do indeed deal with water as their subject matter, they also enact a process of hydromediation, in foregrounding how water becomes 'water' for us humans. In other words, they treat water as a medium, thus offering what could be described as a media-sensitive account of water. Being watery-dependent media (the way all media are, as foregrounded by Cubitt (2016), and perhaps always have been through their dependence on steam power and other forms of water-based energy), they engage with water not just on the level of concept, or theme, but also on the level of substance, interweaving the different aqueous layers while also engulfing the scholar – herself largely made up of water – in both the analysis and its object.

The Pearl Button

Opening with the verse line by Chilean poet Raúl Zurita, "We are all streams from one water", Patricio Guzmán's *The Pearl Button* is a meditative documentary on the role of water in human history. The film starts by positioning water as an interplanetary medium that came to Earth on comets as ice, and is believed to have subsequently contributed to the

Figure 7.2 — Still from *The Pearl Button*.

formation of seas (Figure 7.2). This vast sense of cosmic history is anchored in the specific history of Chile and its people in the film – both its native Western Patagonia inhabitants, who were water nomads, traversing long stretches of the estuary in their canoes, and its modern citizens, who are said to have lost intimacy with the ocean. At the time of the Pinochet regime, inaugurated in 1973 after the US-sponsored coup to challenge the democratically-elected government of Salvador Allende which supported the redistribution of justice to many more Chileans, including its first peoples, the ocean took on the role of a silent witness to the operations of the military dictatorship. This role is slowly revealed in the film, via the grim discovery of the remnants of a decomposed body, brought to the shore by ocean waves. Through the collection of oral testimonies, Guzmán patches together a dark story of the ocean as a burying ground for thousands of victims of the Pinochet rule, with metal rails attached to their packaged corpses intended to ensure their undetectable decomposition on the ocean bed. The director then engages the help of Chile's historians to enact a symbolic burial of the washed-up body, using a stand-in package resembling the original wrapped-up corpse and disposing of it from a helicopter, the way Pinochet's army was said to have done with victims' bodies. He also presents underwater footage from the search for other victims' remnants, in the aftermath of the discovery of the first washed-up body.

One of the objects found by the divers is a button that most likely came off one of the victims' clothing, over which the camera lingers for a prolonged time. In its cosmic circularity and everyday objecthood, the button provides a link between the recent events in Chile and its colonial history. In the course of the film, Guzmán introduces us to the story of Jemmy Button, a native of Tierra del Fuego, who was sold to an English sea captain for a pearl button and taken to England in 1830 to be 'civilized' – only to return several years later a broken man, belonging nowhere. The eponymous button thus serves as a reminder of different forms of violence to which human lives and human bodies have been exposed throughout Chile's history, both under its colonial rule and its military dictatorship.

Some film critics have castigated Guzmán for his "gauche poeticism" (Parkinson 2016), the lack of "trenchancy and restraint" and even for "trying to find resonances where none naturally exist" (Gilbey 2016) – as if the latter was not a *sine qua non* definition of creative editing. Artistic tastes aside, there seem to be two issues that the (predominantly Western) critics, schooled in the European visual formalism or American fast-paced narrativity, seem to be discounting, or overlooking, in their reviews of *The Pearl Button* – which also lead to my two propositions for understanding Guzmán's documentary in terms of hydromedia. The first concerns the possibility of seeing the adoption of fluidity on the director's part as a purposeful methodological trope not just for the narrative but also for the filmic medium itself. The second, related one, is that the film's theoretical sensibility may actually be read as an attempt to decolonize the Western filmic 'mind-(and-eye)-set' by opening up to a different mode of experiencing both media and matter. It is a sensibility we could describe, after Chen, MacLeod and Neimanis, as 'thinking with water', i.e. as a way of bringing water "forward for conscious and careful consideration … in remembrance and recognition of the watery relations without which we could not live" (2013: 3). It is also a way of enabling a different relation with water – and acknowledging this relation as mutually constitutive.

The Pearl Button attempts to rethink geo-history as heritage embodied and embedded in the lives and bodies of the people, and in the land and sea they inhabit. Water serves there as a conceptual connector but also as a narratological and visual medium. And thus, alongside its various sources, both archival and present, photographic and filmic, the film features a montage of high-definition shots of ice-covered rocks, rain, hail, river foam, waterfalls cascading down the cordillera, Patagonia's glacial sea line and water droplets. The director plays with scale, perspective and movement to enact a sense of creative displacement in the viewer. This somewhat vertiginous mode of shooting and editing has some deeper significance, beyond mere visual formalism: it de-anchors the human from his [*sic*] self-awarded position as the pinnacle of the chain of beings, and reconnects him back to the flow of matter across different scales. It also enables a temporary displacement of the standpoint from which theories – and vistas – of what we humans call 'the world' get envisaged and articulated. We could therefore go so far as to suggest that Guzmán's work enacts what I earlier called 'watery filmmaking', with water constituting the material base of both the filmmaker and the medium, the conceptual conduit and the subject matter – but also functioning as an enabler of a new form of 'fluid montage' that does indeed try to 'find resonances where none naturally exist'.

This mode of filmmaking draws on the pre-rational, instinctual form of seeing and making connections. Philosopher Henri Bergson sees the instinctual mode of perception as synonymous with time and movement: it is a way of engaging with the world that renounces any predefined concepts of this world (see 1944: 248–49, 362). Water can become a lesson in reacquainting ourselves with our surroundings and re-experiencing ourselves not just in, but also as part of, the world. As Bergson poignantly highlights in *Creative Evolution*, "He who throws himself into the water, having known only the resistance of the solid earth, will immediately be drowned if he does not struggle against the fluidity of the new environment: he must perforce still cling to that solidity, so to speak, which even water presents. Only on this condition can he get used to the fluid's fluidity. So of our thought, when it has decided to make the leap. But leap it must, that is, leave its own environment" (1944: 211–12). Guzmán's cinematography can therefore perhaps be understood as an attempt to leap beyond the solid conventions of Anglo-American cinema, both in its narratological and essayistic guises, but also beyond the Anglo-American way of writing history, by reaching out to a different mode of knowing, thinking and perceiving.

The film builds on the visual and conceptual method of Guzmán's earlier documentary, *Nostalgia for the Light* (2010), which also looked at Chile's traumatic history from the point of view of cosmic history, but with a focus on the medium of light. One of the most memorable lines from that earlier documentary comes from an interview with astronomer Gaspar Galaz, who in a conversation with the director reveals that all our experiences have already happened and thus belong to the past:

> "The camera I am looking at now is a few meters away and is therefore already several millionths of a second in the past in relation to the time on my watch. The signal takes time to arrive. The light reflected from the camera or from you, reaches me after a moment. A fleeting moment as the speed of light is very fast.... Moonlight reaches us in one second,

sunlight – in eight minutes." "So we don't see things at the very instant we look at them?" "No, that's the trap. The present doesn't exist. It's true. The only present that might exist is the one in my mind. There's always a lapse in time." "Astronomers manipulate the past, just as archaeologists do" (*Nostalgia for the Light* 2010).

In both *Nostalgia for the Light* and *The Pearl Button* the camera thus becomes a device for time travel: it is knowingly incorporated as a framing device, of which the viewer is occasionally made aware both through the framing techniques and through the inclusion of filmmaking artefacts – cameras, cables, green screens – into various shots. It is also a reminder of the fact that all images come to us belatedly, and that they are mediated by light that serves not just as a filter, but also, more importantly, as a vehicle that carries an image through air and time. The presence of the camera as a device for purposeful seeing with a view to obtaining fixed images visualizes this process of temporal deferral in perception, and hence in experiencing, the here and now. Yet *The Pearl Button* also takes some steps towards overcoming this temporal gap by providing a material link between present and past in the form of water, which, in Guzmán's words, is "an intermediary force between the stars and us" (*The Pearl Button* 2015). This explains why his second documentary returns to the astronomic imagery so prominent in *Nostalgia for the Light*, with several shots in the film featuring large telescopes situated in the observatory on the Atacama desert, 'the driest place on Earth'.

Indeed, Guzmán's story of water as outlined in *The Pearl Button* is thus narrated through the lens of not only Chile's history but also its geography, with its estuaries, 4,200 kilometres of rugged coastline, archipelagos, glaciers, streams – and deserts. It could thus be said to offer an indirect challenge to the abstraction of water criticized by Linton as the dominant aquatic narrative of the modern world. Through his multilevel account, Guzmán opens up one of the foundational structures of Western hydro-epistemology, namely the hydrologic cycle, where water is understood in terms of standalone substance which is removed from social relations as well as specific geographical locations (see Linton 2010: 98, 103). Represented in mathematical terms, water in the hydrologic cycle is expressed in the "Rainfall = Evaporation + Runoff" equation (Linton

2010: 133) – which allows for it to be subsequently subjugated to management, quantification and, last but not least, commodification. In his analysis of the construction of 'modern water' as an abstraction, Linton throws light on what he calls "hydrological Orientalism" (2010: 123), i.e. the privileging of Northern geographies and Northern perspectives in the accounts of water's ontology. This attitude is manifested in the "dominant (Western) apprehension of deserts and arid lands as barren, poor, uncivilized places that must be hydraulically re-engineered in order to be made civilized" (ibid). Yet the Atacama desert, in all its aridity, is a pinnacle of technoscientific sophistication and power, with its observatories "funded by the international scientific community, searching the sky for the universe's past, the place from which the very matter from which humans are constructed originates" (Martin-Jones 2013: 712).

Guzmán tells us in a voiceover that "for both the Indigenous people and the astronomers, water is a concept that is inseparable from life itself", with telescopes attempting to bring the universe closer to what we once knew. The film also introduces the cosmological framework developed by the Selk'nam (one of the Indigenous people of the Patagonian region), who believed that, after death, they would turn into stars. The juxtaposition of the images of the starry sky captured by modern telescopes and the photographs of the bodies of the Selk'nam painted with multiple round dots establishes an intriguing cosmology of mediation

Figure 7.3 — Still from *The Pearl Button*.

that poses a challenge to our own, seemingly rather parochial, Western epistemologies (Figure 7.3). Going beyond the hydrologic cycle from which humans are absent, while lending an ear to human suffering in different moments in time, Guzmán eschews an easy humanism by reinscribing humans into the *cosmic* cycle – in the form of stardust. Indeed, all of the carbon, iron and nitrogen, not to mention water's key component, oxygen – chemicals that make up our bodies – were already present in the primordial cloud that appeared as a result of the death of the ancient stars, and that went on to form the solar system. In this sense, to turn to the Indigenous wisdom of 'stardust people' is not to engage in a naïve celebration of cultural difference, with *our* supposed appreciation of the other's way of life and mode of thought only ending up re-confirming our own standpoint as rational (while highlighting the other's viewpoint as interesting). What I am proposing to see as happening here, drawing on Eduardo Viveiros de Castro's theory of perspectivism, is therefore far more than a relativism that accepts the plurality of viewpoints within different cultures, without posing any foundational questions for the Western idea of rationality, relation – or, indeed, 'culture'. As explained by Peter Skafish in the introduction to the Brazilian anthropologist's *Cannibal Metaphysics*, "Viveiros de Castro treated the suppositions of Amerindian cosmology not only as demanding a critique of ostensibly universal Western concepts but also as a possible and actual basis for our own thinking" (2014: 12). Amerindian myths adopt a different perspective towards who counts as 'human' and what counts as 'communication', with various cosmic entities, those we Westerners term animate and nonanimate, being able to communicate with each other. As Viveiros de Castro explains,

> if a subject is an insufficiently analyzed object in the modern naturalist world, the Amerindian epistemological convention follows the inverse principle, which is that an object is an insufficiently interpreted subject… The most common case is the transformation of something that humans regard as a brute fact into another species' artefact or civilized behaviour: what we call blood is beer for a jaguar, what we take for a pool of mud, tapirs experience as a grand ceremonial house, and so on. Such artefacts are ontologically ambiguous: they

are objects, but they necessarily indicate a subject since they are like frozen actions or material incarnations of a nonmaterial intentionality. What one side calls nature, then, very often turns out to be culture for the other (2014: 62).

If all beings do indeed perceive, reach out, communicate and form culture(s) (see Margulis and Sagan 2000: 27; Zylinska 2012: 204), the socio-political designation of some of these beings as occupying a central position in a particular cosmology, and of others as being their servants, inferiors or food, needs to be accounted for. The cosmology of mediation outlined by Guzmán is premised on the role of water as a "shockingly promiscuous medium... that goes and bonds with practically everything" (Linton 2010: 4). Water is a communicative medium in the sense of the term as used by Canadian media theory discussed earlier, but it is also an elementary medium that can work as a liquefying agent for our entrenched socio-cultural concepts and positions. An element that joins, ontologically, not just humans with nonhumans, but also those we twenty-first century Western subjects recognize as humans with our cosmic heritage, it reminds us of the way theories are made, thought is produced and borderlines are drawn. In its different modes of circulation and identification, water as hydromediated by Guzmán and other works of critical anthropology that see non-Western views as more than just "interesting" challenges the positioning of "the subaquatic as the domain of the ethnically Other" (Starosielski 2012: 150) in early environmental cinema – an approach that contained the power of 'native' bodies by equating them with animals while also foreclosing on a deeper examination of the complexities of aqueous geopower. It thus opens up a possibility for "the permanent decolonization of thought" (Viveiros de Castro 2014: 40).

Even the Rain

This possibility has also been explored by another film from the Spanish-speaking world that deals with the problematic – and problem – of water: *Even the Rain* directed by Icíar Bollaín and starring Gael García Bernal (2010). Bernal is Sebastián, an honest young director, who, accompanied by a cynical and world-weary producer Costa (Luis Tosar), is making a film about Christopher Columbus's arrival in the New World. The crew are shooting in Cochabamba in Bolivia, the poorest country in South America, where the rates of pay are so low that the local Indian actors – who, owing to the lingering ironic logic of colonial equivalence, are to impersonate the Taino Indians from Hispaniola (today's Dominican Republic) in the film – can all be paid as if they were mere extras. "Two fucking dollars a day and they feel like kings," announces Costa. However, the production comes to a halt when the actors in Sebastián's film take on a role in a real-life drama: protests against the Bolivian government's attempts to privatize the water networks by selling their management to a multinational corporation. Presented by the authorities as a conflict between modernization and native-like victimhood on the part of 'illiterate' Indians, who have distrust 'embedded in their genes', the protests are based on real-life events that occurred in Bolivia in 2000. For several years the impoverished Bolivian government had been selling the nation's 'commons', i.e. its communication and transportation networks, to foreign investors. Yet its attempts to privatize the water supply met with unexpectedly strong resistance, led by the 'Coordinadora de Defensa del Agua y de la Vida' (Coordinator of Defence of Water and Life), as a result of which the multinational dropped its plans and withdrew from Bolivia. The film's title comes from the impassioned speech made by one of its leads. The Indigenous actor Daniel (Juan Carlos Aduviri), who is also the leader of the Cochabamba protests, cries out at one of the demonstrations that form part of what have become known as the 'water wars': "Against our will they sell off our rivers, our wells, our lakes and even the rain that falls upon us! By law!... What are they going to steal next? The sweat from our brow? All they'll get from me is piss!" On being released from jail, in which he finds himself after one of the protests, but without knowing that the film crew have already struck a deal with the authorities to hand him back as soon as the filming is finished, Daniel pleads with Costa: "You don't understand. Water is life."

Even the Rain contains many scenes of cinematic knowingness, with the violent acts of the *conquista* mirrored in the cultural and economic colonization enacted by the Spanish-led film crew on the Indigenous population. One of the most powerful scenes depicts the forced conversion, prior to their crucifixion and burning, of the Taino rebels in an attempt to ward off further resistance against the Spaniards' attempt to extract precious resources, such as gold, from the local land. On being reassured that 'good Christians go to heaven', Indian chief Hatuey, aka Daniel the leader of the water protests, spits his tormentors in the face and shouts: "Send me to hell!" At another moment, when the Indigenous mothers refuse to simulate the drowning of their babies instead of ceding them to the conquistadores – partly because they fear for their babies' safety, and partly because they cannot accept that "it's what happened" – Daniel poignantly reminds Sebastián, who insists that "it's really important for the film": "Some things are more important than your film". The dramatic arch is reached when Daniel's daughter, who also has a role in the film, is severely injured in the city protests, with Costa – interpellated by Daniel's wife Theresa's appeal to him as *amigo* – abandoning his previous mercurial attitude to his impoverished employees and driving through the violence-ravished city in order to save the girl's life. The film ends with Daniel presenting Costa with a boxed thank-you gift: a small bottle of water, or *yaku* in Quechua, which symbolizes 'life' (Figure 7.4).

While it may be easy to accuse Costa's transformation of being rooted in sentimentalizing Hollywood tropes, the film itself is too knowing about different film histories and genres to allow such an easy dismissal. Costa's moral dilemma and its subsequent resolution can perhaps instead be seen to be a rearticulation of the problematic raised by Viveiros de Castro in the following terms: "what do anthropologists owe, conceptually, to the people they study?" (2014: 39). The question of filmmakers' responsibility not only for the people they film but also for those they

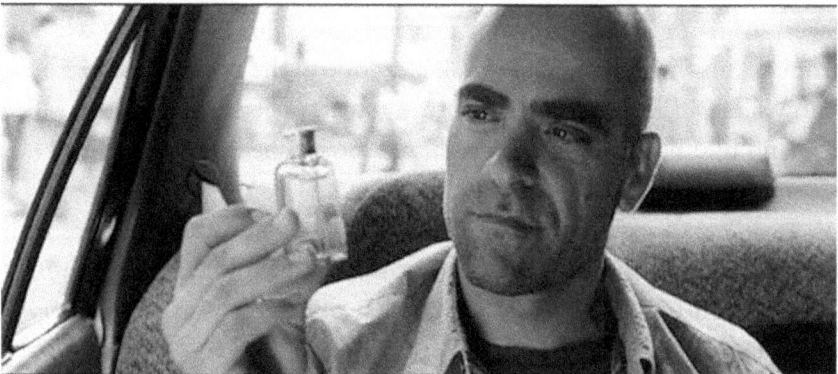

Figure 7.4 — Still from *Even the Rain*.

employ, as actors as well as crew, resonates strongly in *Even the Rain*. The main shooting of the film is accompanied by assistant director María's (Cassandra Ciangherotti) simultaneous shooting of a black-and-white documentary about the film, but also about the socio-political situation unfolding on the ground. As viewers we are therefore presented with multiple instances of 'film within the film', with various other media such as cameras, cables, dollies, props and bits of script making their way into the visual frame.

Like *The Pearl Button*, *Even the Rain* inscribes itself in the logic and structure of hydromedia, I suggest, because it is more than just a film *about* water, where water would be reduced to a resource for human consumption, in either 3D (e.g. as a substance for drinking, washing or construction) or just 2D (as a visual object on screen). In its media awareness, the film therefore mediates water and is mediated by it, turning it into both a narrative device and a conduit for a politico-ethical enquiry. Outlining a cosmology that links first-world historical consciousness with Indigenous knowledges, water becomes an actant here, playing a crucial role alongside the human actors – those starring in Icíar Bolla's, Sebastián's and María's nested films, but also those appearing in the trajectory of what the modern West calls 'history', and in its non-Western 'storied' counterparts. As Viveiros de Castro explains,

> The ethnography of indigenous America is replete with references to a cosmopolitical theory describing a universe inhabited by diverse types of actants or subjective agents, human or otherwise-gods, animals, the dead, plants, meteorological phenomena and often objects or artifacts as well-equipped with the same general ensemble of perceptive, appetitive and cognitive dispositions: with the same kind of soul. This interspecific resemblance includes, to put it a bit performatively, the same mode of apperception: animals and other nonhumans having a soul "see themselves as persons" and therefore "are persons": intentional, double-sided (visible and invisible) objects constituted by social relations and existing under a double, at once reflexive and reciprocal – which is to say collective-pronominal mode. What these persons see and thus are as persons, however, constitutes

the very philosophical problem posed by and for indigenous thought (2014: 56).

The theory of hydromedia therefore allows us to read water as an actant and a person, one among many others. As "each watery entity has been somewhere, sometime before, cycling through its various articulations through millennia" (Chandler and Neimanis 2013: 74), water can be said to trouble Western linear historicity by opening onto a different time and a different knowledge. Water also enacts a rupture in the modernist logic of human development and progress by breaking the naturalness of the flow of capital associated with modernity. Unvesseled water challenges the corporate logic of containment while also denaturalizing the modernist water imagery which, to cite Janine MacLeod, implies "the movements of capital" (2013: 42). Through this, we can glimpse the fact that "water is profoundly shared among the living, the dead and the unborn" (MacLeod 2013: 51) – a statement that echoes the Indigenous cosmologies which make their way into *The Pearl Button*.

The ethics of hydromedia: a conclusion

While the Western way of thinking normally deduces politics from ethics, we could perhaps suggest, in conclusion, that the cosmopolitical theory of watery entanglements described by Viveiros de Castro, where the polis includes all sorts of human and nonhuman 'persons', opens onto ethics. Given that all beings in the world are made of water, and that "Our watery milieu are enfolded into our bodies, repeating our ancestors differently", Chandler and Neimanis go so far as to suggest that "water constitutes a proto-ethical material phenomenon", or even that water actually "makes ethics possible" (2013: 62). Water is an ethical medium because it foregrounds the fact that no being is a self-enclosed entity that can 'encyst' itself from others. Indeed, it *is* this fact. Water thus literally liquidates the liberal stand-alone subject of ethics. Yet for an ethical event to occur, and to be acknowledged as such by watery humans, a cut is needed in the aqueous flow (see Zylinska 2014: 38–44, 98–100). This will need to take the form of an account on the part of those who call themselves human to recognize and take responsibility for the multiple forms of watery enfolding, where one entity's flourishing can signal another's demise.

The ethics of hydromedia can thus be seen as an enactment of what Kember and I have called "an ethics of mediation", which names "these processes of agential resolution that carry a human inflection: they are processes of 'differential cutting', of making pragmatic in-cisions into the flow that also have the force of ethical de-cisions" (2012: 171). To bring back the human into the cosmology of multiple 'persons' as enabled by the texts and practices discussed in this chapter is not to reinstate humanism, or express preference for the *human* modes of flourishing and facilitation. The interrogation of the latter is precisely what consti-tutes the primary task of such ethics. Yet it does involve the recognition of the human capacity for story-ing, and the need to turn such an account into an ethical interpellation. Because, as Donna Haraway has poignantly highlighted, "It matters what thoughts think thoughts. It matters what knowledges know knowledges. It matters what relations relate relations. It matters what worlds world worlds. It matters what stories tell stories" (2016: 35). In other words, it matters what we see, show and tell, about whom, with what and why. Film is a medium that is capable of taking on this task.

References

Bergson, H. 1944. *Creative Evolution*. Trans. Arthur Mitchell. New York: Random House, The Modern Library.

Bruno, G. 2014. *Surface: Matters of Aesthetics, Materiality, and Media*. Chicago, IL: University of Chicago Press.

Chandler, M. and A. Neimanis. 2013. Water and Gestationality: What Flows beneath Ethics. In *Thinking with Water*, edited by C. Chen, J. MacLeod and A. Neimanis, 61–83. Montreal and Kingston: McGill-Queen's University Press.

Chen, C., J. MacLeod and A. Neimanis. 2013. Introduction. In *Thinking with Water*, edited by C. Chen, J. MacLeod and A. Neimanis, 3–22. Montreal and Kingston: McGill-Queen's University Press.

Cubitt, S. 2005. *EcoMedia*. Amsterdam and New York: Rodopi. Kindle version.

Cubitt, S. 2016. *Finite Media: Environmental Implications of Digital Technologies*. Durham, NC: Duke University Press. Kindle edition.

Deleuze, G. 1986. *Cinema 1: The Movement Image*. Trans. Hugh Tomlinson and Barbara Habberjam. Minneapolis, MN: University of Minnesota Press.

Gabrys, J. 2011. *Digital Rubbish: A Natural History of Electronics*. Ann Arbor, MI: University of Michigan Press.

Gilbey, R. 2016. The Pearl Button seeks to reveal Chile's suffering – but doesn't quite hold together. *New Statesman*, March 17. https://www.newstatesman.com/culture/film/2016/03/pearl-button-seeks-reveal-chiles-suffering-doesnt-quite-hold-together.

Haiven, M. 2013. The Dammed of the Earth: Reading the Mega-Dam for the Political Unconscious of Globalization. In *Thinking with Water*, edited by C. Chen, J. MacLeod and A Neimanis, 213–31. Montreal and Kingston: McGill-Queen's University Press.

Haraway, D. J. 2016. *Staying with the Trouble: Making Kin in the Chthulucene*. Durham, NC: Duke University Press.

Hjorth, L. et al. 2016. *Screen Ecologies: Art, Media, and the Environment in the Asia-Pacific Region*. Cambridge, MA: MIT Press.

Illich, I. 1985. *H_2O and the Waters of Forgetfulness: Reflections on the Historicity of "Stuff."* London: Boyars.

Kember, S. and J. Zylinska. 2012. *Life after New Media: Mediation as a Vital Process*. Cambridge: MIT Press.

Kittler, F. 1999. *Gramophone, Film, Typewriter*. Stanford, CA: Stanford University Press.

Linton, J. 2010. *What Is Water? A History of a Modern Abstraction*. Vancouver: UBC Press.

MacLeod, J. 2013. Water and the Material Imagination: Reading the Sea of Memory against the Flows of Capital. In *Thinking with Water*, edited by C. Chen, J. MacLeod and A. Neimanis, 40–60. Montreal and Kingston: McGill-Queen's University Press.

Margulis, L. and D. Sagan. 2000. *What Is Life?* Berkeley: University of California Press.

Martin-Jones, D. 2013. Archival Landscapes and a Non-Anthropocentric 'Universe Memory.' *Third Text* 27(6): 707–22.

Maxwell, R. and T. Miller. 2012. *Greening the Media*. Oxford: Oxford University Press.

Parikka, J. 2015. *A Geology of Media*. Minneapolis, MN: University of Minnesota Press.

Parkinson, D. 2016. The Pearl Button Review. *Empire Online*, March 14. https://www.empireonline.com/movies/pearl-button/review/.

Peters, J. D. 2015. *The Marvelous Clouds: Toward a Philosophy of Elemental Media*. Chicago, IL: University of Chicago Press.

Skafish, P. 2014. Introduction. In E. Viveiros de Castro, *Cannibal Metaphysics*, 9–33. Minneapolis, MN: Univocal.

Starosielski, N. 2012. Beyond Fluidity: A Cultural History of Cinema under Water. In *Ecocinema Theory and Practice*, edited by S. Rust, S. Monani and S. Cubitt, 149–68. New York; London: Routledge.

Viveiros de Castro, E. 2014. *Cannibal Metaphysics*. Trans. Peter Skafish. Minneapolis, MN: Univocal.

Zylinska, J. 2012. Bioethics Otherwise, or, How to Live with Machines, Humans, and Other Animals." In *Telemorphosis: Theory in the Era of Climate Change*, vol. 1, edited by T. Cohen, 203–22. Ann Arbor, MI: Open Humanities Press.

Zylinska, J. 2014. *Minimal Ethics for the Anthropocene*. Ann Arbor, MI: Open Humanities Press.

Filmography

Even the Rain. 2010. Directed by Icíar Bollaín. Film. DVD.

H2O. 1929. Directed by Ralph Steiner. Film. YouTube.

Nostalgia for the Light. 2010. Directed by Patricio Guzmán. Film. DVD.

The Pearl Button. 2015. Directed by Patricio Guzmán. Film. DVD.

Chapter 8

Reclamation Legacies

Denis Byrne

Coastal reclamations

Coastal reclamations extend human terrestrial habitat into the sea by the mechanism of draining wetlands and mudflats or dumping soil, rock and dredged marine sediments into coastal waters to expand coastal terra firma or create artificial islands. The 're-' in the word 'reclaiming' implies a prior ownership, an implied claim that all space is incipiently human habitat. The privileging of human occupancy over that of other species, which has led to intertidal and marine space occupied by fish, crustaceans and seagrass being classified as waste (van Dooren 2014: 77), is mirrored in colonial practices of classifying Indigenous lands as waste or barren land, as seen in the colonial mapping of parts of the upland rainforest of Borneo (Peluso 1995; Tsing 1993) where the 'barren' classification has been a preliminary to the issuing of logging and mining licences. While the reclaiming of the littoral zone for agriculture has been proceeding for about 1,000 years in Asia and Europe, the acceleration seen in the rate of reclamation (for agricultural, industrial, infrastructural and residential purposes) over the last two to three centuries has been driven by the growth economics of capitalism. Whether it is Indigenous land or the littoral zone which is being colonized, the rapidly expanding rates of commodity consumption over that time have produced a corresponding expansion of the human ecological footprint, measured in terms of "the area of land or sea needed to produce the resources consumed by a given population and absorb its waste" (Bonneuil and Fressoz 2016: 245; see also Moore 2015). Coastal reclamations are part of that expansion, forming an element of capitalism's "second nature" (Bonneuil and Fressoz 2016: 22).

Coastal reclamations are a signature landform of the Anthropocene, an epoch that for the reasons given above might better be termed the 'Capitalocene' (Moore 2015). Their expansion over the last century or so is exemplified in the fact that 11,000 kilometres of the coast of China is now under some form of reclamation, half that country's coastal wetlands having been destroyed by reclamation between 1950 and the year 2000 (Ma et al. 2014). In Tokyo Bay, when you stand on one of the newer artificial islands that are still mostly bare soil, it can seem as if this land, which is devoid of vegetation, has been newly exposed by a withdrawing sea, evoking the retreat of the sea that occurred 115,000 years ago at the onset of the last glaciation. Actually, of course, the sea is currently rising, not shrinking, not withdrawing. The sea is growing in volume, as it did at the *end* of the last glaciation, about 11,000 years ago, only this time it is a result of thermal expansion of seawater as an effect of anthropogenic global warming. The sea is now pushing back against the walls of our reclamations, creating waterlines that are a zone of nervousness and stress. This zone is a good place to think about territory, territorialization, loss and the future.

Reclamation counterfactual

I want to draw here on Claire Colebrook's (2017) concept of an 'Anthropocene counterfactual', which she laid out in her chapter in the volume *Anthropocene Feminism*. There she identifies one effect of the Anthropocene as being a new self-consciousness among us of our role in imposing difference on the world, this coming after an era of feminist and queer work devoted to dismantling the idea of intrinsic difference – notably gender and sexual difference – in favour of a kind of *indifference*. Reclamations can stand as examples of what we now see as absolute difference-making by humans: hard-edged platforms of new land are superimposed on what in many cases were complex intertidal ecotones. But with her "Anthropocene counterfactual," Colebrook (2017: 5) sees the world possessing a "complexity that will always exceed any differences we read into the world" – it possesses, in this sense, a kind of "indifference" to human difference-making.

One element of Colebrook's argument is that we should pay more attention to the points at which we have imposed difference and ask – and this is where the counterfactual comes in – what would have happened if we had not inscribed this difference?

> If the Anthropocene is the return of difference – because humans are once again exceptional, but now in their destructive and inscriptive impact – it might be worth asking how such difference operates. What might it mean to think a counterfactual scenario where humans had *not* inflicted the difference of the Anthropocene on the planet? (Colebrook 2017: 5).

Three years ago, I began a study of the history of coastal reclamations in the Pearl River Delta in Southern China, a region where land began to be reclaimed for agriculture around a thousand years ago (Byrne 2018). I am now pushed to ask what difference it is that these reclamations have inscribed. The earliest reclamations in the Delta involved the simple expedient of placing a row of rocks out in the mudflats, parallel to the shore. At high tide, delta waters bearing a heavy sediment load would sweep over the rock boundary and with the retreating tide leave behind enough of that sediment that over the space of a few years a deposit of sediment would build up inside the boundary of stones. This deposit would eventually become a new rice field. This process mimicked the natural process of delta formation (Bianchi 2016) in which riverine waters slow as they approach the sea, causing them to drop much of their sediment, so forming mudflats which, with repeated floods, become perennially dry land. The farmers and the lineage trusts which since Ming (1368–1644) times took a leading role in reclamation projects (see Faure 2007) can be seen to have emulated this natural fluvial process in order to accelerate land creation and to direct it to locations most favourable to them. You might say they have acted within the delta ecology to redirect it slightly. The reclamations grew almost by themselves, following a careful and subtle aligning of human intentions with the intentionality or force of natural processes. While conventionally we think of farming communities growing crops, here they were in a sense also growing land, or growing land in order to grow crops.

In retrospect, the *difference* made by these agricultural reclamations was less consequential for other-than-human lifeforms than the new kind of reclamation – I will term it 'mechanical reclamation' – which began to be seen in the Pearl River Delta from the late nineteenth century. The paddy fields and fishponds that occupy the old reclamations provide habitat for a broad spectrum of wildlife, including migratory birds, reptiles and small mammals (Lou et al. 2014). By contrast, the new reclamations, which have been created to serve as platforms for industrial estates, container ports, residential development and airports, tend to be hostile to non-human life. They contain large expanses of concrete and other forms of Anthropocene rock that form surfaces which are more-or-less impervious and are more easily 'dedicated' to exclusive human use than is the case with agricultural reclamations.

While the fluvial sediments that comprise the agricultural reclamations were transported there by the gravity-impelled flow of rivers (the Pearl River Delta is fed by two main river systems), the 'substance' of the new reclamations, which includes demolition debris, urban waste, quarried rock and dredged sediment, is transported to the reclamation site mechanically. Fossil fuel burning earth-moving machinery (including graders and bulldozers, trucks and dredges) have allowed a vast redistribution of matter across space. This form of mechanical redistribution is characteristic of the Anthropocene. Whereas agricultural reclamations in the delta, relying on the motive power of water, could only be situated where this power was operative, fossil fuel enabled a reordering of the location of reclamations. In the words of Christophe Bonneuil and Jean-Baptiste Fressoz (2016: 203), fossil fuel "conferred on the capitalist the freedom to store energy and to mobilize it at a desired moment in the degree needed". Referring more specifically to coal power – and it will be remembered that the first mechanical reclamations, in the nineteenth century, were enabled by steam shovels and steam dredges – they observe that, "The steam engine made it possible to homogenize space, to ignore location, watercourses and gradients" (ibid). Finally, it will not escape notice that the purpose of some of the delta's largest new reclamations (the container ports and airports of Hong Kong and Shenzhen), formed by mechanical redistribution of matter over local and regional space, is to facilitate the mobility and redistribution of people, commodities and manufactured goods over transnational and global space.

Thinking counterfactually, as Colebrook (2017) encourages us to, means asking not just how the difference of reclamations operates but what would have happened if the difference of agricultural reclamations and then mechanical reclamations had not been inscribed on the natural-cultural space of the Pearl River Delta in the first place. The most obvious answer is that delta shorelines would today be radically different. Prior to the inscription of reclamations there existed a complex intertidal transition zone between land (as in dry land, or terra firma) and the sea, principally taking the form of mudflats and the mangrove ecologies they support. Reclamations, whether of the agricultural or mechanical type, collapse this transition. They emplace bunds and sea-walls which draw a hard line between land (the new land of the reclamation) and sea, a line that tends to be drawn straight rather than curvilinear. Apart from their impact on intertidal ecologies, one of their other effects is that for many millions of humans, their most common contact with the sea now occurs at these sites of sharply drawn land-sea interface.

Figure 8.1 — Reclamations on the south side of Weiyuan Island, Dongguan City, in the Pearl River Delta. The buildings on the left were constructed on a mid-twentieth-century reclamation; the fields on the right are part of a late-twentieth- to early-twenty-first-century reclamation. (Photograph by Denis Byrne, 2018).

The production of flatness

In 2018, 55% of the world's population lived in urban settings.[1] There they spent much of their lives on anthropogenic surfaces, including on the formed surfaces of roads and carparks and on the horizontal floor surfaces of buildings which rest on levelled sites. Typically, these surfaces are flat, as are the surfaces of the great majority of coastal reclamations. While coastal reclamations function to expand the amount of terrain available for human terrestrial activities, a key element of this is their provision of flatness. Barry Higman (2017), in his book on the subject of flatness, maintains that the anthropogenic expansion of flatness has been fundamental to the Anthropocene.

Since deep in our hominid past, it seems to have suited us to eat, sleep and socialize on flat surfaces. The landscape architecture scholar, David Leatherbarrow (1999, 2002), has been particularly concerned with the platforms (or slabs) we create for houses to rest on and the platforms and the terraces we create for gardens, ponds, swimming pools and so on. He stresses the way we have privileged flatness as human habitat, and he describes the projects of levelling required to produce flat terrain in situations where it does not naturally occur. In the Pearl River Delta today – for example in Zhongshan prefecture to the north of Macao – everywhere one sees the raw scars where the sides of hills have been cut away to create flat terrain for factories and residential developments. In some cases, hills have been entirely excised from the face of the landscape for this reason. Much of the rock and soil, the material substance of these hills, has ended up being used to create new coastal reclamations or to extend old ones, meaning that flatness is achieved at both ends of the process. The impression you get in traveling through such areas is of a brutal reshuffling of earth materials, roughly emulating natural processes of erosion that would take millions of years to achieve something like the same levelling.

The human footprint on the Earth expanded slowly through the millennia when we were hunters and gatherers, more rapidly after we began domesticating plants and animals, about 11,500 years ago, more rapidly still from the time of the Industrial Revolution, beginning in the mid-1700s, and very rapidly in the period since World War II. Matt Edgeworth (2014, 2016) has introduced the concept of the *archaeosphere* as a useful way of drawing attention to the fact that vast areas of the Earth are now covered by the modified soils and terraced hillslopes of agriculture, the concrete and asphalt paving of roads, airports and container ports, the underground infrastructure of tunnels, pipes and wiring below our cities, the burgeoning landfill sites and the reclamations which extend coastal terrain out into the sea. The archaeosphere is a layer of varying thickness that has expanded at an accelerating rate and nowhere is this more apparent than along the world's coastlines. On Japan's main island of Honshu, for instance, 60% of the coastline is now classified as 'artificial', which is to say that for the most part it is concrete.[2] Honshu has swapped much of its pre-existing coastline of beaches, dune fields and wetlands for reclamations that constitute an ocean of concrete forming a flat platform for the enactment of contemporary life – forklifts drive over it, kids bounce balls on it. Some of Japan's skateboard parks are located in this 'second nature' requiring concrete surfaces to be sculpted into the hills and hollows which 'concrete disciples' favour.[3]

Most of this concrete coast dates from the time of Japan's post-war 'economic miracle', beginning in the mid-1950s and representative of a surge in the pouring of concrete surfaces that began at that time in many parts of the world and has gathered pace since then. As an Anthropocene marker, this concrete, and the archaeosphere more generally, is likely to be much easier for most people to apprehend than the plutonium traces which fell to Earth following nuclear testing in the 1950s (Waters et al. 2017). This seems a good reason to find new ways to draw attention to coastal reclamations: whether they are characterized by concrete surfaces or not, they help make the Anthropocene visible in a tangible, graspable way. The ability of people to grasp the Anthropocene as a material reality seems a crucial prerequisite for any widespread popular mobilization against the dark future which it portends. But coastal reclamations can only serve this role once people understand that they are in fact reclamations and not natural landscapes. The problem here is that, once created, many of them, including those that support waterfront parks (Byrne 2017), come to seem natural in their own right and the point where they were sutured to the pre-existing coastline can be difficult to detect. For each new generation born into this habitat, the reclamation is "given in its sensuous certainty," to borrow the words of Sara Ahmed (2010: 241) who writes that, "What passes through history is not only the work done by generations but the 'sedimentation' of that work as the condition of arrival for future generations". Trees sprout from the anthropogenic soil of many reclamations and their fallen leaves form humus that helps create the conditions for earthworms and other organisms to flourish in that soil. Plants, buildings and infrastructure spread across these new landforms, exaggerating the reclamation's appearance of longevity and making it more difficult for us to recall or imagine the beach or marshland it displaced. At the same time, perhaps we always 'know' reclamations for what they are. Thom van Dooren (2016: 200), in reflecting on the massive reclamations that house much of the Port of Rotterdam, has been inspired by Michel Serres's writing to think of such topographic alterations as *markings*, in some ways not unlike the territorial markings of other species.

I suggest that in order to "think a counterfactual scenario" (Colebrook 2017: 5) in relation to reclamations and to ask what would have happened if we had not inscribed the difference they represent, we first need to 'excavate' them and thus to unwind them historically. Against the tendency to assimilate them as natural phenomena, we need to give them a history, in other words, to 'unwind' back to the moment of their creation (or one might say 'inscription') and then go back to the moment before that in order to think how the world was, and how we were, before them. I agree with Timothy Morton's view (2017: 147) that an awareness of the present's *continuity* with the past is important to "the question of what kind of world we want to inhabit". This form of observation is, for him, "futuristic because thinking the contours of this continuity is part of how to exit from it: you have to figure out what form of prison you are in before you can escape" (2017: 147).

Sea-level rise and the walk back

When the plane touches down on the runway of Kansai airport in Osaka Bay, you register the thump of several hundred tonnes of aircraft, people and luggage reconnecting with terra firma. You have landed, but what kind of 'land' is this? The two rectangular islands on which Kansai airport rests were created in two stages between 1987 and 2007 by depositing

430 million cubic metres of quarried fill material on the clay floor of
the bay (Mesri and Asce 2015). But however massive and imposed the
Kansai reclamation is, it is, like all reclamations, contingent on a host of
environmental elements remaining favourable to its continued existence.
Many reclamations rely on storms not growing in intensity over time
until they reach the point where they weaken or carry away the polders
and seawalls built to prevent the sea reclaiming the space of the reclamations. Coastal reclamations are contingent on sea levels not rising beyond
the point where it is feasible to keep them dry.

In the case of Kansai airport, the analyses carried out by marine
engineers in the pre-construction phase failed to predict how quickly
and to what extent the Holocene and Pleistocene alluvial clay layers in
the seabed would compress, or 'settle', under the load of the reclamation. By 2012 the rate of settlement of the two airport islands had been
measured at between 17 and 30 metres, leading to a situation in which,
when typhoon Jebi struck on September 4, 2018, with an accompanying 3.29 metre storm surge, the islands were flooded by the sea, trapping
over 3,000 passengers and airport staff. Photographs circulated globally showed Renzo Piano's iconic terminal building seeming to be afloat
in Osaka Bay.[4]

Kansai airport's 2018 submersion was a 'reminder' of the effect of
previous coastal inundations, particularly that which occurred when the
sea rose approximately 120 metres following the end of the last glaciation 11,700 years ago (reaching its present level around 2,000 years ago).
Archaeologists have made the point that for most of the 200,000 years
that modern humans have existed, sea levels have been significantly lower
than they are today and thus the global territory available to us as a species was some 20 million square kilometres greater than today (Harff et
al. 2016). This extent of habitat is, you might thus say, what we are used
to as a species. It is also true, of course, that the experience of witnessing rising sea levels and coastal recession is not new to us. The travellers
in Kansai's terminal who looked down through the glass walls at the sea
as it advanced over the runways and aprons were re-enacting, in a sense,
what our coast-dwelling forebears experienced during the early Holocene
coastal recession. In areas of gently sloping terrain, some effects of sea-
level rise in the early Holocene "would have been readily apparent within

the lifetime of individuals or within living memory, and, perhaps, were dramatically so. Hence, we may expect that sea-level change would have affected past systems of belief and cosmology, as well as more practical matters of subsistence and social interaction" (Harff et al. 2016: 2). In the lowest-gradient parts of the Sunda Shelf (which joins the landmasses of the Malay peninsula and present-day Indonesia to mainland Southeast Asia) the sea would have moved inland tens of kilometres over the course of a person's lifetime (Wurster and Bird 2016). It is not, then, the phenomenon of sea-level rise itself which is novel in human experience; what is novel is the situation of rising sea levels precipitated by actions of our own, not least among them the discharge into the atmosphere of fossil emissions during air travel.

Applied to coastal reclamations, the kind of counterfactual thinking proposed by Colebrook (2017) would require us to mentally accommodate the geologic timescale of Pleistocene and Holocene sea-level oscillations. One of the more interesting affordances of these reclamations is that they invite us to experience time in an unusually physical way. In some parts of the world, including the Zhongshan area of the Pearl River Delta and the south side of Hangzhou Bay (also in China), coastal reclamations have been extended seaward in a serial manner. Over the course of several centuries, new 'bands' of agricultural reclamation have been added to the outer or 'leading' edge of earlier ones to form a pattern which, when seen from the air, looks not unlike a series of tree rings (Byrne 2018). Archaeologically, the passage of time is typically marked by the vertical accumulation of occupation layers and other anthropogenic, vertically ordered strata so that, in excavating down through these stratigraphic sequences one proceeds vertically back/down through time. With 'serial' reclamations, however, such as those of Zhongshan and Hangzhou, the temporal order is horizontal, meaning that in walking (or bicycling, or driving etc) out across these reclamations we move laterally out through time. Similarly, in Tokyo Bay, one can walk out across a series of adjacent artificial islands, linked by bridges, that date from the 1910s, 1920s, 1960s and later. The experience here is of walking, in a kind of reverse archaeology, out through the stratigraphy of the twentieth century ('reverse' in the sense that here one proceeds from the earliest to the latest, opposite to the way one proceeds in an archaeological excavation).

Figure 8.2 — 'Walking out.' A bridge linking the artificial islands on the
west side of Tokyo Bay. (Photograph by Denis Byrne, 2016).

Having walked to the end of these reclamation sequences one turns
and walks back inland. As the sea rises over the next centuries – the current conservative estimate is a 65cm rise by 2100 (Weeman and Lynch
2018) – people will be required to make the 'walk back' in front of an
expanding sea. Although we routinely speak of 'defending' present coastlines against sea-level rise, there is nothing aggressive about the sea's
expansion. David Leatherbarrow (1999: 172) offers this depiction of
water's agency:

> Water has the virtue of unselfish willingness to sacrifice its
> present form for the shape of its next container, doing this
> continually and insistently, as if this act of humility were its
> lifelong task and higher purpose – as if its charge were to fill
> up every space it enters the way sound does a room, pressing
> everything other than itself out of its new container.

Natural coastlines are in a dialogical relation with the sea: they allow
incoming tides to enter them, to fill up embayments and submerge mudflats; they allow the sea to erode their headlands, carve out coves, lay
down and take away beaches. Equally, as seen in the case of river deltas,

coastlines expand into the sea. Rachel Carson's wonderful books on marine ecology at the water's edge, including *Under the Sea Wind* (1941) and *The Edge of the Sea* (1955), memorably depict these mobile spaces of interpenetration. The 'difference' a reclamation makes is to inscribe a land-sea boundary that in its straightness and hardness is designed to negate and resist interpenetration. It installs an artificial terra firma designed never to sacrifice its own form, never to give up its substance to the sea. In this respect, we must give due attention to seawalls: what they are, what they do and what they teach.

Figure 8.3 — Ifugao rice terraces in the Cordillera of Luzon, Philippines. (Photograph by Frank George, taken between 1890 and 1923. Collection of the Library of Congress).

We might begin by turning inland. Among our first large-scale efforts at levelling land were the terraced fields we inscribed on hillslopes which turned hillsides into cascading sequences of platforms that in many cases are supported by stone walls. In Asia, terracing expanded with the advent of wet rice agriculture and its requirement for dead-flat fields which are effectively ponds in the early part of the crop cycle. The technology of terracing has allowed groups like the Ifugao of the Luzon highlands to occupy steeply dissected terrain in which an economy based on wet rice agriculture would otherwise be impossible (Conklin 1980). What agricultural terracing does is push flat land up the sides of hills whereas coastal reclamations push it out to sea. But the walls that mark the outer edge of these platforms become lines of tension. They require constant, laborious maintenance to counter the effects of erosion. The *difference* made by the generations who built the terraces becomes a legacy of labour for subsequent generations. The same can be said of those who have built the seawalls that support and protect coastal reclamations.

The seawall

The outer edges of coastal reclamations are almost always marked by some form of hard barrier, typically in the form of masonry seawalls (mostly stone or concrete or the two in combination) and revetments. While seawalls 'take' the force of incoming waves, revetments (sloping structures formed of materials such as wood, rock or concrete tetrapods) are designed to absorb and dissipate wave energy. Put simply, reclamations displace the sea, pushing it out of space that it has formerly occupied and thus setting up a line of tension between the mass of the reclamation and the mass of the sea. A reclamation's sea barrier may have to contend with the force of waves, storm wave setup (the effect of a storm front in pushing up the sea level at the coast) and the force of ocean currents, but it also has to contend with the ordinary hydrostatic pressure of the body of water that is the sea. If a reclamation can be thought of as pushing the sea out of a certain space, then it can also be thought to exist in defiance of the ongoing push of the sea to reoccupy that space.

Figure 8.4 — The reclamation and seawall at Elizabeth Bay on
Sydney Harbour. (Photograph by Denis Byrne, 2017).

Were a reclamation suddenly to vanish, the sea would flood back
into the space it formerly occupied. Free of the limits imposed on it by
the reclamation's sea barrier it would, in Leatherbarrow's (1999: 172)
terms, reshape itself in accord with the shape of the beaches, inlets and
headlands of the former coastline. But even where the reclamation and
its sea barrier remain in place, the sea works against or 'worries' the barrier, seeking out its weaknesses, testing its resolve. In an earlier discussion
of the nineteenth-century sandstone seawall at Elizabeth Bay, in Sydney
Harbour (Byrne 2017: 53), I point to the way the sea and the sea spray
oxidize the minerals in the sandstone, causing the stone blocks of the wall
to erode relatively quickly (compared, say, to granite seawalls). When the
tide is out, a remnant sliver of beach is exposed at the base of the wall,
allowing me to examine the effects of this erosion close-up. On one occa-
sion, standing on the beach facing the wall, it occurred to me that as the
sandstone blocks eroded, they allowed the sea to advance inland a mil-
limetre at a time back towards where it was prior to the construction of
the reclamation in the 1880s. In doing so, the sandstone appeared to be
at least as amenable to the sea's impetus as it was to our intention for it to
defend the reclamation against the sea. If we concede that the harbour

waves hitting the wall 'intend' not to end there but to run up the sandy surface of the former beach, just as they did in the days before the wall was built, then the eroding wall is responding to the sea's intentions at the same time as it temporarily serves our purpose of keeping the sea out of the reclamation.

Some would say that to speak of the sea's 'intention' to reclaim the space of the reclamation is to fall into the error of anthropomorphizing the sea. I could, alternatively, speak of the sea as having an *impetus* or momentum which would see it reoccupy this space. But I agree with Bruno Latour (2017: 52–54), who, in discussing the measures taken by the US Army Corps of Engineers to stop the Mississippi River spilling over into the bed of the Atchafalaya and hence flooding New Orleans, points to the way the Corps itself has anthropomorphized the Mississippi delta, deploying "the vocabulary of battle" against a "dangerous" river. The whole discourse of coastal engineering posits the sea – particularly the now-rising sea – as a threat that has to be defended against.

Thinking counterfactually, the 'danger' of the Mississippi being captured by the bed of the Atchafalaya is only a danger in relation to the fact that New Orleans and other settlements and infrastructure have been positioned in a way that leaves them vulnerable to the consequences of that capture. Similarly, what makes us particularly vulnerable to sea-level rise is that we have concentrated so much human settlement and infrastructure so close to the coast that we are now deeply committed to sea level remaining static. It is estimated that 450 million people and over 4,000 settlements are located within 20 km horizontal distance and 20 metres elevation of the coastline (Small and Nicholls 2003: 595), and in many countries the proportion of population in near-coastal locations is increasing. To dramatize the situation, people are rushing to the coast and the sea is rising to meet them. I maintain that to ascribe intentions to the sea in Sydney Harbour simply recognizes the nature of the relationship we entered into with the sea when we carried out our reclamations, whereas to think of ourselves as under attack by the sea – or by a noncompliant nature more generally – is to refuse to own our past actions and turn our back on who and what we are. If we are now at heightened risk from the sea it is because we have, in a sense, pushed ourselves *into* the sea.

Conclusion: Paul Virilio goes to the beach

In the language of coastal engineering, the Elizabeth Bay seawall is a 'hard defence'. By way of concluding, I would like to draw a comparison between that seawall and Atlantikwall, the system of concrete bunkers and other fortifications built by Nazi Germany between 1942 and 1944 along the French littoral to defend against an Allied invasion of Europe. With the breaching of the Atlantikwall during the Normandy landings of June 1944, the Allies, like a flood of water, filled up the space of France and pushed German forces back into Germany.

A year later, Paul Virilio, at the tender age of thirteen, took a train to the Normandy coast and had his first encounter with the sea, access to the Atlantic littoral having been forbidden during the German occupation. What he saw when he left the train station at his destination and began walking to the beach was so novel and marvellous to him that he temporarily lost his bearings (Virilio 1994: 10). This is how he described the experience when looking back as an adult:

> Advancing in the midst of houses with gaping windows, I was anxious to be done with the obstacles between myself and the Atlantic horizon; in fact, I was anxious to set foot on my first beach. As I approached Ocean Boulevard, the water level began to rise between the pines and the villas; the ocean was getting larger, taking up more and more space in my angle of vision. Finally, while crossing the avenue parallel to the shore, the earth line seemed to have plunged into the undertow, leaving everything smooth, no waves and little noise. Yet another element was before me: the hydrosphere (1994: 12).

Virilio (1994: 9) describes his discovery of the sea as a "precious experience" and "an event in consciousness of underestimated consequences". He became a great fan of the beach and a frequent visitor to the Atlantic coast. In the course of these visits he naturally became familiar with the Atlantikwall bunkers, now lying derelict. As a young man, he used one of them as a cabana and in 1958 began a photographic survey of them. "I would hunt these grey forms," he tells us, "until they would transmit to me part of their mystery" (1994: 11).

In all, there were 15,000 bunkers making up the 'wall'. Virilio would come to describe them as "funerary monuments of the German dream" (1994: 29) because they represented a fallback of defensive strategy by a military whose initial success had been based on the strategy of offensive speed, or *blitzkrieg*. He would quote a statement made by Mao Tse-tung in 1942: "If Hitler is obliged to resort to strategic defence, fascism is over and done with" (1994: 28).

Virilio's survey of the bunkers became the subject of a 1975–76 exhibition at the Museum of Decorative Arts in Paris which featured his superb black and white bunker photographs, the catalogue for that exhibition (including text by Virilio) forming the basis of the book, *Bunker Archaeology* (1994). But the survey had also been seminal to Virilio's book, *Speed and Politics* (1986), in which he charted the accumulating role of speed in the history of warfare, tracing a sequence that leads from the Medieval fortress whose power lay in its immobile, static resistance through to the "lightning warfare" of the Third Reich in which "stasis is death" (1986: 67).

In seeing a resemblance between the Third Reich's fantasy of installing an impermeable 'wall' across the French littoral and present-day

Figure 8.5 — The seawall recently added to the top of the quay on Honmura Island in Japan's Seto Inland Sea. (Photograph by Denis Byrne, 2016).

projects to build seawalls to protect us against sea-level rise, I am particularly taken with the idea that although Germany instinctively rejected the principle of stasis which the bunkers represented – Virilio (1994: 29) notes the fact the Hitler repeatedly refused ever to visit the wall – the massive solidity of the bunkers provided some sense of security. These "littoral boundary stones" (1994: 11) might be said to have provided a reassurance that this edge of Europe could be rendered as a hard boundary, a concrete frontier, even when fighter planes and bombers had already rendered the idea of such boundaries meaningless. For Virilio (1994: 12), the history of speed – in this case in the form of aircraft – had 'shipwrecked' the Atlantikwall: "These concrete bunkers were in fact the final throw-offs of the history of frontiers" (ibid).

We stand at the end of a 2000-year history of relative sea-level stability, a temporal interval that is negligible geologically but nevertheless seems sufficient for us to cling to the waterline we know rather than begin the 'walk back'. In trying to hold what we have, we confront the fact that much of 'what we have' is not just the coastline bequeathed to us 2,000 years ago when the Holocene marine incursion stalled but also all the territory gained by reclamation. In the battle to hold this expanded version of the mid- to late Holocene coastline, we are beginning to deploy steel-reinforced concrete seawalls, the rough equivalent to the Third Reich's reinforced concrete bunkers. Coincidentally, the concrete tetrahedrons deployed by the German army engineers on the Normandy littoral to trap Allied landing craft and amphibious tanks (Virilio 1994: 27) find their equivalent in the concrete tetrapods typically piled up in front of seawalls to dissipate, or 'trap', wave energy.[5]

Staying with Virilio, I return to the issue of flatness which I maintain is essential to understanding the proliferation of coastal reclamations. In an essay on the place of the Atlantikwall bunkers in Virilio's work, the sociologist Mike Gane (2000) mentions Virilio's discovery that some of the bunkers he examined had toppled over or tilted as the sand dunes they were sited on eroded. This meant that the horizontal plane of their floors was now inclined at an angle. Collaborating with the architect Claude Parent, Virilio began working on designs for 'oblique' buildings and urban precincts in which flat living surfaces were replaced by sloping ones or where flat surfaces were linked together in assemblages by sloping surfaces (Parent and Virilio 2004) in the form of ramps. Parent and Virilio designed the church of Sainte-Bernadette du Banlay in Nevers, a suburb of Paris, a structure that has inclined floors and looks strikingly like a bunker.

Neither the public nor the architectural profession were receptive to the architecture of the 'oblique function' (Parent and Virilio 2004) and, apart from Sainte-Bernadette du Banlay church, it never achieved physical form. What interests me, however, is that in breaking from the 'static' space of horizontality, to which humanity has been condemned, this architecture, by disorienting the body and demanding of it the effort to walk up and down slopes, was intended to force users to become self-conscious of the way their lives were ruled by the conventional architecture of horizontality or – in my terms, flatness – and to question those conventions. The experience would, in Gane's (2000: 87) words, add "alienation to alienation".

In the case of coastal reclamations, I see no point in seeking to alienate people from the reclamations they inhabit or use. Many of them have no choice but to keep on occupying reclamations and when the time comes that many reclamations are inundated, the resettlement of inhabitants and relocation of industries will no doubt in itself inflict environmental destruction elsewhere.[6] However, as I mentioned earlier, I believe there is value in working to give greater visibility to reclamations and to encourage self-consciousness of what they are and how they got there. Giving them this visibility would contribute to the larger task of giving visibility and tangibility to the Anthropocene, against the agenda of those interests (the oil industry, for example) which seek to make it invisible.

Acknowledgements

I wish to thank Andrea Connor for alerting me to new thinking on flatness. I am grateful also for the comments provided by Jennifer Hamilton, Gay Hawkins and Astrida Niemanis on an earlier version of this chapter.

Notes

1. United Nations, Department of Economic and Social Affairs, 2018, https://www.un.org/development/desa/en/news/population/2018-revision-of-world-urbanization-prospects.html.

2. According to a 1996 survey by the Ministry of Land, Infrastructure and Transport, the 'natural coastline' in Japan as a whole totals 17,660 km, the 'semi-natural' coastal areas make up 4,358 km, and 'artificial' coastal areas total 11,212 km (Hesse 2007).

3. See https://www.concretedisciples.com/skatepark-directory/skateparks/japan_c97/.

4. Ibid. The airport authority maintains that its use of 'sand drain' technology has now slowed the rate of subsidence; see Kansai International Airport Land Co., 'Technical Information, Approach to Settlement, Condition of Settlement', http://www.kiac.co.jp/en/tech/sink/sink3/index.html.

5. So ubiquitous have tetrapods become in the coastal landscape of Japan they have become subjects of cult interest (Hesse 2007).

6. I am indebted to Paul James (Western Sydney University) for raising this point in a discussion in February 2019.

References

Ahmed, S. 2010. Orientations Matter. In *New Materialisms: Ontology, Agency, and Politics*, edited by D. Coole and S. Frost, 234–7. Durham, NC: Duke University Press.

Bianchi, T. S. 2016. *Deltas and Humans: A Long Relationship Now Threatened by Global Change*. Oxford: Oxford University Press.

Bonneuil, C. and J.-B. Fressoz. 2016. *The Shock of the Anthropocene: The Earth, History, and Us*. Translated from the French by D. Fernbach. London: Verso.

Byrne, D. 2017. Remembering the Elizabeth Bay Reclamation and the Holocene Sunset in Sydney Harbour. *Environmental Humanities* 9(1): 40–59.

Byrne, D. 2018. Reclaiming Landscape: Coastal Reclamations Before and During the Anthropocene. In *The Routledge Companion to Landscape Studies*, 2nd ed., edited by P. Howard, I Thompson, E. Waterton and M. Atha, 277–87. London: Routledge.

Carson, R. 1941. *Under the Sea Wind*. New York: Simon and Schuster.

Carson, R. 1955. *The Edge of the Sea*. New York: Houghton Mifflin Harcourt.

Colebrook, C. 2017. We Have Always Been Post-Anthropocene: The Anthropocene Counterfactual. In *Anthropocene Feminism*, edited by R. Grusin, 1–20. Minneapolis, MN: Minnesota University Press.

Conklin, H. C. 1980. *Ethnographic Atlas of Ifugao*. New Haven, CT: Yale University Press.

Edgeworth, M. 2014. Archaeology in the Anthropocene: Introduction. *Journal of Contemporary Archaeology* 1(1): 73–77.

Edgeworth, M. 2016.Grounded Objects: Archaeology and Speculative Realism. *Archaeological Dialogues* 23(1): 93–113.

Faure, D. 2007. *Emperor and Ancestor: State and Lineage in South China*. Stanford, CA: Stanford University Press.

Gane, M. 2000. Paul Virilio's Bunker Theorizing. In *Paul Virilio: From Modernism to Hypermodernism and Beyond*, edited by M. Featherstone, 86–102. London: Sage.

Harff, J., G. N. Bailey and F. Lüth. 2016. Geology and Archaeology: Submerged Landscapes of the Continental Shelf: An Introduction. In *Geology and Archaeology: Submerged Landscapes of the Continental Shelf*, edited by J. Harff, G. N. Bailey and F. Lüth, 1–8. London: Geological Society.

Hesse, S. 2007. Tetrapods. *Japan Times*, July 22.

Higman, B. W. 2017. *Flatness*. London: Reaktion.

Latour, B. 2017. *Facing Gaia: Eight Lectures on the New Climatic Regime*. Translated from the French by C. Porter. Cambridge: Cambridge University Press.

Leatherbarrow, D. 1999. Leveling the Land. In *Recovering Landscape: Essays in Contemporary Landscape Architecture*, 171–84. New York: Princeton Architectural Press.

Leatherbarrow, D. 2002. *Uncommon Ground: Architecture, Technology and Topography*. Cambridge, MA: MIT Press.

Lou, Y., H. Fu and S. Traore. 2014. Biodiversity Conservation and Rice Paddies in China: Toward Ecological Sustainability. *Sustainability* 6: 6107–24.

Ma, Z., D. S. Melville, J. Liu, H. Yang, W. Ren, Z. Zhang, T. Piersma and B. Li. 2014. Rethinking China's New Great Wall. *Science* 346 (6212): 912–14.

Mesri, G. and S. M. Asce. 2015. Settlement of the Kansai International Airport Islands. *Journal of Geotechnical and Geoenvironmental Engineering* 141(2): 04014102/1–04014102/16. https://www.scribd.com/document/308958732/ASCE2014-Online-Kansai-Settlement-Mesri

Moore, J. 2015. *Capitalism in the Web of Life: Ecology and the Accumulation of Capital*. New York: Verso.

Morton, T. 2017. *Humankind: Solidarity with Nonhuman People*. London: Verso.

Parent, C. and P. Virilio. 2004. *The Function of the Oblique: The Architecture of Claude Parent and Paul Virilio*. London: AA Publications.

Peluso, N. L. 1995. Whose Woods are These: Countermapping Forest Territories in Kalimantan, Indonesia. *Antipode* 27(4): 383–406.

Small, C. and R. J. Nicholls. 2003. A Global Analysis of Human Settlement in Coastal Zones. *Journal of Coastal Research* 19(3): 584–99.

Tsing, A. L. 1993. *In the Realm of the Diamond Queen*. Princeton, NJ: Princeton University Press.

van Dooren, T. 2014. *Flightways: Life and Loss at the Edge of Extinction*. New York: Columbia University Press.

van Dooren, T. 2016. The Unwelcome Crows: Hospitality in the Anthropocene. *Angelaki* 21(2): 193–212.

Virilio, P. 1986. *Speed and Politics: An Essay of Dromology*. Translated from the French by Mark Polizzzotti. New York: Semiotext(e).

Virilio, P. 1994. *Bunker Archaeology*. Translated from the French by George Collins. New York: MIT Press.

Waters, C. N., J. Zalasiewicz, C Summerhayes, I. J.Fairchild, N. L. Rose, N. J. Loader, W. Shotyk, A Cearreta, M. J.Head, J. M. P. Syvitski, M. Williams, M. Wagreich, A. D. Barnosky, A. Zhisheng, R. Leinfelder, C. Jeandel, A. Gałuszkao, J. A. Ivar do Sul, F. Gradstein, W. Steffen, J. R. McNeill, S. Wing, C. Poirier and M. Edgeworth. 2017. Global Boundary Stratotype Section and Point (GSSP) of the Anthropocene Series: Where and How to Look for Potential Candidates. *Earth-Science Reviews* 178: 379–429.

Weeman, K. and P. Lynch. 2018. New Study Finds Sea Level Rise Accelerating. *NASA Global Climate Change*, February 13, 2018. https://climate.nasa.gov/news/2680/new-study-finds-sea-level-rise-accelerating/

Wurster, C. M. and M. I. Bird. 2016. Barriers and Bridges: Early Human Dispersals in Equatorial SE Asia. In *Geology and Archaeology: Submerged Landscapes of the Continental Shelf*, edited by J. Harff, G. N. Bailey and F. Lüth, 235–50. London: Geological Society of London.

Chapter 9

Human-Nature Offspringing:
Indigenous Thoughts on Posthuman Heritage

J. Kelechi Ugwuanyi

Introduction

Katherine Hayles suggests in her book *How We Became Posthuman* (1999) that the term posthuman both incites terror and excites pleasure. Explaining the terror it incites, she posits that 'post', with its dual connotation of superseding the human and coming after it, hints that the days of "'the human' may be numbered" as intelligent machines come to dominate humans on the planet (1999: 283). On the other hand, the pleasure it incites is that posthumanism opens up new ways of understanding what being human means today. In the words of Cary Wolfe, "the human occupies a new place in the universe, a universe now populated by what I am prepared to call nonhuman subjects" (2009: 47). The 'numbered days' alluded to by Hayles are also found in posthuman discourses in the arts, humanities and social sciences. Some of the reasons for embracing posthumanism in recent times include potential domination by intelligent machines, anthropocentric exaltation and the neglect of nonhuman species in the environment, the glaring consequences of a dying ecological system caused by rapid industrialization and urbanization, and the fear of 'doomsday' that now stares humanity in the face. The term posthumanism has thus been used in three principal ways: "as a world after humanity; as forms of body modification and transhumanist 'uplift'; and, used here, in the sense of a world comprised of the more-than-human" (Cudworth and Hobden 2018: 5).

Most explanations of posthumanism attempt to overcome a duality that has separated the human from the nonhuman. Building on (and sometimes appropriating) Indigenous knowledge systems that highlight human–nature relational ontologies (see also Cole 1998; Cajete 2000), and 'ecological connectivity' (Rose et al. 2003), "posthumanist texts enact universalizing claims and, as a consequence, reproduce colonial ways of knowing and being by further subordinating other ontologies" (Sundberg 2013: 42). Elsewhere, Sundberg (2011: 321) contests that posthumanism is "a relational ontological approach framing the human and nonhuman as mutually constituted in and through social relations" (see also Castree 2003). In this sense, it is formed through "complex knowledge systems wherein animals, plants and spirits are understood as beings who participate in the everyday practices that bring worlds into being" (Sundberg 2013: 35). This posthuman position reinforces the theorization of classical animism, whose tenets present a duality of human and nonhuman (Tylor 1913 [1871]; Spencer 1889), a division that Indigenous knowledge reconciles.

If "heritage is used to construct, reconstruct and negotiate a range of identities, social and cultural values and meanings in the present" (Smith 2006: 3), this relational analytic of human and nonhuman (and their mutuality) in the ecosystem is very important to critical heritage studies. Posthumanist discourse resonates with critical heritage studies, which aims to "critically engage with the proposition that heritage studies need to be rebuilt from the ground up, which requires the 'ruthless criticism of everything existing'" (Smith 2012: 534), as well as "tackle the thorny issues those in the conservation profession are often reluctant to acknowledge" (Winter 2013: 533). Recognizing the new geological epoch of the Anthropocene, Harrison (2015) draws attention to an urgent need to rethink the future of heritage, proposing a model of "connectivity ontologies" between human and nonhuman, culture and nature (2015: 27). Such connectivity ontologies, he explains, are "modalities of becoming in which life and place combine to bind time and living beings into generations of continuities that work collaboratively to keep the past alive in the present and for the future" (ibid). In a similar vein, a group of archaeologists and heritage scholars reacted to the announcement of the Anthropocene age by remarking that heritage ontologies and epistemologies will have to be renegotiated to accommodate the complexities associated with the new age of the Earth (Solli et al. 2011). To enable this renegotiation, the anthropocentric focus of both traditional and critical heritage studies discourses, which focus on heritage as a product of social and cultural process (Smith 2006) or an outcome of narrative (Ankersmith 2009; Partner 2009), must be reconsidered.

With combined insights from animism, posthumanism and critical heritage studies, this chapter examines the essence of a surviving practice of human-nature relationships. Specifically, it uses the example of a human-tree connection found among the Igbo of Nigeria to address the questions: What is the lifeline between human and 'non-human' agency in processes of human existence and recreation? Could it be an attitude of value creation? Is this understanding within or beyond religion? Consequently, the chapter examines how posthumanism and animism overlap and contradict each other, what this might mean for future heritage initiatives that attempt to grapple with posthumanist thinking, and the implications of this thinking on notions of territoriality in the age of the Anthropocene.

This work is part of a longer-term study exploring the negotiation between global heritage discourse and existing beliefs and value systems in the context of the village arena (or 'square') among the Igbo of Nigeria (Ugwuanyi 2019). The study employed ethnographic techniques of field observation, in-depth interviews and focus group discussions to collect data. Seven villages in Nsukka Igbo were selected and studied: Useh Aku, Umu-Obira Nkporogu, Ogor Ikem, Amokpu Uhunowerre, Amegu Umundu, Ebor Eha-alumona and Onicha Enugwu-Ezike. Nine months were spent collecting data, supplementing more than three decades of the author's life as a member of the culture. Following the ethics employed for the research, interview references are coded.

Posthumanism in the colonial construct of the 'Other' (animism?)

The peak period of posthuman discourse coincides with the recognition of a new geological epoch, the Anthropocene, announced in the first decade of the twenty-first century (see Crutzen and Stoermer 2000; Zalasiewicz et al. 2008). Posthumanism unsettles the anthropocentric orientation of the world with the view that the attribution of 'human' might exist beyond the Anthropocene. Miah, for example, has argued that "the 'post' of posthumanism need not imply moving beyond human-ness in some biological or evolutionary manner. Rather, the starting point should be an attempt to understand what has been omitted from an anthropocentric worldview" (2008: 72). This omission produced a tension between anthropocentric and non-anthropocentric ontologies. In contrast, many Indigenous peoples in Africa have long recognized the need for a relational ontology to ensure survival between species, believ-ing that "without such understanding, the universe may appear incom-plete" (Braidotti 2002; 2016). This relational understanding was dis-missed and seen as a primitive form of religion by anthropologists who termed it 'animism' and/or 'totemism'. E. B. Taylor promoted the theory of animism in his 1871 (1913) book *Primitive Culture*, dismissing it as a primitive 'belief in spiritual beings', which, according to him, was the beginning of all religions. Emile Durkheim expanded on Taylor's posi-tion, arguing that the emergence of animism "was due to the particular mentality of the primitive, who, like an infant, cannot distinguish the ani-mate and the inanimate" (in Garuba 2012: 1). More recently, Harvey has suggested that "contemporary animists do not offer assertions about the origins, development and true nature of all religion, but a focused dis-cussion about particular ways of being related to the world" (2005: 83). Whereas E. B. Taylor and other anthropologists viewed the reverence 'humans' share with 'non-human' species as merely a religious attitude of 'belief in spiritual beings', contemporary scholarship argues that 'ani-mism' may be helpful in "drawing attention to ontologies and episte-mologies in which life is encountered in a wide community of persons only some of whom are human" (Harvey 2005: 81). This conclusion was reached after further research among Indigenous communities chal-lenged the views of classical animism (e.g. Bird-David 1999; Sillar 2009; Brightman et al. 2012).

'Indigenous people' is a term construed here to mean "groups with ancestral ties/claims to particular lands prior to colonization by out-side powers" (Sundberg 2013: 34), "whose nations remain submerged within the states created by those powers" (Shaw, Herman and Dobbs 2006: 268). The propositions of classical animism, contemporary ani-mism and posthumanism are to some extent entangled and complemen-tary. Indeed, the three share some crucial characteristics: (a) the human is identified with life; (b) nature and culture and human and nonhuman are disembodied and independent; (c) the 'Other' attributes life to the nonhuman. At the same time, contemporary animism is classified in my reading as an enhancement of animistic ideas rather than the outright rejection of classical animism. For classical animists, attitudes that attribute human agency to nonhumans were considered 'primitive' and 'religious' (see Tylor 1913 [1871]; Spencer 1889; Durkheim 1912 [1995]). Contemporary animists on the other hand perceive this to mean 'habituation' (see Harvey 2005; Bird-David 1999; Morrison 2013), while posthumanism sees it as 'mediation' (see Braidotti 2016; Sundberg 2013). However, all recognize the community of life, the duality of human and nonhuman, and the interactive exchange that exists among them. Whereas classical animism sees this interaction as an elevation of nonhuman to human status, contemporary animism and posthumanism view the interaction as an interchange that promotes survival. What is the future of heritage in the context of these theorizations?

It could be said that some practices of mediation between human and nonhuman beings in Africa may qualify for a classical animistic classification. At the same time, there are more cultural values that overlap with religious value than animism can accommodate. Here we might note Mbiti's (1969) contestation that religion permeates every aspect of life in Africa. Should we also say that posthumanism is a religious attitude or maybe a form of (post)religious thinking? If posthumanism is understood to mean "a new way of combining ethical values with the well-being of an enlarged sense of community, which includes one's territorial environmental inter-connections" (Braidotti 2016: 26), then posthumanists are simply asking for a return to or a recognition of the very roots of 'animistic' interests that prevailed before, during and after colonialism. It could also mean that posthumanism is a way for the 'West' to appropriate what they have condemned in the past in order to make it acceptable in a 'global' arena, especially with the geological and geographical realities of the current age. By universalizing relational ontologies through posthumanism, the territorial particularity of cultures is again questioned, thus reinventing a knowledge pattern that 'provincializes the West' (see Chakrabarty 2000) in a multicultural and multispecies universe.

As we continue to make a case for posthumanism that is grounded in a desire that both human and nonhuman species are granted equality and ethical treatment, there are clear links to an expansion of animism which seeks to revive ways of life that are similar to that of 'primitive' times. If we are to interrogate the cultural attitude that produced 'animism' through the lens of posthumanism, it has immediate implications for critical heritage studies. First, animism assumes that the Other recognizes nonhuman life, a point posthumanism also makes when it ascribes human characteristics to nonhumans or highlights the relational co-existence between such entities. There are important links here with ongoing work in archaeology and heritage focused on the 'life' ascribed to certain objects (see Sørensen 2013). Symmetrical archaeology, for example, echoes some of the thinking that animism and posthumanism provoke (see Witmore 2007; Shanks 2007). The key insight of this conception of the archaeological record is that human and nonhuman objects are less distinct than previously thought, but also that the life we imagine for certain nonhuman things in the world cannot be expressed other than the way (professional) humans recognize. Could we then say that all heritage has a 'life' that is similar to or different from that of human? Should we, in other words, continue to see heritage as a static-moribund phenomenon that does not change in the Anthropocene age?

If heritage is premised on the past, present and future being connected by a certain continuity of human experience, then Solli et al.'s (2011: 42) question of "whether global warming may cause environmental change of such a magnitude that sudden cultural ruptures are unavoidable" raises crucial concerns for the future of the field. Against this backdrop, there is a need to champion new ways of thinking about heritage in the age of the Anthropocene. It is the contention of this chapter that examples of human and nonhuman relational ontologies across posthumanism and/ or animism have significant implications for heritage futures. Our hope is to seek a clearer understanding of these connections with specific reference to the knowledge systems and historic practices of the Igbo villages in Nigeria.

Human-tree relationships in Igbo villages

Among the Igbo, for example, such trees as the *Iroko; Udara* (*Udala*); *Uburu* (*Uvuru*); *Oroma* (*Oloma* or *Orange*); *Oji* (kola tree); *Ogirisi* (*Ogilishi*); *Ngwu; Ogbu; Ofo; Abosi* (*Avosi*); *Akpu* (*Apu*)... have some recognized mystic cultural symbolisms which sanctify where they stand or grow... the *Udara*, the *Uburu*, the *Orange, Ube* (the pear tree), the *Ube Osa*, with abundant fruits are often associated with the mystic science of life, procreation (*omumu*) and maternity (fertility)... The *Ogirisi* tree (*Origishi*) is reputed to possess talisman qualities capable of counteracting evil charms... the *Ngwu* tree, the *Ofo* tree and the *Abosi* (*Avosi*) are the sacred trees par excellence, a ritual symbol of mystic knowledge and power (Okolie 1992: 16).

Even though Okolie aptly captures the relationship between trees and humans in Igbo cosmology in his study of Igbo villages, he uses an animistic analogy to explain their connectivity. Trees are "cosmic pillars" (Eliade 1965 in Okolie 1992: 15) in this reading, creating a link between the sky, the Earth and the world below, visible as the "Tree of Life". Shelton (1965, 1971) identified specific 'life trees' among the Nsukka Igbo, including *Ogbu* or *Alagbaa* (Ficus elastica) and *Echikeri* – supposedly pronounced *Echikara* (*Spondias monbin*). Such "life trees" "can be cut down and chopped into pieces, then tossed on the earth, and the pieces will sprout and become trees" (1971: 65–8).[1] In tracing the ancestry of one of the villages he studied, Shelton (1971: 93) found that Ezeocha, eldest son of Eze Owuru, is symbolized by an *ube* or 'pear' tree in the village arena. Similarly, Talbot (1926) had earlier observed that Chukwu's shrine (God's altar) is symbolized by *Ogbu, Akpu* or *Awha* trees. With reference to other sacred trees among the Ibibio, he asserts that "should a branch from this fall in the direction of any of the inhabitants, it is thought to be a sign of coming misfortune to that family" (1926: 114). Opata and Apeh's (2016: 130–31) finding in *Otobo Dunoka* – a village arena – in the ancient iron smelting site of Lejja shows that three *Ojirioshi/Ogirishi* (*Newbouldia beauv*) trees planted at the entrance to the arena symbolize the three brothers who are the progenitors of three quarters of Lejja town.

Considering these relational ontologies between humans and trees, and speaking outside the animistic context within which his analysis was carried out, Talbot (1926: 113) held the view that "nothing shows more clearly how near are these people to the heart of Nature than the close ties between Earth's human children and their brothers, the great trees." Despite modernity, such relational ontologies survive today, as my findings from many Igbo villages show.

In Useh Aku, an age-long *Ube* tree (*Canarium schweinfurthii*) in *Otobo Useh* – the village arena – is believed to be one of the routes through which children enter the village (Figure 9.1). In the past, a rite was performed whenever a new child was born in the village, to thank the tree for bringing forth a child unto the Earth. Moreover, it is told that children are often around the tree, mostly to pick the seeds, and there is hardly a time one will not find children under that particular *Ube* tree. Because a branch has never fallen on any of them, they have strong feelings that

Figure 9.1 — *Ube* Otobo, Useh Aku. (Photograph by J. Kelechi Ugwuanyi).

the tree loves and protects children. Consequently, women who are seeking to have a child are sometimes told in divination to go to pray under the tree. There are many testimonies about those who have gone to pray for children and have received them. As an adult member of the village informed me,

> in the days of yore, before the coming of Oyibo [this could mean 'modernity' or 'White man'], when a child is born, the parents take a day old chick to the arena and tie it on the Ube tree with a tender palm frond. Cooked food and wine are also taken along to the Otobo for prayers, thanksgiving and to have a communal meal. I performed this rite for this, my child [she points to her son of about 11 years]. I am not sure that people still do it today (interview, December 16, 2016).

The same woman also revealed that there were sacred Ujuru (*Irvingia gabonensis*) and Akpaka (*Pentaclethra macrophylla*) trees (both now dead) in the arena in the past. However, those who received revelation through Afa divination go to pray under the trees.

An Ụdara tree (*Chrysophyllum albidum*) in Otobo Amegu Umundu is said to represent the lives of the members of the village (Figure 9.2). According to many years of observation, an Amegu villager (interview, January 26, 2017) said that "each time a branch of the tree falls, it's a signal for the death of a chief, a leader, a famous person or a philanthropist in the village". Past Onyishi Amegu (ruler/leader, the oldest man in the village) have resisted attempts to cut down the tree. The most recent attempt was made in c.2004 during a rural electrification project in the community. A demand was made to cut down the tree on the side where the wiring was to traverse, but the Onyishi refused to approve it and reassured his people that such 'evil' (referring to cutting down the tree) against their ancestors would never happen during his reign. Another fearful deity in the Amegu village is Igbudọcha, whose shrine is a simple cluster of trees and grass.

In Otobo Ogwu, in the arena of the great Ogwudinama deity in Umu-Obira Nkporogu, a tree (unidentified species) is used to symbolize Dimgbokwe and another – an Akpụ tree – Odiọkara.[2] What Dimgbokwu and Odiọkara represent is explained in the following interview from an

Figure 9.2 — *Ụdara* Otobo, Amegu Umundu. (Photograph by J. Kelechi Ugwuanyi).

Obira villager about how the people received the Ogwu deity and the knowledge of the Igbo calendar (see also Ugwuanyi and Schofield 2018),

> ... *Diugwu Egbune consulted a great dibia* [medicine man] *by name, Dimgbokwe from Obosi (in the present Anambra state) to prepare for him a medicine with which to identify the name of the woman when next she visits. Dimgbokwe came to Umu-Obira, prepared a Ọgwụ* [medicine] *called Odiọkara and planted Akpụ* [silk cotton tree – *Ceiba pentandra*] *where the medicine was kept* (Interview, February 11, 2017; see also Ugwuanyi 2017).

In the full narrative, some women visited one of the founding fathers of Umu-Obira (Diugwu Egbunne) and challenged him that if his people could identify the visitors' names, then they would be allowed to live among them. When it became impossible for the people of Umu-Obira to identify the visitors, they left with a promise to visit again. Diugwu Egbunne consulted a *dibia* (Dimgbokwe) whose efforts helped reveal the names of the visitors at their next visit. After the encounter, two different trees were used to symbolize Dimgbokwe – the *dibia* and Odiọkara – the medicine he made. Noting Okolie's (1992) excerpt above, it is common to find in all the village arenas in Umu-Obira an *Ọgbụ Otobo*, a tree in the village arena that symbolizes the sacredness of the *Otobo* as a politi-cal, cultural, religious and social space in Igbo culture (Ugwuanyi and Schofield 2018).

What is the meaning of this human-tree relationship in the three vil-lages? Could it be traced to an ancestral relationship with the Mother Earth? Is it a relational attitude to other ecological species or a cultural/ religious practice? Is it possible that trees conceived in this sense could help humanity to survive? The answer to the last question is 'Yes' and 'No'. It is 'No' because trees cannot (to the best of my knowledge) birth a human in a biological sense. It is 'Yes' because of the inevitable impor-tance of trees for human survival. I will return to this point. Worthy of examination is the synchronized relationship between human and non-human in the Igbo cosmos within which the tree-human relationship is closely connected. The following discussion examines this 'connectivity ontology' in the agnostic sense of what 'life' – birth, death and rebirth – could mean in Igbo cosmology. Also, how territoriality, belongingness and mutuality bring together the Earth's contents into a united 'life' – a community of living.

'Life' entanglements in the Igbo universe

In the Igbo cosmos, *Ani, Ala, Ana* or *Al'* (depending on dialectal differ-ences) is an entanglement of all the things that exist; the Nsukka Igbo call it *Al'*, the term to be used here. It is the Igbo word for 'land' or 'ground', which is cosmologically the Earth goddess, the mother of all things, 'the Mother Earth' (Uchendu 1965; Ifesieh 1989; 1994; Cole 1982); her sanctity is a responsibility of all species – human, animal, plant – or land-forms and water bodies. Altering any one of these could affect the healthy survival of the whole. Based on this ontology, the Igbo believe that higher animal-human beings must respect and recognize the uniqueness and systemic relationships between all things that exist in the environment to achieve healthy living.

Corroborating ethnographic information with literature indicates that moral values, ethics of behaviour, and laws/policies of the land are believed to dwell in the cosmos of the 'Mother Earth', who also controls fertility. She could be good or bad based on people's ability to keep to the laws of the land. Violation/desecration, what the Igbo call *aru, nsọ Al'* or *imeru Al'*, meaning 'taboo' or 'altering of the universe', spells doom for the people and the environment, which require cleansing or ritual placation to restore order (see Oriji 2007; Uchendu 1965; Meek 1937). Being on good terms with *Al'* symbolizes being in peace with procreation and living in a healthy society. Two villagers from Onicha Enugu argue that,

> Any year we prayed in Ọnụ Al' and open it,[3] members of the village that do evil, especially those that unjustly killed a man or woman, die [see also Meek 1937: 25], economic crops produce well, we have a bumper harvest, too many good things happen in this land. But, many Onyishi in recent times have refused to do this right thing because they are afraid of chains of deaths that will follow and because they are not righteous themselves (Interview, March 17, 2017).

No one can order the opening of *Ọnụ Al'* in any village without the approval of *Onyishi* (the ruler/leader, the oldest man in a village) who sits in-between the living, the (dead) ancestors and gods. Describing the position of the *Onyishi*, Aja (2002: 30) makes the point that "the living eldest man, by divine law, is the sustaining link of life binding the ancestors and their descendants. It is he who reinforces the life-force of his people and that of other inferior forces: animals, vegetables and things".

In places like Onicha Enugu, many members of *Ọha* (council of elders) have been in conflict with past and present *Onyishi* for not praying and opening the *Ọnụ Al'* as had been done previously in order to heal the land, repair the environment and bring prosperity to the people. Some *Onyishi* claimed they are now Christian converts who should not do any such thing. On the other hand, many elders believe that the degenerating values and dwindling agricultural output experienced in the region today are a result of disobedience, misbehaviour and evil doings against the ethics of life sanctioned by *Al'*, the procreator, 'the Mother Earth'. Every village and town has an *Ihu/Ifu/Iru Al'* or *Ọnụ Al'* (meaning 'the face of the Earth', 'the beginning of the Earth' or 'the altar of the Earth goddess'). Even though there is no physical demarcation of the earth except that made by a body of water, each independent village refers to the Earth goddess as their own, for example *Al'* Useh, *Al'* Umu-Obira, *Al'* Amegu and so on. This is not a physical division; it is a cosmic demarcation of the territorial features and conceptualization of the Earth and the species that relate to the people in the community of the living.

One significant feature of *Ọnụ Al'* is a small grove with a cluster of tree species (Figure 9.3). The trees, according to Ifesieh (1989), are planted during the founding of *Ihu/Ifu/Iru Al'* or *Ọnụ Al'*. The connection between this tree of the *Ihu/Ifu/Iru Al'* or *Ọnụ Al'* and the human-tree relationship can also be linked to the practice of *Ili Elo*, which literally means 'burying of the umbilical cord'. Each time a child is born, the umbilical cord is cared for and monitored; when it falls off, the parents pick it up and bury it under a sapling tree, normally one with economic significance: "palm tree, local pear tree (*ube*), breadfruit tree, local apple

Figure 9.3 — A typical feature of *Ọnụ Al'*. (Photograph by J. Kelechi Ugwuanyi).

(*udara*) tree or plantain or banana tree" (Uchegbue 2010: 158). In the words of Uchendu, "the Igbo who cannot point to the burial place of his navel cord is not a *diala* – freeborn. A child whose navel cord was not buried is denied citizenship" (1965: 59). Territoriality and belonging-ness are manifest in the process of *Ili Elo*, which enmeshes bodies in the community of the living and shows that "human beings share life with *Ala* (Earth), that we have our nature which is partly made of earth, and that our substance comes from the fruits of the earth and at death we rejoin in our bodies, the composition of earth whilst our soul joins our ancestors, who with the authority of *Ala* rule and govern the Earth" (Ileogu 1974: 23). It expresses an early oath-taking on behalf of a child to abide by the moral values, ethics and laws sanctioned by the *Al'*. It also symbolizes an early dedication of the child to "the goddess of *Ala*, to the ancestors and to the community, and symbolically admitted or intro-duced to the *Omenala* (tradition) of his people" (Uchegbue 2010: 159). In this ontology, the tree under which the umbilical cord is buried becomes a living witness to this oath and dedication as a member of the wider universe, where human and nonhuman are 'equal'. They are equal because they all pass through birth, death and rebirth or reincarnation, a process that returns them to *Al'*, the Mother Earth. I will expand on this point and how it affects heritage in the following section. Meanwhile, the idea of human-nature offspringing is becoming more comprehensible. What lessons can we learn here about relational ontology between humans and nonhumans? What are the implications for heritage in the age of the Anthropocene?

The 'life' in heritage: Why posthumanism matters in the Anthropocene

It is strongly hinted above that posthumanism is another way of express-ing animism. What differentiates both is the time and context of appli-cation. The concept of 'animism' emerged when the 'West' was develop-ing knowledge about the long history of humanity through analogy with the living conditions of the 'Other'. Posthumanism on the other hand has gained traction at a time when humanity is looking to counteract the looming effects of climate change and technological encroachment. As a result, it revisits the mutual living conditions among species that observers of animism dismissed as 'primitive religion'. In their mutualism, not only trees (a living thing in this case) participate in the community of living; *Al'*, which simply connotes 'land' or 'ground' but transcends into 'Earth goddess' or 'Mother Earth' in the cosmic sense, is the centre of fertility for all that exists. More important to decipher is the human commitment to birth, death and rebirth with trees standing as witnesses. It is safe now to infer that the *Ube* Otobo and *Ụdara* Otobo in Useh and Amegu villages respectively are either trees under which the umbilical cord of the people's founding fathers were buried or those that were instituted at their death to symbolize their ancestral position. Nonetheless, human agency is given to nonhumans with comparable functioning and capacities. This connectivity is an act of posthumanism (or animism), and – at the same time – a heritage of the people.

The manner in which processes of decay and ruination are played out in the conservation narratives of Odiọkara and its symbol – the dead *Akpụ* tree – is also noteworthy. The tree was allowed to die; another species of tree replaced it, in order to continue the Odiọkara narrative. This explains the progression in the life cycle of the Odiọkara that the *Akpụ* tree symbolizes. It moved from birth to living to death, decay and decomposition, and was then reborn – appearing in another form of life (the sapling *Ọgbụ* tree). There was no need to keep the *Akpụ* tree alive to continue to hold the 'authentic' narrative of Odiọkara. It was simply allowed to die to complete its life cycle to benefit the Earth; as Shanks writes in another context, "decay and ruin reveal the symmetry of people and things" (1998: 22). Presenting his work on "the life of an artefact," Shanks argued that "raw material is taken and transformed according to conception of design, an artefact produced, distributed or exchanged, used, consumed and lost or discarded. It may be recycled, given new life" (1998: 16).

Following on from the mutual living ontology of posthumanism, the empirical case studies above and Shanks' thoughts on 'the life of an artefact', I would suggest that heritage has a 'life': a life that is not biological, a life that is not professionally induced, a life that has its own kind of consciousness different from that of the human – a 'utilitarian life'. Every heritage has a birthing or production mission, which is the original purpose of its coming to life in the first place. The utilitarian life of heritage is its usefulness to self, to nature and to culture through the stages of birth, living, death and rebirth – a fulfilment of the birthing or production mission (Figure 9.4). How can heritage be useful to self, nature and culture? Answers to this question bring back the 'yes' and 'no' response to the earlier question: Is it possible that trees conceived in this sense could help humanity to survive? Trees provide oxygen for respiration, give us food, improve water quality, control soil erosion, enhance wildlife habitat and so on. Trees live, die and decay to enhance other 'life'. This usefulness serves the living community in which the tree is but one life amongst many. It provides a service considered advantageous to all – itself, nature and culture or humans (or people). Many heritage resources also serve their community, living or dead. What is important is that the posthumanist mediation or 'connectivity ontology' makes such living progressive and continuous.

At birth, heritage could die or proceed to the living – the utilitarian stage. While in the community of living, it could spoil, deteriorate or be abandoned; from there, it would either die or be repaired or revived. Death at birth or during a living stage could see the heritage resurrected

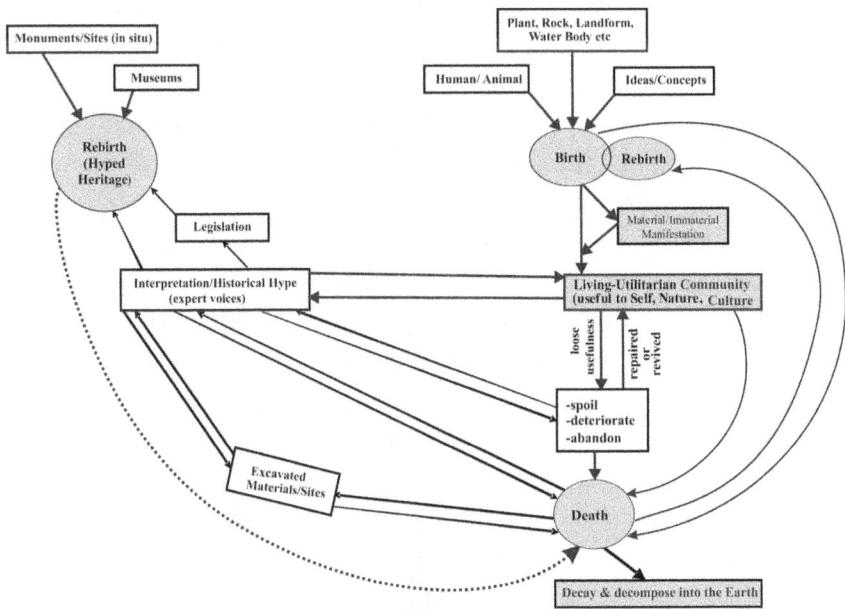

Figure 9.4 — Diagram illustrating the 'life-cycle' of a heritage. (Drawn by J. Kelechi Ugwuanyi).

by replacement, as in the case of the *Akpụ* tree that symbolizes Odiọkara in Umu-Obira, or through a dead idea coming alive again via manifestation.

The posthumanist tenet of mutual living relates more to the utilitarian stage, where all things that exist share some kind of ethical responsibility to life. At death, they decay and decompose to regenerate the Earth, thus helping to complete the biocyle. However, the current conservation approach delays this natural occurrence. In our bid to protect and conserve heritage for the future, we make heritage bypass the stages of the life cycle and delay processes of death, decay and decomposition. In other contexts, we may refuse heritage its living-utility, or (in the case of archaeological excavations) open up others that have already begun to 'decompose'. We use our expert voice – interpretation that I might call 'historical hyping' – along with legislation to give heritage another kind of life, making it a 'hyped heritage'. The new life we give to heritage could also mean 'rebirth'. We divert heritage from moving through the life cycle and force things to 'die' in other ways by making them dwell in the past, stopping them from going through the process of decay and decomposition, instead 'rebirthing' them to live another life in the present. By so doing, we also deny the Earth the power of regenerating beings, one of the major reasons why the ethics that posthumanism proposes is very important in the Anthropocene.

At the 'rebirth' stage, heritage is removed from the community of the living and 'saved' for the future. Museums and conserved monuments, sites or landscapes are examples of heritage with a new or alternative life. At rebirth, heritage may serve another purpose that is different or closely related to its original birthing mission. I have always wondered how the future will receive these things; as a gift from us – the good custodians of history – or as evidence of the calamity 'we' caused with our actions. That is, if anything we are keeping survives the Anthropocene. Even some of the collections we hold dear for the future are in our time becoming 'un-inherited' (Sinamai 2019); this reality forces us to think seriously about the number of things we could hold in trust to survive into the future. Our attitude delays the natural processes of the biocycle and strongly contributes towards the Anthropocene. To borrow from

Shanks (1998: 18), "Death may be delayed. But immortality cannot be achieved" for any heritage.

Concluding thoughts

Drawing on posthumanist thinking, I want to conclude by arguing that, in the Anthropocene epoch, conservation might have to pay more attention towards caring for a utilitarian heritage in the community of life than to saving a (past) dying one. Equally, current realities might require us to allow many heritages, especially those that are already on the brink of collapse, to die, decay, decompose and regenerate the Earth. "Unlimited accumulation, and keeping everything forever, are being called into question" (Morgan and Macdonald 2018: 1), giving rise to 'the crisis of accumulation' (Harrison 2013) that might in the end prove there is 'no future in archaeological heritage management' (Högberg et al. 2018). The posthuman/animist examples in this chapter are based on a model of living heritage where things are allowed to die, decay and decompose while the narratives that go with them are held onto in their new form. I refer to this elsewhere as an 'in-use' model of heritage (see also Ugwuanyi 2018; 2019). Although this idea is still being developed, it involves a negotiated ethics that conserves forms of heritage in their living or utilitarian context to make them survive longer before death; the success of this would be dependent on the combined efforts of experts and non-experts, with the crucial aim of ensuring that heritage is not separated from living communities.

Posthumanism suggests that the anthropocentric bias of heritage being just for 'humankind' is no longer tenable. Heritage is of the Earth, living among the community of beings and should belong to all. A re-conceptualization that sees heritage as a member of a living community, where human and heritage share the same faith of birth, living, death and rebirth, is required. This is just one of the many ways we could think about managing heritage 'profusion' from a posthuman perspective. Other scholars have made similar suggestions: 'de-growing museum collections' (Morgan and Macdonald 2018); 'curating decay' (DeSilvey 2017); and 'avoiding loss aversion' (Holtorf 2015).

It is difficult to let go of the many beautiful and desirable things we have made and continue to make. It is however unrealistic to think that we can save every past and present experience in the future. We have to let go of heritage in the same way we let go of loved ones when they die. Posthumanism makes us know that our shared responsibilities in the community of living (where heritage is a member) would mean accepting the life cycle of birth, mediated living, death (and eventual decay and decomposition) and rebirth, to regenerate the Earth and keep the cycle going. Thinking about heritage in this sense would help us to make informed decisions for the future of heritage, especially in the Anthropocene epoch.

Acknowledgements

This work was supported by the Overseas Research Scholarship, University of York; Tweedie Exploration Fellowship, University of Edinburgh; and Gilchrist Educational Trust. I very much appreciate members of the villages that participated in this research. My thanks to Dr Elizabeth Currie, Dr Paul Edward Montgomery Ramírez and Joëlla van Donkersgoed for reading through the original draft and making useful suggestions. The essay was first presented at the International Conference of the Association of Critical Heritage Studies (ACHS) held in China in 2018; I thank the participants for their helpful comments.

Notes

1. Agbedo (2015) found the local name to be *ejirooshi*.

2. The *Akpụ* tree lived, died, decayed and decomposed into the Earth to regenerate life. In this process of regenerating life, a sapling *Ogbụ* tree sprouted in the same location, and the people used it to replace the dead *Akpụ*, thus bestowing on it the narratives of Odiọkara that the dead *Akpụ* was carrying whilst alive.

3. In the *Ọnụ Al'* (the altar of *Al'*), there is a buried and covered pot that needs to be opened for one month (of the local calendar) after offering prayers. The buried pot is taken to be the *Ọnụ Al'*, therefore, opening it is construed as opening the *Ọnụ Al'*.

References

Agbedo, C. U. 2015. Deities and Spirits in Ọlido. In *Ọlido in Historical and Contemporary Times*, edited by C. U. Agbedo, 157–68. Nsukka, Nigeria: A KUMCEE-Ntaeshe Book.

Aja, E. 2002. Philosophy and Human Values: An African Perspective. *Journal of Liberal Studies* 10(2): 28–66.

Ankersmith, F. 2009. Narrative, an Introduction. In *Re-figuring Hayden White*, edited by F. Ankersmit, E. Domanska and H. Kellner, 77–80. Stanford, CA: Stanford University Press.

Bird-David, N. 1999. 'Animism' Revisited: Personhood, Environment, and Relational Epistemology. *Current Anthropology* 40: 67–92.

Braidotti, R. 2002. *Metamorphoses: Towards a Material Theory of Becoming*. Cambridge: Polity Press.

Braidotti, R. 2016. Posthuman Critical Theory. In *Critical Posthumanism and Planetary Futures*, edited by D. Banerij and M. R. Paranjape, 13–32. New York: Springer.

Brightman, M., V. E. Gratti and O. Ulturgasheva (eds). 2012. *Animism in Rainforest and Tundra: Personhood, Animals, Plants and Things in Contemporary Amazonia and Siberia*. New York: Berghahn.

Cajete, G. 2000. *Native Science: Natural Laws of Interdependence*. Santa Fe, NM: Clear Light Publishers.

Castree, N. 2003. Environmental Issues: Relational Ontologies and Hybrid Politics. *Progress in Human Geography* 27(2): 203–11.

Chakrabarty, D. 2000. *Provincializing Europe: Postcolonial Thought and Historical Difference*. Princeton, NJ: Princeton University Press.

Cole, H. M. 1982. *Mbari: Art and Life among the Owerri Igbo*. Bloomington: Indiana University Press.

Cole, P. 1998. An Academic Take on Indigenous Traditions and Ecology. *Canadian Journal of Environmental Education* 3: 100–115.

Crutzen, P. J. and E. F. Stoermer. 2000. The Anthropocene. *Global Change Newsletter* 41: 17–18.

Cudworth, E. and S. Hobden. 2018. *The Emancipatory Project of Posthumanism*. London: Routledge.

DeSilvey, C. 2017. *Curated Decay: Heritage Beyond Saving*. Minneapolis: University of Minnesota Press.

Durkheim, E. 1912 [1995]. *The Elementary Forms of Religious Life*. Trans. K. E. Fields. New York: The Free Press.

Garuba, H. 2012. On Animism, Modernity/Colonialism, and the African Order of Knowledge: Provisional Reflections. *E-flux Journal* 36: 1–9.

Harrison, R. 2013. Forgetting to Remember, Remembering to Forget: Late Modern Heritage Practices, Sustainability and the 'Crisis' of the Accumulation of the Past. *International Journal of Heritage Studies* 19(6): 579–95.

Harrison, R. 2015. Beyond 'Nature' and 'Culture' Heritage: Toward an Ontological Politics of Heritage in the Age of Anthropocene. *Heritage and Society* 8(1): 24–42.

Harvey, G. 2005. Animism. In *A Contemporary Encyclopedia of Religion and Nature*, 81–83. London: Thoemmes Continuum.

Hayles, N. K. 1999. *How We Became Posthuman: Virtual Bodies in Cybernetics, Literature, and Informatics*. Chicago, IL: University of Chicago Press.

Högberg, A., C. Holtorf, S. May and G. Wollentz. 2018. No Future in Archaeological Heritage Management? *World Archaeology* 49(5): 639–47.

Holtorf, C. 2015. Averting Loss Aversion in Cultural Heritage. *International Journal of Heritage Studies* 21(4): 405–421.

Ifesieh, E. I. 1989. *Religion at the Grassroots: Studies in Igbo Religion*. Enugu, Nigeria: SAAP Press.

Ifesieh, E. I. 1994. Power and Religion: Implications and Illustrations. *Nsukka Journal of the Humanities* 7: 112–25.

Ileogu, E. 1974. *Christianity and Ibo Culture*. Leiden: E. J. Brill.

Mbiti, J. 1969. *African Religion and Philosophy*. London: Heinemann.

Meek, C. K. 1937. *Law and Authority in a Nigeria Tribe*. Oxford: Oxford University Press.

Miah, A. 2008. A Critical History of Posthumanism. In *Medical Enhancement and Posthumanity*, edited by B. Gordijn and R. Chadwick, 71–94. New York: Springer.

Morgan, J. and S. Macdonald. 2018. De-growing Museum Collections for New Heritage Futures. *International Journal of Heritage Studies*. DOI: 10.1080/13527258.2018.1530289.

Morrison, K. M. 2013. Animism and a Proposal for a Post-Cartesian Anthropology. In *A Handbook of Contemporary Animism*, edited by G. Harvey, 38–52. London: Routledge.

Okolie, M. A. E. 1992. The Village Square in Igbo Cosmology. *Nigerian Heritage* 1: 11–19.

Opata, C. C. and A. A. Apeh. 2016. Ụzọ mma: Pathway to Intangible Cultural Heritage in Otobo Ugwu Dunoka Lejja, South-eastern Nigeria. *International Journal of Intangible Heritage* 11: 128–39.

Oriji, J. 2007. The End of Sacred Authority and the Genesis of Amorality and Disorder in Igbo Mini States. *Dialectical Anthropology* 31, nos 1/3: 263–88.

Partner, N. 2009. Narrative Persistence: The Postmodern Life of Narrative Theory. In *Re-figuring Hayden White*, edited by F. Ankersmit, E. Domanska and H. Kellner, 81–104. Stanford, CA: Stanford University Press.

Rose, D., D. James and C. Watson. 2003. *Indigenous Kinship with the Nature World*. Sydney: NSW National Parks and Wildlife Services.

Shanks, M. 1998. The Life of an Artefact in an Interpretive Archaeology. *Fennoscandia Archaeologica* XV: 15–30.

Shanks, M. 2007. Symmetrical Archaeology. *World Archaeology* 39(4): 589–96.

Shaw, W. S., R. D. K. Herman and G. R. Dobbs. 2006. Encountering Indigeneity: Re-imagining and Decolonizing Geography. *Geografiska Annaler* 88(3): 267–76.

Shelton, A. J. 1965. The Presence of the 'Withdrawn' High God in North Ibo Religious Belief and Worship. *Man* 65: 15–18.

Shelton, A. J. 1971. *The Igbo-Igala Borderland*. New York: State University of New York Press.

Sillar, B. 2009. The Social Agency of Things? Animism and Materiality in the Andes. *Cambridge Archaeological Journal* 19: 367–77.

Sinamai, A. 2019. *Memory and Cultural Landscape at the Khami World Heritage Site, Zimbabwe: An Un-inherited Past*. London: Routledge.

Smith, L. 2006. *Uses of Heritage*. London: Routledge.

Smith, L. 2012. Editorial. *International Journal of Heritage Studies* 18(6): 533–40.

Solli, B., M. Burström, E. Domanska, M. Edgeworth, A. González-Ruibal, C. Holtorf, G. Lucas, T. Oestigaard, L. Smith and C. Witmore. 2011. Some Reflections on Heritage and Archaeology in the Anthropocene. *Norwegian Archaeological Review* 44(1): 40–88.

Sørensen, T. F. 2013. We Have Never Been Latourian: Archaeological Ethics and the Posthuman Condition. *Norwegian Archaeological Review* 46(1): 1–18.

Spencer, H. 1889. *Principles of Sociology*. New York: D. Appleton.

Sundberg, J. 2011. Diabolic *Caminos* in the Desert and Cat Fights on the Rio: A Posthumanist Political Ecology of Boundary Enforcement in the United States–Mexico Borderlands. *Annals of the Association of American Geographers* 101(2): 318–36.

Sundberg, J. 2013. Decolonizing Posthumanist Geographies. *Cultural Geography* 21(1): 33–47.

Talbot, A. 1926. *The Peoples of Southern Nigeria: A Sketch of their History, Ethnology and Languages*, vol. 4. London: Frank Cass.

Tylor, E. B. 1871 [1913]. *Primitive Culture*. London: J. Murray.

Uchegbue, C. O. 2010. Infancy Rites among the Igbo of Nigeria. *Research Journal of International Studies* 7: 157–64.

Uchendu, V. C. 1965. *The Igbo of Southeast Nigeria*. New York: Holt, Rinehart and Winston.

Ugwuanyi, J. K. 2017. 'We Live in the Past': Heritage, 'Igbo Calendar' and Variegations among the Igbo of Nigeria. A paper presentation at the International Conference for Ph.D Candidates in Humanities and Cultural Studies on 'Non-Contemporaneity' organized by the International Promovieren in Wupetal (IPIW) and Bergischen Universitat Wupertal, Germany, September 26–29.

Ugwuanyi, J. K. 2018. Hegemonic Heritage and Public Exclusion in Nigeria: A Search for Inclusive and Sustainable Alternatives. *West African Journal of Archaeology* 49: 71-93.

Ugwuanyi, J. K. 2019. The Igbo Village Square, Heritage and Museum Discourse: An Ethnographic Investigation. Ph.D. thesis, University of York.

Ugwuanyi, J. K. and J. Schofield. 2018. Permanence, Temporality and the Rhythms of Life: Exploring Significance of the Village Arena in Igbo Culture. *World Archaeology* 50(1): 7–22.

Winter, T. 2013. Clarifying the Critical Heritage Studies. *International Journal of Heritage Studies* 19(6): 532–45.

Witmore, C. L. 2007. Symmetrical Archaeology: Excerpts of a Manifesto. *World Archaeology* 39(4): 546–62.

Wolfe, C. 2009. *What is Posthumanism?* Minneapolis: University of Minneapolis Press.

Zalasiewicz, J., M. Williams, A. Smith, T. L. Barry, A. L. Coe, P. R. Bown, P. Brenchley, D. Cantrill, A. Gale, P. Gibbard, F. J. Gregory, M. W. Hounslow, A. C. Kerr, P. Pearson, R. Knox, J. Powell, C. Waters, J. Marshall, M. Oates, P. Rawson and P. Stone. 2008. Are We Now Living in the Anthropocene? *GSA Today* 18, no. 2: 4–8.

Chapter 10

Ruderal Heritage

CAITLIN DESILVEY

All change

It is now clear that we live in a world increasingly defined by the rapid and unpredictable transformation of social, ecological and geophysical systems. While transformation, as such, is nothing new, the speed and the scale of the changes we, and other species, are experiencing, is unusual (if not entirely unprecedented). The relevance of the concept of 'heritage' in this moment is not immediately obvious – given its association with conservative and preservative instincts, and its fixation on the past as its point of reference and locus of value. In critical heritage studies, however, the wider recognition of inevitable transformative change has been paralleled by the emergence of new theoretical approaches, which understand heritage as a socially-embedded, future-oriented process through which the past is brought into the present to shape novel environments and practices. In this recent work, heritage significance is framed as an emergent, relational property – not an intrinsic quality linked to the preservation of certain material states (Harrison 2013; Pétursdóttir and Olsen 2014; DeSilvey 2017; Harrison et al. 2020). These alternative approaches see change and transformation as an integral element of heritage, with the potential to generate new connections between people, the past and the future. They also recognize that with our human imprint now penetrating deep into global ecologies and geologies, the distinction between nature and culture – and natural and cultural heritage – is an artefact of a world we no longer inhabit (Harrison 2015).

For the past several years, I have been working at this interface with a team of researchers,[1] seeking to understand how the practice of heritage-making is sustained (or enhanced) in relation to materials and landscapes caught up in active transformative processes. In the places we studied, the making of future heritage is not about conserving objects or artefacts as stable entities but about maintaining continuity with the past through processes of change and innovation. In our research, we sought to understand how transformations that could be interpreted as loss on one register could also provide opportunities for the emergence of other relational configurations and trajectories. These places provide a glimpse of how we might find our way in what I will describe as a 'ruderal' future, where disturbance is the norm, and where our strategies for survival will depend on making alliances with more-than-human agents and entities. In the following section I introduce each of the three landscapes we explored (all undergoing transformation at a scale and a velocity more pronounced than the background condition of change characteristic of comparable landscapes) before returning to discussion of the concept of the ruderal in the context of heritage practice.

Sites and synergies

In mid-Cornwall's china clay country, north of the town of St. Austell, more than two centuries of industrial extraction, on an increasingly large scale, have produced a patchwork, punctured landscape mosaic. Imerys, a multi-national company based in France, continues to extract deposits of kaolin (decomposed granite) in massive open pits, reliant on heavy machinery and a substantial processing infrastructure of pipes, roads, plants, tunnels, tracks and tanks. Other areas are now 'post-operational' and held in limbo, awaiting redevelopment or reuse. In the meantime, these post-operational spaces undergo renaturalization, either intentionally (through replanting of heathland and forest) or passively (through the emergence of rogue plantations of rhododendron and other ostensibly 'invasive' species). Around the edges and in isolated pockets, elements of the industry are conserved and presented as artefacts of industrial heritage, as at the Wheal Martyn Clay Works

north of the town of St. Austell. Many more structures and infrastructures are left to their own devices: the landscape is scattered with redundant rail lines, pyramidal waste tips, disused industrial buildings and massive concrete settling tanks. No one really knows what to do with these remainders, although some of them have accrued value and are celebrated as icons of local heritage. 'Preservation' of such features is problematic, however, and regional heritage bodies, which recognize the continual change brought about by evolving industrial process *as* the heritage of this landscape, have struggled to find a clear way forward (Kirkham 2014).

Several hundred miles away, on the east coast of England, Orford Ness is a 15km-long spit composed of loose stone, or 'shingle'. The Ministry of Defence (MoD) occupied the site for most of the twentieth century, for use as an airfield and then for classified research into bomb ballistics, aerial warfare and atomic weapons testing. In 1993 the National Trust acquired the former MoD property on the Ness (citing the nature conservation value of the vegetated shingle habitat) and applied a policy of 'continued ruination' to selected structures associated with the tenure of the Atomic Weapons Research Establishment (AWRE) (DeSilvey 2017). This unusual cultural heritage management philosophy is set against a backdrop of rapid coastal change. Processes of erosion and longshore drift continually rearrange the coastline, and in recent years the Coastal and Intertidal Zone Archaeological Network (CITiZAN) has worked to record archaeological features as they are undermined and erased. Our research explored the dynamic and fluid nature of this place and tried to understand how this quality is celebrated by local managers and creative practitioners (Bartolini and DeSilvey 2020a). But we also discovered nodes of attempted durability and fixity that work against this embrace of change, including a high-profile attempt by a private owner to protect the historic Orford Ness lighthouse, adjacent to the National Trust property, from inevitable erosion. Some of the MoD structures were listed decades ago, but the AWRE facility was only designated as a Scheduled Monument in 2014. Historic England holds that the new designation does not trigger a presumption of protection, but accept they have no control over how other people interpret the significance of such designations.

In the Côa Valley, in north-east Portugal, a concentration of prehistoric rock art animal figures shares a landscape with a rewilding pilot, led by a local organization, Associação Transumância e Natureza (ATN), and supported by Rewilding Europe. Here, the landscape is changing rapidly and has been for decades as the region's population gradually declines: the villages have hollowed out and fallen into partial ruin; the olive groves and arable fields are untended and overgrown. ATN sees the widespread land abandonment as an opportunity, and wants to return semi-wild horses and back-bred cattle (and, eventually, ibex and other large herbivore species) to the landscape, to allow their grazing to transform the landscape into something that resembles (but does not replicate) what it was in the past, 20,000 years ago, when representations of horses and other animals were carved into the stone along the river (DeSilvey 2019). In the long term, the hope is that the introduced animals will once again become prey in an expanded range for existing predator species, such as lynx and wolves. But this transformation is uneven and gradual. The introduced animals still need care and management; traditional practices of animal husbandry intersect with the rewilding initiative in curious ways; ruined houses are restored to accommodate ecotourists (DeSilvey and Bartolini 2018; Bartolini and DeSilvey 2019). The "new natural heritage" (Jepson and Schepers 2016: 2) produced by rewilding catalyzes landscape reengagement and reconnection, but the depopulation trend continues.

Ruderal thinking

In these open and uncertain landscapes, the people who are responsible for steering them into the future must continually accommodate and negotiate ambiguity, instability and emergent processes. One concept that we have found useful when thinking about these places is that of the 'ruderal.' A word with its roots in the Latin word for 'rubble', ruderal is an ecological term used to describe opportunistic plant species adapted to take root in disturbed environments. In my old Webster's Dictionary, the first definition assigns a clear disruptive agent: a ruderal plant is found 'growing where the natural vegetational cover has been disturbed by man' [sic]. The second definition offers a more passive and indirect

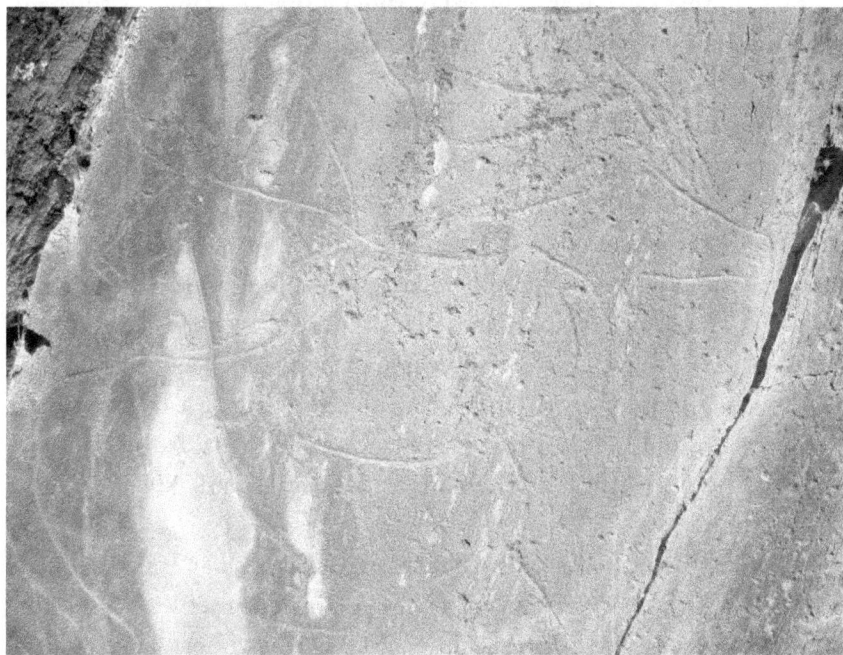

causation: 'a weedy and commonly introduced plant growing where the vegetational cover has been interrupted'. In this second sense, the driver of disturbance is not specified. This ambiguity echoes a refrain in public debate around the concept of the Anthropocene; disturbance is evident, but there is a persistent desire in some quarters to identify the causal agent as something or someone other than us, to shift blame onto epic Earth processes or some equally indifferent inhuman force (Clark 2011). The concept of the 'ruderal' contains within it both sets of possible meanings, and in this sense it collapses any functional distinction between 'natural' and 'cultural' environments. It may apply to plants that colonize environments scoured by fire or by flood, but it can also describe an adaptive response to conditions produced through processes of industrial extraction and construction, war and other forms of human wasting.

In both inflections of the term, disruption creates the conditions of possibility for the emergence of new (plant) communities, made up of individual interlopers often described as 'weeds' or 'invasive species', aggressive and indiscriminate. But the term also contains associations of renewal, in that the advance guard of ruderal species stabilizes soils and supports the emergence of conditions that allow other less aggressive species to succeed them. Central to the concept of the ruderal, however, is a recognition that the ecosystems that emerge in the wake of disturbance are novel and non-analogue – they have never existed before. The reference to the 'interrupted' may suggest a break in successive progression, but in the places where ruderal process holds sway, there is no clear path back to a historical baseline ecological state (Jackson and Hobbs 2009). As such, ruderal thinking aligns with the awareness in heritage studies that we cannot restore an imagined, authentic version of the 'the past': we can only borrow scraps from the available past to assemble a viable future, and in this sense disturbance can be seen as the opportunity for the emergence of alternative trajectories and novel narrative configurations.

The term 'ruderal' has caught the imagination of a handful of scholars outside ecology, who are drawn, as Sarah Cowles observes, to the way in which "ruderal species perform ecological, metaphoric and cultural work" (2017: 1). Bettina Stoetzer, in her research on the unruly edges of the city of Berlin, uses the ruderal as a tool for "rethinking cultural migra-tion, human-nonhuman relations and unintended ecologies" (Stoetzer 2018: 308). She writes,

> [A] ruderal analytic shifts attention to heterogeneous and unexpected life amid rubble. Ruderal ecologies grow in the inhospitable environments created by war and exclusion; they emerge by chance and entail illegal border crossings... the per-spective of ruderal ecologies accounts for the ways in which biological life, cultural identifications and strategies of survival are never authentic or pure but always situated within histo-ries of disturbance (ibid).

Other scholars do not use the term directly but are pushing their thinking into sympathetic territory. Stephanie Wakefield has writ-ten about the cultural relevance of the ecological concept of the 'back loop', which asserts that systems do not remain in a steady state, but experience continual phases of collapse and unravelling, followed by creative phases of 'release and reconfiguration' (Wakefield 2018). "In the release phase," she writes, "energies and elements previously cap-tured in the conservation stage are set free" (2018: 79). It is possible to extend this thinking to heritage contexts, to explore how 'energies and elements' that had been stabilized through conservation can be released through processes of decay and disintegration, and through embrace of weedy natural-cultural combinations (DeSilvey 2017). In a recent essay on 'auto-wilding', Anna Tsing writes, "So many of us are Anthropocene weeds. Weeds are creatures of disturbance; we make use of opportunities, climb over others and form collaborations with those who allow us to proliferate. The key task is to figure out which kinds of weediness allow landscapes of more-than-human liveabil-ity" (2017: 17).

Ruderal heritage research, then, is orientated to ongoing instances of both destruction and renewal, and focused on the opportunities that emerge from inhabiting disturbed substrates and sensibilities. In explo-ration of this theme at our three case study sites, I draw out the ecologi-cal and cultural resonance of ruderal thinking by attending to stories of specific species and situations. Through these stories, I explore how a focus on heritage transformation, rather than preservation, unravels the boundary between categories of natural and cultural heritage management and opens out opportunities to salvage meaning from apparent loss and disintegration.

Of moss and mountains

In the clay country, past disturbance by industrial excavation has created pockets of dense and diverse plant life, most noticeably where landscapes have been allowed to revegetate on their own over decades. At Lansalson Pit, part of the complex of redundant clay workings that became the Wheal Martyn Clay Works, the banks surrounding the blue-green pool are crowded with common ruderal species – buddleia, bramble, bracken – but also more exotic rhododendrons, escaped from local gardens. These pioneer plants, adapted to colonize nutrient-poor exposed soils, have an ambivalent status, oscillating between saviour and scourge. In one sense, they stabilize the scarred landscape, and act as a literal ecological place-holder until more formal plans for remediation and redevelopment emerge. But certain species, such as the rhododendron, are recast as alien invasives, threatening to destabilize remnant indigenous ecosystems.

While the species mentioned above are ubiquitous and unescapable in the clay country, the region is also home to several ruderal species that are valued for their rarity, and carefully monitored and managed. One bryophyte (moss) species unique to Cornwall, the Western rustwort (*Marsupella profunda*), colonizes unshaded or lightly shaded clay and granitic rocks, and appears to have evolved to prefer the open, exposed conditions created by ongoing industrial activity in the clay country. In the 1990s the moss was identified at several sites in and around both dormant and active pits, and some of these sites were subsequently protected with SSSI and SAC designations. Because the necessary disturbance was absent, however, the moss became shaded by encroaching gorse and bramble. The plants are now largely extinct in the designated areas, and the species has been categorized as 'Vulnerable' on the International Union for Conservation of Nature (IUCN) Red List (Hodgetts 2011). Natural England has identified 'refuge' sites for species

translocation, and entered into a collaboration with Imerys to help replicate the heavy industrial activity that will maintain the unique ecological
conditions required to maintain viable populations (Callaghan 2014).[2]
In this paradoxical instance of ruderal heritage in action, the conservation of natural heritage is only possible through the "periodic large-scale
disturbance" brought about by extractive intervention (2014: 7).

The largest recorded population of Western rustwort occurs in the
area surrounding the West Carclaze Sky Tip (also known as Great
Treverbyn Sky Tip), where a social analogue of ruderal process has
played out over the last several years. After its initial deposit as a waste
by-product of clay extraction, the artificial sand-mountain was left to
its own devices for sixty years, gradually accruing significance to local
people. In 2014 a perceived threat that a proposed redevelopment
scheme would level the tip led local residents to submit a listing appli-
cation to English Heritage. As one observer commented, change (or
the spectre of change) 'drew out value' and forced people to articulate
their desire to save the feature as part of their heritage – disturbance,
in this instance, elevated significance. Despite recognition of its status
as an 'iconic local landmark', English Heritage judged that the tip did
not have a sufficiently 'high level of historic importance' to merit designation (English Heritage 2014). The Sky Tip is now integrated into
the marketing of the West Carclaze Garden Village development, however, as a value-added piece of local history, which will be retained as
"the centre feature" of China Clay Heritage Park, and be allowed to
"naturally weather and erode... producing its dramatic sculptural
shapes". The promotional materials for the new scheme promise: "Our
plans are to let nature take its course and allow it to evolve as the cen-
trepiece of an extensive recreational and wildlife habitat, which will
bring pleasure and enjoyment to residents and visitors alike" (Eco-Bos
2018). The Sky Tip provides an example of ad hoc management of a
hybrid natural/cultural heritage feature, where the developers,
through a process of conflict and negotiation, have agreed to accommodate inherent instability – although their commitment to this
agreement in principle has not yet been tested in practice (Bartolini
and DeSilvey 2020b).

Poppies and police towers

On Orford Ness, we have been working with partners to understand how heritage is made (and unmade) in this unstable environment, where disturbance – by bomb-testing and wave-action alike – is the historical norm. The Ness harbours many ruderal species, as the mobile shingle and the ruined structures provide the exposed substrates needed for their establishment and survival. Some more common species are considered to be invasive weeds or innocuous interlopers, while others, like the yellow-horned poppy (*Glaucium flavum*), are deemed to be 'native' and thus worthy of conservation, whether they are growing on the shingle ridges or in the concrete foundations of a derelict building (DeSilvey 2017). The management challenges here, however, are not dissimilar to those posed by the management of the clay country moss. I photographed a yellow-horned poppy on the raised beach south of the lighthouse in March 2012, and then six years later came upon a poppy uprooted by a recent storm in more or less the same location, its roots exposed and the plant toppling down the beach crest. The toppled poppy could be seen as evidence of the destruction caused by accelerated, anthropogenic climate change, sea-level rise and increasing storm intensity. But the poppy thrives on disruption, and its seeds are adapted to take root in newly accreted shingle ridges. Once it has established, however, it is vulnerable to trampling by careless walkers – though the warnings of unexploded ordnance tend to keep people on the marked paths at Orford Ness.

We can find a loose cultural analogue to the poppy in a derelict wooden police tower, which used to stand a few hundred yards south of the Orford Ness lighthouse. The tower was built in 1956 as an observation post for the AWRE security police, located inside a defensive perimeter fence. In a 2009 National Trust survey the tower was still secure, but by 2012 its foundation was very close to the beach crest. By the time of the first CITiZAN survey in 2015, it was gone; in 2016 the survey team measured and recorded the foundation slab where it had tumbled down the beach, reduced to broken blocks of rubble. I returned to the site with CITiZAN in March 2018, and all that was left was a fragment of concrete jutting out of the steep beach face, and a few smaller fragments scattered down the beach. The mood

of the survey team was not mournful, but curious, forensic: the loss of this feature, in a sense, justified the labour spent surveying, measuring, documenting and recording. It also became clear that only three years after the collapse of the tower, people no longer agreed on exactly where it had stood. One of the National Trust employees claimed the feature recorded by CITiZAN was not the police tower base, but another eroding concrete foundation. We discussed the tower over breakfast in the Orford Ness bunkhouse, looking at old maps and photos, comparing and considering. One of the CITIZAN staff shared a 1951 image she had found of a similar tower located north of the lighthouse and posited that the tower in question was not built in 1956 but was relocated from the other site. Uncertainty, in this instance, created openings for dialogue and deliberation; history frayed and had to be woven back together. A lively dialogue about the past in place was generated out of disruption and erosion of evidence, in a collaborative process arguably more productive and generative than a passive encounter with the static tower as a piece of 'heritage'. Here, we found a heritage practice that was not trying to hold back change, but was working with it, and finding opportunities for engagement and (re)connection.

Fire and friction

In the Côa, management of ruderality is central to both the rewilding effort and to wider landscape concerns, which come into focus in relation to particular practices. In this rugged and exposed landscape, local shepherds and farmers have a well-established tradition of seasonal burning; they have used fire for centuries as a management tool to maintain the vegetation in a state of early succession, and encourage new growth of forage-friendly plant species. As farming practices have been gradually withdrawn from the landscape, however, pioneer species such as broom have begun to grow in greater densities, and fires have become more intense and destructive (occasionally threatening the remnant cork and olive trees, which can withstand moderate wildfires but may be killed by high-intensity blazes). On our first visit to the area in 2015, ATN staff told us about the local perception of the unworked landscape

as 'unclean' and dirty, and their concern about fire-setting by one of the shepherds in a local village. A few years prior they had taken the shepherd to court for violating the new restrictions on fire in the reserve (Leuvenick 2013: 20).

As part of the rewilding of the area, ATN aims to disrupt the cycle of burning and the continual reversion to first-stage succession to allow woody tree species to become established, and to eventually recreate a semi-forested landscape mosaic. The grazing and browsing of the reintroduced horses and cattle are intended to keep the ruderal species in check and make the landscape less vulnerable to damaging fires, as other species gradually move in and the system 'gains resilience'.[3] Following years of conflict and disruption, local residents are now becoming more accepting of ATN's strategy, as they witness the effect of the new management practices. One couple with property in a local village remarked on the frequent fires in the past, and their perception that in the years following the release of the horses, fires within the Faia Brava reserve had become very rare. In this landscape, fire is both an expression of intangible cultural heritage and a contested ecological agent, managed by conflicting interests. But the conflict and disturbance have catalyzed a tentative transition into a future state in which the entanglement of natural and cultural heritage becomes the basis for a new, shared understanding of landscape dynamics.

Last words

I hope these brief landscape forays have shown how ruderal thinking can offer a productive conceptual tool for a critical heritage practice oriented to latency and release, instability and emergence. In such an orientation, memory and materiality are unhitched from the pursuit of permanence to instead work through change and disturbance; in the process new attachments are formed, and old ones are reimagined. The trick is to look out for the seeds waiting to germinate in the rubble. As poet Gary Snyder reminds us, "manzinita seeds will only open/after a fire passes through/ or once passed through a bear" (Snyder 1974: 19).

Notes

1. This chapter is based on research carried out as part of 'Heritage Futures', an Arts and Humanities Research Council (AHRC) 'Care for the Future: Thinking Forward through the Past' Theme Large Grant (AH/M004376/1), awarded to researchers at University College London, University of Exeter, University of York and Linnaeus University (Sweden). I am grateful to Nadia Bartolini, Heritage Futures Transformation theme Research Associate, and Antony Lyons, Senior Creative Fellow on the Heritage Futures project, who shared fieldwork experiences in the UK and Portugal and who informed this work through ongoing intellectual and creative collaborations. Additional thanks are owed to our project partners and the individuals who participated in our research and generously offered their insights and expertise. Material developed for this chapter also appears in the co-authored volume, *Heritage Futures: Comparative Approaches to Natural and Cultural Heritage Practices*, written by R. Harrison, C. DeSilvey, C. Holtorf, S. Macdonald, N. Bartolini, E. Breithoff, H. Fredheim, A. Lyons, S. May, J. Morgan and S. Penrose (with contributions by A. Högberg and G. Wollentz), UCL Press, 2020. Chapters by N. Bartolini and A. Lyons in that volume provide more indepth analysis of the three sites introduced in this chapter. See further information at www.heritage-futures.org

2. I am grateful to David Hazlehurst, Natural England, for sharing the story of the Western rustwort in our conversation on June 4, 2015 in Penryn, Cornwall.

3. Personal communication with ATN coordinator, December 12, 2017.

References

Bartolini, N. and C. DeSilvey. 2019. Rewilding as Heritage-making: New Natural Heritage and Renewed Memories in Portugal. In Routledge Handbook of Memory and Place, edited by S. de Nardi, H. Orange, E. K. Koivisto, D. Drozdzewski and S. High, 305-314. Abingdon and New York: Routledge.

Bartolini, N. and C. DeSilvey. 2020a. Recording Loss: Film as Method and the Spirit of Orford Ness. *International Journal of Heritage Studies* 26(1): 19-36

Bartolini, N. and C. DeSilvey. 2020b. Making space for hybridity: Industrial heritage naturecultures at West Carclaze Garden Village, Cornwall. *Geoforum* 113: 39-49.

Callaghan, D. 2014. Survey of Western Rustwort (Marsupella profunda) and other Bryophytes at Land at West Carclaze, Cornwall. Cornwall Council, PA14-12186. West Carclaze eco-community.

Clark, N. 2011. *Inhuman Nature: Sociable Life on a Dynamic Planet*. London: Sage.

Cowles, S. 2017. Ruderal Aesthetics. http://www.ruderal.com/articles/

DeSilvey, C. 2017. *Curated Decay: Heritage Beyond Saving*. Minneapolis: University of Minnesota Press.

DeSilvey, C. 2019. Rewilding Time in the Vale do Côa. In *Rethinking Historical Time: New Approaches to Presentism*, edited by M. Tamm and L. Olivier, 193-206. London and New York: Bloomsbury Academic.

DeSilvey, C. and N. Bartolini. 2018. Where Horses Run Free? Autonomy, Temporality and Rewilding in the Côa Valley, Portugal. *Transactions of the Institute of British Geographers* 44(1): 94–109.

Eco-Bos. 2018. Heritage: Respecting the Past. West Carclaze Garden Village website. http://www.westcarclaze.co.uk/heritage.html

English Heritage. 2014. Great Treverbyn Tip. Reject at Initial Assessment Report, August 14, 2014. Report number: 485801.

Harrison, R. 2013. *Heritage: Critical Approaches*. Abingdon and New York: Routledge.

Harrison, R. 2015. Beyond 'Natural' and 'Cultural' Heritage: Toward an Ontological Politics of Heritage in the Age of Anthropocene. *Heritage & Society* 8(1): 24–42.

Harrison, R., C. DeSilvey, C. Holtorf, S. Macdonald, N. Bartolini, E. Breithoff, H. Fredheim, A. Lyons, S. May, J. Morgan and S. Penrose. 2020. *Heritage Futures: Comparative Approaches to Natural and Cultural Heritage Practices*. London: UCL Press.

Hodgetts, N. G. 2011. A Revised Red List of Bryophytes in Britain. *Field Bryology* 103: 40–49.

Jackson, S. T. and R. J. Hobbs. 2009. Ecological Restoration in the Light of Ecological History. *Science* 325 (5940): 567–9.

Jepson, P. and F. Schepers. 2016. *Making Space for Rewilding: Creating an Enabling Policy Environment. Policy Brief.* Oxford and Nijmegen: Rewilding Europe.

Kirkham, G. 2014. *United Kingdom China Clay Bearing Grounds: Mineral Resource Archaeological Assessment*. Report commissioned by English Heritage for Cornwall Council.

Leuvenink, A. 2013. Facilitating Social Learning to Increase Levels of Local Involvement: The Case of Associação Transumância e Natureza in Portugal. Master's thesis dissertation, Wageningen University.

Pétursdóttir, Þ. and B. Olsen. 2014. Imaging Modern Decay: The Aesthetics of Ruin Photography. *Journal of Contemporary Archaeology* 1(1): 7–23.

Snyder, G. 1974. *Turtle Island*. New York: New Directions.

Stoetzer, B. 2018. Ruderal Ecologies: Rethinking Nature, Migration, and the Urban Landscape in Berlin. *Cultural Anthropology* 33(2): 295–323.

Tsing, A. 2017. The Buck, the Bull and the Dream of the Stag: Some Unexpected Weeds of the Anthropocene. *Suomen Antropologi* 42(1): 3–21.

Wakefield, S. 2018. Inhabiting the Anthropocene Back Loop. *Resilience: International Politics, Practices and Discourses* 6(2): 77–94.

Chapter 11

Extracted Frontiers: A Call from the North

ANATOLIJS VENOVCEVS

From microchips to smartphones to electric cars, humanity's dreams of techno-salvation are built on the crude materiality of extracted metals and minerals. This extraction conveniently avoids large population centres in affluent Western democracies and instead clusters around the world's social peripheries. This slam poem, first presented as a spoken performance at the 8th Winter School of the Estonian Graduate School of Culture Studies and Arts in Tallinn, represents a call from the north – one of the largest frontiers for modern mining activities. By drawing on a few examples of past and present extractive landscapes, it aims to highlight the Arctic's physical, environmental and social costs for our technological transcendence. New ways of understanding humanity need to be rooted in the real material costs and consequences of our new and future technologies.

Hello everyone and thanks for your time,
For my part I will be slamming in rhyme.
I'm Anatolijs from UiT;
I am here to tell you some history.

My work is in Canada, Labrador –
And its tiny towns that mine iron ore.
They were assembled in order to feed
The hunger for steel and industry greed.

It was the fifties and we won the war
But in so doing we ran out of ore.
The new query as we raced for the stars –
"Where to get stuff for tanks, buildings and cars?"

Labrador, Canada, cold and remote,
A tundra traversed by foot or by boat.
Snubbing the Innu who lived there before,
A railroad was built in this quest for ore.

The valleys were filled and mountains were moved,
An engineering feat the railway proved.
Into the forests from north shore Quebec,
For four hundred miles the path made its trek.

Forest fires were started, chemicals spilt,
Rivers poisoned as this marvel was built;
But at the railroad ends, new towns emerged,
Car-based suburbs where no highways converged.

And mine ore they did in open-pit mines,
Blasting vast craters in thin northern pines.
Thus millions of tons was pulled from the north
By folks who came to the Labrador Trough.

As a colony this venture was seen –
Mines for settlement as farming had been.
Some people got rich and goods were attained.
Material gains, material drained.

Despite hard work, the glories were fleeting.
A crash in the price left the towns bleeding.
Some mines were closed while the rest were downsized;
In a free market, precarity's prized.

However the mining heritage stays,
The past continues in multiple ways:
The paths, the craters, the buildings remain,
The waters polluted, the caribou slain.

Yet memories of wealth and small town pride
Make some past transgressions easy to hide,
And it could be good news that prices rebound
And more land is set as extractive ground.

There's new mines now too, not just iron ore –
There's nickel and cobalt in Labrador.
Rare-earth elements can also be found,
With new technology demands abound.

But now things are different, Labrador's changed,
Towns barely survive by new work arranged.
No settlements grow, the future's in doubt.
Most workers fly in, most workers fly out.

How this relates to the things we explore –
New futures, Anthropocene and much more?
My point is simple and comes from the north –
Remember the waste as we venture forth.

For instance energy, how green can it be,
When there's only waste for people like me?
In lands that get flooded, mined out and burned,
For carbon-free life so desperately yearned.

Or take something that we all have at hand,
Smart gadgets that meet our every demand.
But what can be said on this conception
When my friends do not have cell reception?

Thus gets constructed Anthropocene's dream
Through outpouring waste from central regime.
To build and sustain a tech-future charm
While outsourcing the material harm.

We get all the waste and reap few rewards.
Material culture outlasts all words.
And in far futures when we are all gone,
Our toxic legacies will linger on.

Thus to build the new post-humanities
We can't just ignore externalities.
I am talking pollution, destruction,
Suicide, drugs and social dysfunction,

Violence, poverty, the boom and the bust,
Resettled, removed towns turning to rust,
Waste in the north for the southern demand,
Wealth built on stolen Indigenous land.

This is a call and my message to you –
Heed the material legacies too!
Waste is unequal, our lives aren't alike,
Tech-futures are suspect, so drop the mic.

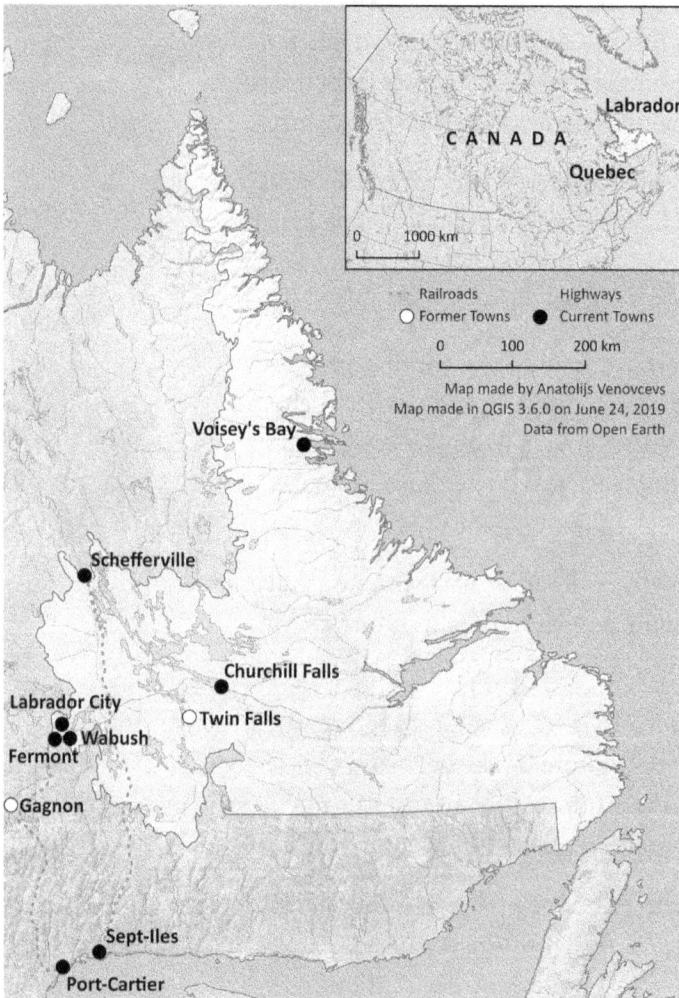

Figure 11.1 — Map of Labrador showing its relationship to Canada along with
related towns and infrastructure. (Map by Anatolijs Venovcevs).

Figure 11.2 — Trailer home subdivision, Labrador City, Labrador, an example of fast-built modernity. (Photograph by Anatolijs Venovcevs).

Figure 11.3 — Carol Lake mine, Labrador City, Labrador. The trucks in the photo are 7.7 metres high and typically carry 30 tons of rock per load. (Photograph by Anatolijs Venovcevs).

Figure 11.4 — Closed grocery store Labrador City, Labrador. (Photograph by Anatolijs Venovcevs).

Figure 11.5 — An abandoned rail line, Wabush, Labrador. (Photograph by Anatolijs Venovcevs).

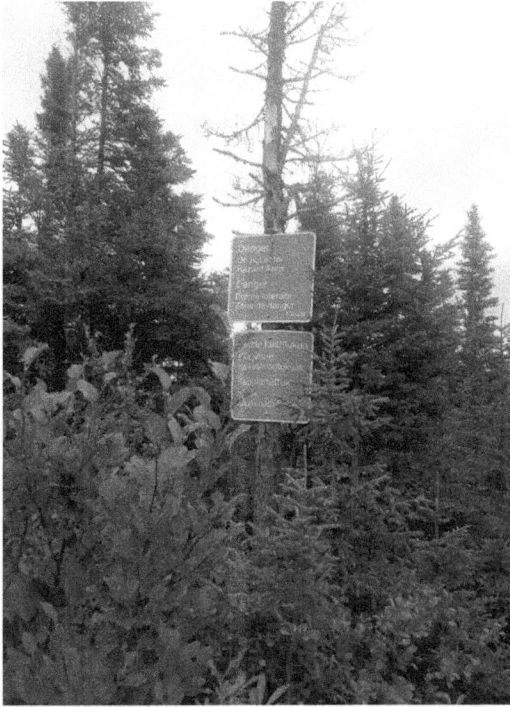

Figure 11.6 — Keep-out sign written in four different languages, Happy Valley-
Goose Bay, Labrador. (Photograph by Anatolijs Venovcevs).

Acknowledgements

The author would like to thank the editors for the opportunity to publish this poem. Additionally, very special thanks to Esther Breithoff, for seeing the value in the poem and putting the author in touch with the editors. Finally, the team of *Unruly Heritage: An Archaeology of the Anthropocene* project at The Arctic University of Norway deserve gratitude for their feedback and commentary on the early draft of this poem and for allowing the author to study the tangled legacies of mining in Labrador.

Chapter 12

When We Have Left the Nuclear Territories

ANNA STORM

Humans have begun to withdraw from nuclear territories. Out of the world's more than 600 reactors, about 170 are already permanently shut down and many more are in a limbo state of temporary shutdown (IAEA 2019).[1] Military and civilian nuclear plants are decommissioned and dismantled, and contaminated lands are evacuated. At the same time as nuclear territories in this way are unmade and abandoned, new enclosed terrains are created to store the radioactive remains; sometimes adjacent to the former nuclear facility, sometimes elsewhere. Eventually, these storage sites are also expected to be left behind, before the radioactive danger has ceased.

Since it is not possible to detect radiation with unaided human senses, the visualization of its effects and management is critical and politically contested. Atomic utopias, but foremost atomic dystopias, are themes abounding in literature, photo, film and artwork, and to some extent present within heritage practices. The "merging of beauty and terror" (Pratt 2017: G172) forming a sublime apocalypse is also how the human movement of leaving nuclearized land is often described. There are, however, also other articulations and visualizations of this movement, neither drawing on sublimity, nor employing "the artistic principle of estrangement" (ibid), but rather conveying reassuring notions that everything is normal and under control (Storm 2018).

For instance, the process by which humans are leaving nuclear territories is interlinked with new roles and accountabilities being attached to the nonhuman. In physical and imaginative ways, animals and vegetation, bedrock and clay, are attributed the role of guardians of radioactive remains. Controlled dismantling of nuclear plants is envisioned to result in neat and tidy green lawns, hazardous nuclear waste is suggested to be safely contained by copper, clay and bedrock, and contaminated areas are depicted as spontaneously rewilding into a pristine character as a consequence of human absence. These new roles and attachments – I will argue in this chapter – make up parts of how the human heritage of nuclear activities is projected to be cared for, guarded and, perhaps, reconciled in the future.

Radioactivity has been described as a "powerful ghost" from the past (Gan et al. 2017: G8) and contamination as a "tracer" of relations (Swanson et al. 2017: M8). Yet, a more explicit heritage approach reveals an expansive palette of understandings, articulations and movements of nuclear 'things' into or out of sight. While some aspects of the nuclear past are formally canonized within institutionalized heritage practices, some are not, and it is toward these practices which sit primarily outside of official heritage processes I will primarily steer my view, by scrutinizing visualizations of closed-down nuclear power plants being dismantled, and descriptions of radioactive waste being regulated and disposed of. I will demonstrate how these material and symbolic remnants of human activity so significant for defining, designating and understanding the Anthropocene are successively – and in striking contrast to institutionalized heritage processes – made invisible in various ways, through the ongoing deterritorialization of nuclear lands. Nonetheless, I also acknowledge how the haunting 'ghost' of radioactivity will interfere with this process of making-invisible, accompanied by, among others, a pheasant, a bison, sacred springs, bentonite clay and human myth.

Nuclear territories, nuclear heritage

The majority of nuclear territories are industrial and social enclaves, located along a waterfront in a countryside setting. Although small experimental reactors of the 1940s and 50s were located in urban areas, in the physicists' own university campus' geography (Storm, Krohn Andersson and Rindzevičiūtė, 2019), during the following decades, secret nuclear military facilities and large-scale commercial nuclear power plants were commonly established further away from population centres, in rural

areas (Blowers 2017). Nuclear enterprises took shape in close entanglement with a purposely built mono-industrial community, where the
social contract between the plant management and its locally based work-
force was tight (Storm and Kasperski, forthcoming). The 'nuclear way of
living' in these workers' communities (Wendland 2015) was marked, on
the one hand, by high status and privilege, and on the other hand, by fear
and ignorance of actual risks (Brown 2013; Masco 2006; Šliavaitė 2005).

The need for uninterrupted access to large amounts of water to cool
the atomic fission process directed the nuclear territorialization to sea-
shores, lakesides and riverbanks. Due to the level of secrecy and levels of
calculated risk, nuclear plants were enclosed within well-defined 'security
buffer zones' with special restrictions. This resulted in a co-production of
industrial land surrounded by a cultivated or uncultivated landscape of
some kilometres' radius that was not further developed by human settlement or other industries. Hence, the initial countryside localization was
often reinforced by an adherent protection of existing biological habitats. In addition, many of these non-exploitable areas within the security
buffer zones were given the status of nature reserves. As Joseph Masco
describes the buffer zones of military nuclear sites in the US: "most of
these areas were fenced off... and isolated from human contact [and
were consequently among] the most heavily fortified wilderness areas
in the world" (Masco 2006: 313) – areas now transformed into wildlife
reserves through large schemes. Already during operations, nuclear territories were thus inhabited by comparatively undisturbed nonhuman
biota, living on the grounds of the security buffer zones, but also in the
water outside and flowing through the plant (Storm 2018), and in the air
surrounding it.

While a number of small urban reactors were dismantled decades ago
and thereby provided grounds for experimentation and learning within
nuclear decommissioning (Laraia 2018), the first generations of bigger
plants are now reaching the end of their technical life and face decommissioning and dismantling at a large scale. Together with evacuated
contaminated areas around disastrous plants such as Chernobyl and
Fukushima Daiichi, and new nuclear facilities built to store radioactive
waste, we are therefore currently witnessing a "movement by which one

leaves a territory... [and] simultaneously extends the territory in new ways" (Sterling and Harrison, this volume: 24).

Nuclear heritage – or as it is often articulated with reference to earlier terminology, atomic heritage – is institutionalized particularly in two empirical domains: in terms of nationalized scientific achievements of early experiments, and in terms of globalized nuclear disasters of World War II (Storm, Krohn Andersson and Rindzevičiūtė 2019). Since heritage is "generally invoked as a positive quality" (Harrison 2013: 7), contributing to positive-affirmative identities and belongings in the present, the first empirical domain highlighting scientific achievements easily fits into an established heritage logic. Heritage practice does also address difficult, dissonant and dark pasts (cf. Logan and Reeves 2009; Macdonald 2009; Tunbridge and Ashworth 1996) which then closely resonates with the second empirical domain of atomic heritage, highlighting nuclear disaster and warfare. As Sharon Macdonald has shown, difficult heritage is often incorporated in the overarching positive-affirmative logic. Like 'positive' pasts, difficult pasts can work as "bolstering senses of national togetherness and pride" (Macdonald 2015: 9), but they can also perform a contrast to contemporary progress (Jönsson and Svensson 2005) and pass on warnings not to repeat what once went horribly wrong, and in that sense contribute to an affirmation of the present. Recently, however, there have been attempts to politically address difficult pasts by means of transparency, apology and morality (Macdonald 2015), thus making heritage a potential tool not only for bolstering, but for reconciliation (Storm, forthcoming).

The processes where humans are abandoning nuclear territories, where nuclear plants undergo dismantling, where workers' communities negotiate their identity and radioactive waste is relocated to new storage sites – the focus of this chapter – are not to any significant extent part of institutionalized heritage concerns (see though Tafvelin Heldner, Dahlström Rittsél and Lundgren 2008). Instead, suggestions about such measures have until recently been framed as a joke or met with hearty laughs (Storm 2014: 70, 96), perhaps partly depending on the ambiguous character of these industrial and social movements, which do not easily fit any of the established and more straight-forward 'positive' or 'difficult' categories roughly outlined above.

The nuclear industry itself has begun to explore issues of preserving memory and information to future generations about storage sites for long-lived hazardous nuclear waste (OECD/NEA 2015; see also Holtorf and Högberg 2014; Harrison et al. 2020), but is generally not engaging in broader heritage discussions (see though Dounreay Site Restoration Ltd 2010; NDA 2007). Instead, the most responsible action and desired outcome from a professional nuclear decommissioning point of view is a complete erasure of any trace of the nuclear activity leading to "unrestricted release" (Laraia 2018: 110) of the nuclear territory.

Nonetheless, as pointed out by Gan et al. with reference to the work of Lesley Stern, a landscape "remembers the movement of things" (Gan et al. 2017: G3) and the clean and controlled withdrawal envisioned by the nuclear industry might be destabilized by human as well as nonhuman reterritorializing engagements. As the case studies in this chapter show, the nuclear territory humans are abandoning is partly described as rewilding into a refuge, symbolically healed from the consequences of a "violently uneven modernity" (ibid) into a 'scar' one can live with (Storm, forthcoming; Storm 2014). Partly, and conversely, abandoned nuclear territory is envisaged as being transformed into a controlled natural environment confirming and extending the "fable of Progress" (Swanson et al. 2017: M2) in which the nuclear industry is already a key actor. This controlled environment, epitomized by a green lawn, is nevertheless disturbed by the radioactive remains which refuse to entirely disappear. At this point, we are finally recommended to rely on stones as nonhuman guardians with long-lasting inscriptions of warning, and on bedrock that will hopefully keep our unwanted heritages safely stored for the deep future.

This chapter focusses on three case studies, each of which represent distinct and detailed articulations of processes of leaving and recreating nuclear territories. I make a close reading of, first, an eight-minute semi-official film documenting the dismantling of the Yankee Rowe nuclear power plant in the US; second, a lavish book describing the situation in the countryside areas of Belarus contaminated by radioactive fallout from Chernobyl, twenty years after the disastrous event; and third, an exhibition connected to the Barsebäck nuclear power plant in Sweden in two versions, one set up before the plant was inaugurated, one set up after its closure. While the film is easily available online and the book is available in print, the reading of the exhibition is based on several visits between 2010 and 2018.

The readings are guided by the various meanings of 'territory' and its verbal derivations (Cambridge Dictionary 2019; The Free Dictionary 2019), highlighting connotations of potential conflict and threat, of socio-political disadvantage and of borders to protect, involving humans as well as nonhumans. Together, the three cases, which involve visuals, text, sound, texture and space, suggest futures that are, on the one hand, technically and geographically precise, and, on the other hand, utterly indistinct, uncertain and open, even mythological.

Collapsing buildings and a flying pheasant at Yankee Rowe, US – a film

The short film *Yankee Rowe Demolition Video* starts with a dedication: To the 'outstanding Yankee Rowe Decommissioning Team. You're the best!' (Yankee Rowe 2018). Then, without further ado, the viewer is thrown into a visual story of how this US nuclear power plant was physically dismantled during four years in the 2000s. To the soundtrack of the blues-rock song 'The House is Rockin', the first footage shows a sequence of collapsing buildings. It is a snowy winter and the light is grey. Heavy machinery takes big bites out of built structures, a long emergency staircase comes sliding down along a sphere-shaped white building, and men in full protective white clothing work with blowtorches on a platform high up a chimney. The flames are intense in colour and the chorus repeats: "well the house is a rockin', don't bother knockin'".

The almost joyful energy of these first minutes of the eight-minute film alters into a tone of addressing challenges, as the theme from the action movie *Mission: Impossible* starts playing. The change of music is accompanied by a text visual cockily stating: 'Mission: Possible', and then we are back to collapsing buildings and men with cutting blowtorches. However, all of a sudden we are up in the air, flying over the white spherical building. The aerial photo moves and we also see parts of the surroundings. It is now summertime. The plant is located at the shores of a dammed river reservoir, enclosed by wooded hills and with no other buildings or infrastructures discernible. Coming back to the ground, we face, not the *Mission: Impossible* star Tom Cruise, but a woman with long brown hair under her hardhat, in short sleeves and yellow gloves, cleaning a removed piece of the spherical building's outer walls.

The third soundtrack of the film is a streamlined country rock version of 'I Can See Clearly Now'. The camera pans to offer a distant aerial view of the vast wooded landscape and the water reservoir, approaching the nuclear power plant which gradually comes into sight. When reaching a position directly above the plant, a time-lapse technique makes the many buildings on the industrial site fade away one by one, leaving an empty flat brownish ground (Figure 12.1). The nuclear power plant is gone. The surrounding forests are now sparkling in autumn colours, and still images depict a bald eagle sitting on a branch, a black bear walking in grass and low bushes and a heron on a branch. A pheasant stands on paved ground with trucks and other vehicles in the background, followed by a picture where it takes off flying towards the sky. As a contrasting interlude, the footage shows a dark interior with nuclear fuel assemblies followed by a sunny exterior with sixteen white cylindrical canisters within a fenced, well-ordered area. The song repeats: "I can see clearly now... It's gonna be a bright/bright sunshiny day/look all around, there's nothing but blues skies/look straight ahead, there's nothing but blue skies".

The final credits for the film, and a cartoon of a pheasant running across the screen, are accompanied by Vera Lynn singing the bittersweet World War II song 'We'll Meet Again'. At the time, the famous lyrics "we'll meet again/don't know where/don't know when/but I know we'll meet again some sunny day" was comforting soldiers away from their families and loved ones. To later listeners, however, the song also has other connotations, not the least in relation to the final scene in Stanley Kubrick's 1964 film *Dr. Strangelove* where it accompanies a long series of mushroom clouds emerging from nuclear bombings.

This short film, made in close connection with the decommissioning team in the first decade of the 2000s, and available on a semi-official website of the Yankee Rowe nuclear plant (Yankee Rowe 2018), conveys an insider's documentary account of the efforts of the team who carried out the dismantling work. Nothing is explained to an ignorant spectator, there is no speaker's voice, just a visual sequence of the material

Figure 12.1 — The Yankee Rowe nuclear power plant in the US disappeared during four years in the 2000s. In a documentary video, time-lapse technique makes the many buildings on the industrial site fade away one by one, leaving an empty flat brownish ground. Stills from the Yankee Rowe Demolition Video (2018).

disappearance of non-named built structures. As such, the film is not an official archive, nor is it an insider's account in a personal sense, for example through the identification of recognizable individuals. There are no faces shown, no group pictures of the workers and no activities such as lunches, breaks, comings and goings included. From this perspective, the film is simply a moving photographic album showing how the plant physically disappeared.

However, the film also shows the envisioned future of the site. As a result of the dismantling, an empty flat ground comes into sight; it emerges repeatedly from different viewpoints through time-lapse sequences and successively the ground also becomes greener, turning into a lawn. By zooming out from the industrial site, the aerial photo situates it in a vast wooded landscape intersected by the river. The depictions of wild animals emphasize a future where the former industrial site is no longer populated primarily by humans but is instead rewilded. This harmonious vision of a return to a 'pristine' natural setting is, however, slightly disturbed by the interlude about spent nuclear fuel – the most hazardous radioactive remains left after nuclear power production. While the very plant disappears through the dismantling work, the spent fuel is transferred into a new storage construction, shining bright and with reassuring safety features like a high fence and camera surveillance, but still forming an obstacle to the overall imagined prospects of human disappearance and a return to nature.

The film does not indicate for how long the fuel in the white canisters will stay there, nor for how long a human presence will be required to guard the hazardous material. In addition, the debris from the many buildings torn down disappears in the film like evaporating water. It seems just gone; but of course it is not, it has been moved somewhere else. In the main story of the film, about leaving the nuclear site by undoing it, the radioactive waste therefore remains an exception and a slightly unsettling element as it partly lingers and is partly relocated. Furthermore, it is difficult not to hear a double message in the choice to play 'We'll Meet Again' as the final song of the film. Certainly, it could refer to the dedicated Yankee Rowe decommissioning team, perhaps meeting again at some other nuclear power plant that is to be dismantled. But, to some listeners, it might rather say in an ironic and at the same time worrying way, that this waste is not at all gone, it will stay with us, 'we'll meet again' – and perhaps in forms close to the horrors of the mushroom clouds which are also evoked by the music.

Sacred springs and European bison in the Chernobyl Zone, Belarus – a book

Twenty years after the Chernobyl disaster, a group of well-known Belarusian photographers and writers published a lavish book on the consequences of the catastrophe for the Belarusian countryside entitled *Charnobyl'-Chernobyl-Tschernobyl* (Ramaniuk, Kliashchuk and Byshniou 2006). The nuclear power plant is located in Ukraine, but the border is roughly only ten kilometres away and most of the radioactive fallout was concentrated on hundreds of rural villages, cropland, meadows and forests of south-east Belarus. This radioactive geography has long been attributed to wind directions at the time back in April 1986, but lately it was revealed that Soviet air force artificially also trigged rainfall in some parts of this area before the toxic clouds could reach Moscow and other big cities (Brown 2019: 52–53). After the devastating event, people and livestock were evacuated from a zone of 30 km radius around the plant, including the atomic city of Pripyat with ca. 50,000 inhabitants. While many villages were physically demolished in order to bury the contaminated structures underground, some were just abandoned.

One of the first photos in the book shows two white storks, one of them sitting in a nest on the top of a post, the other caught in flight (Ramaniuk, Kliashchuk and Byshniou 2006: 16). Close by is a floodplain and behind the water are green forests. The landscape is flat. Two seemingly well-kept rowing boats lie on the shore. In the foreground is a blossoming small tree, probably a blackthorn. The caption says: 'Storks are leaving the Zone. Since the dawn of time they settled close to humans and now, when people are gone, they feel unprotected in the abandoned villages. Stork nests decay and fall down just like human houses. The storks do not come back'.

Another photo shows a heavy, muscular European bison gazing into the camera (Figure 12.2), surrounded by bushy forest (Ramaniuk, Kliashchuk and Byshniou 2006: 17). On the ground, just in front of the

Figure 12.2 — The European bison once populated the forests of Belarus. Today it is reintroduced as part of the envisioned, but contested, rewilding of the radioactively contaminated zone around the Chernobyl nuclear power plant. (Photograph by I. Byshniou, reproduced with permission from Ramaniuk, Kliashchuk and Byshniou 2006: 17).

bison, stands what might be a metal tray for cattle feed. In the foreground is a wooden signpost, slightly out of focus, with a warning message in Russian: 'Radioactive contamination. Entry is PROHIBITED'. The caption describes how humans previously hunted the wild European bison to extinction and later restored the stock through big efforts and, finally, in 1995, how a group of bison were brought to the Chernobyl Zone in order to recreate a wild herd there.

Turning yet another page, we meet the eye of a grown-up wolf (Ramaniuk, Kliashchuk and Byshniou 2006: 19). It lies indoors on a floor with its head raised and turned towards the photographer, teeth shown. The floor is covered with pieces of cloth, broken bricks, gravel and pieces of paper. The interior wall behind shows ragged generations of wallpaper. Two windows are shining white in backlight and a metal headboard is leaning against the wall. The caption says, in the voice of the photographer: 'Wolves take eye contact as a challenge. But I avoid it, because he is the master here, not I'.

In the many following photos, captions and running text, the reader meets wild boar, herons, a bear cub, eagle owls, hedgehogs, mice, elk, deer, tortoises, swallows and black storks, all seemingly prospering and in charge of their situation. The Zone – here denoting not only the evacuated and closed-off territory around the Chernobyl plant but the whole area in Belarus suffering from radiation – has in these depictions become a natural wilderness. The swallows assemble on barbed wires in the evenings, the hedgehogs hunt on overgrown village streets and the tortoises find rest in broken, left-behind pots – all signs of human withdrawal, and of animal and vegetation taking back the territory.

In the book, the Zone is described alternately as a refuge, a zoo and a paradise (Ramaniuk, Kliashchuk and Byshniou 2006: 28, 30, 35). One of the authors hears the swallows twittering as if to say: "'Tomorrow will be fine and peaceful here'... A life in which human beings are superfluous" (2006: 35). He also observes how the deer are roaming differently in the Zone compared to the rest of the Belarusian countryside: "they are less cautious here. For here neither a hunter nor a mushroom-gatherer can frighten them" (ibid). Instead, together with foxes, martens and raccoon-dogs, the deer are said to enjoy "gifts from humankind" (2006: 45) such as the crop of apples and pears in abandoned gardens. Depictions of a thriving nature in the Zone, resembling the imagery of this book, have been widely spread in international media but are certainly also questioned, especially when it comes to observed dysfunctional microbes, worms, larvae and insects – key to the process of breaking down organic matter and to pollination (Brown 2019: 138–145).

According to the book, the only animal species that seems less content are the white storks – those who fictionally deliver human babies – who prefer to live close to human settlement and therefore leave the Zone along with the humans, as well as the dogs and cats that are brought along by the closed territory's checkpoint staff. The pets keep disappearing, probably due to the wolves (Ramaniuk, Kliashchuk and Byshniou 2006: 33). Apparently, the rewilding Zone is not a friendly place for too human-dependent species. In passing, the authors also note that, against expectations, no "three-headed snakes, two-tailed wolves and other mutants" (2006: 39) were ever sighted. Thus, not even in terms of radiation-caused biological disturbance is the human visible in the Zone – at least not in this part of the story.

The book conveys an account of spontaneous rewilding within the Zone, due to human absence. The spontaneous processes are furthermore reinforced by the conscious reintroduction of the European bison into this new wilderness – a species that has the status of being a highly valued national animal of Belarus. In his vivid account of centuries of bison lore connected to the Białowieża forest, which straddles the border of today's south-west Belarus and Poland, Simon Schama notes how the bison "became a talisman of survival. For as long as the beast and its succouring forest habitat endured, it was implied, so would the nation's martial vigour" (Schama 1995: 41).

The story about 'Nature' is however only the first part of the book Charnobyl'-Chernobyl-Tschernobyl. It is followed by sections on 'Folk', 'History' and 'Individuals'. When changing perspective, the reader realizes that the Zone is not all that empty of human activity, even though the extent and character of these activities have certainly changed dramatically. Indeed, numerous forced migrants regularly come back to the Zone and a few elderly people have actually resettled in their home villages. Twice a year, honey from the many beehives in the Zone is harvested (Ramaniuk, Kliashchuk and Byshniou 2006: 86). On special days, traditional ceremonies are performed including drinking and washing oneself with water from local sacred springs (2006: 120–1). Cemeteries are carefully maintained and many people now living far away still choose to have their grave in the village of their childhood, now within the Zone (2006: 112–119). In springtime, former village neighbours meet on the orthodox commemoration day Radunica, or Radunitsa, to remember their departed and to eat and drink together at the place where they used to live. The almost idyllic depictions show colourful and sunny gatherings of young and old (2006: 112, 116–17). In parallel, the authors remind the reader that "for us these places have become deadly" and "we shall never live in this home again" (2006: 35, 28).

All the photos, including those depicting solely animals and natural scenery without any trace of human activity whatsoever, are all carefully located in geography. Places have names: Aravichy*, Chapajeuka*, Viazhyshcha*, Dubrova*, Novaja Jelnia*, but at the same time an attached asterisk (*) indicates that these particular villages 'no longer exist' (Ramaniuk, Kliashchuk and Byshniou 2006: 4). The places are thus simultaneously present and absent; this used to be a human settlement with a human name, but now the humans have left and the village is no more, or at least it is changing into something else. To the geographical specification of a no longer existing village or place, strict contamination data is added, such as, "Cs-137 contamination – 84,09 Ci/km^2, 2102, 25 exceeds normal" (2006: 8) and sometimes the distance to the disastrous reactor is also pointed out: "39 km to the reactor" (2006: 4).

In the Belarussian countryside, 485 villages were lost due to the Chernobyl catastrophe, along with fifty-four collective and state farms and 2,500 square kilometres of cultivated land (Ramaniuk, Kliashchuk and Byshniou 2006: 45, 30). The river Dniapro, or Dnieper, passes through the most heavily contaminated areas of Belarus and is here widened into a large floodplain or wetland, which has for a long time been of great significance as a breeding habitat and migration stopover. The wildlife in the area therefore has a history, and since 1988, two years after the catastrophic reactor meltdown, the Belarussian state established the Palessie State Radio-Ecological Reserve in the country's part of the closed-off territory around the Chernobyl plant. The authors of the book do not really address the paradoxes of the establishment of this natural reserve and the fact that a vast majority, about 80%, of the radioactive fallout was captured by its forests, or how the contamination might affect the ecosystems. Instead, it is simply noted that the "most perfect retainers of radiation are the oaks" (2006: 30).

This landscape close to a nuclear power plant was once inhabited by farmers and their cultivated crops, livestock and pets, and the white storks. Although ambiguous, the Zone and the forced and painful evacuation of people and domesticated animals is portrayed in the book as partly healed by the return of the wildlife. After two decades, the people return occasionally for remembrance and for semi-wild harvests such as honey. But the territory now belongs to the wild animals and the non-cultivated vegetation – superficially reminiscent of an imagined pristine paradise. The seemingly prospering wildlife speaks about some kind of reconciliation, but the mourning of the forever lost homes nevertheless still permeates the Zone.

A rune stone and a green lawn at Barsebäck,
Sweden – an exhibition

The Barsebäck nuclear power plant in Sweden was expected to become a major tourist attraction even before it was completed and, as a result, an exhibition pavilion opened in 1972 at the fringe of the plant construction site, three years before the first reactor went into operation. The so-called Expo building is a dark green wooden structure with a sharply pitched roof, located on the meadows almost directly on the shore of the Swedish south-west coast, and was described at the time as "architecturally daring" (quoted in Krohn Andersson 2012: 145). The windy, open, flat landscape offers an uninterrupted view over the Sound towards Denmark, only twenty kilometres away as the crow flies. The Expo housed an auditorium, a small café and a showroom with plates about nuclear technology, the long history of the area and also a number of archaeological objects that were found during excavations preceding the construction work.

One of the displays shows a colourful and illuminated outline of a nuclear reactor. A panel with two buttons invites the visitor to move control rods up and down inside the reactor vessel by pushing the buttons and thereby affect the energy output. In the schematic depiction, the reactor is connected to a turbine which in turn is connected to a generator, to a transformer and finally to the grid. The turbine is also linked to a condenser using cooling water from the sea and a pump that closes the loop back to the reactor.

Another lit plate shows the contours of a male human body surrounded by a number of small photos depicting, for example, a television set, a plate with food, an aircraft, a miner, a city skyline, a team of surgeons at work around an operating table and, in the low corner, the nuclear power plant framed by trees and an open field. The plate informs that the dose of radiation exposure for a pilot is 7.0 mSv/year, for a miner 20.0 mSv/year and for the general public having medical treatment 0.8 mSv/year, and so forth. By comparison, the exposure caused by the nuclear power plant to nearby residents is no more than 0.0007 mSv/year.

At the time of my first visit, a human-size 'rune stone' made with Styrofoam is mothballed in a storage room of the Expo. The inscription in red faux-runic letters says: 'SKB's [the Swedish nuclear fuel and waste management company] pointing device. Point and you will get an answer. Facts about the waste of nuclear power production'. In the middle of the rune stone is an empty space where a digital screen once was, and below this void is an 'engraving' of a snake or dragon that almost bites itself in its tail and so creates a loop. Snakes and dragons are common in Nordic mythology and often depicted on rune stones, for example the dragon Fáfnir who greedily guards a cursed gold treasure or the Midgard Serpent which embraces the universe.

The two reactors at Barsebäck went into operation in 1975 and 1977 respectively. They were finally shut down in 1999 and 2005, after almost three decades of electricity production. The Barsebäck plant was the most politicized nuclear facility in Sweden, and the closedown was highly contested both in the local context, nationally within Sweden and between Sweden and Denmark. As one part of the reorientation after the closure, a new information centre opened in 2008, replacing the old Expo. Like its forerunner, the new 'Infocentre' or 'Barsebäck's energy and environmental exhibition' houses a small auditorium/cinema, showrooms, a café and a corner with books, crayons and paper where kids are encouraged to make drawings to pin to the wall. The new building is low, white and directly connected to, but still outside, the fenced and guarded entrance to the plant area. The walls are glazed, which emphasizes openness and transparency. As regards contents, the themes from the old Expo are still very much present, although in a modernized visual language. In addition, the task to decommission and dismantle a nuclear power plant and how to take care of the radioactive waste are new features.

The explanations of how the technology works are still there, as are the radiation exposure comparisons, in new form. In the Infocentre, the visitor encounters a number of dioramas – three-dimensional miniature models in glass showcases – where female, long-haired Barbie dolls collect their suitcases upon entering the airport, put laundry into the washing machine, sunbathe by the sea and, finally, work at a nuclear power plant dressed in white protective clothing. Again, the figures of exposure to normal background levels of radiation in air travel, homes and sunlight are, with comforting results, matched with exposures for a nuclear industry worker.

Close by is a sequence of three photos answering the question 'What will it look like afterwards?', that is, after decommissioning and dismantling the nuclear power plant. The photos are aerial views of the dismantling of the Connecticut Yankee nuclear power plant in the US (located about 200 km from the Yankee Rowe plant described above in the decommissioning video). The first photo from 2003 shows the Connecticut Yankee plant with numerous industrial buildings, cooling water intakes and parking lots located in between a waterfront and a forest. In the second photo, from 2005, the reactor building and some other buildings are gone. In the third photo, from 2007, the flat space is empty, beige in colour. The place where the reactor once was, stands out as a green oval. The message of these photos in the Barsebäck Infocentre seems to be clear: there are role models to learn from, and there will be nothing left of the Barsebäck plant when we are done here.

In a separate room without windows, the walls imitate granite bedrock which makes the space dark. Here the management of nuclear waste is addressed. A large photo of the Sound between Sweden and Denmark is on the wall. The Öresund Bridge is in the background, an impressive engineering masterpiece arching over small sailing boats discernible in the water. In front of the photo is a free-hanging transparent glass plate featuring a photo of a red and white transport vessel. The caption says that it is MS Sigyn, a specially constructed vessel that transports radioactive waste from Swedish nuclear power plants to the two nuclear waste facilities in the country.

Next to this combination of two photos is a mock-up of a copper canister (Figure 12.3) with deep rectangular compartments to house the most highly radioactive spent fuel assemblies. A caption explains about the thickness and weight of copper and steel in the construction, indicating its ability to resist corrosion and external pressure. Around the shining cylindrical mock-up is a soft bench covered in bright-coloured cloth representing a buffer layer of bentonite clay, on which visitors can sit down and take a rest.

While the archaeological treasures and the rune stone were not brought along from the first to the second version of the exhibition, ancient times and mythological dimensions are present in other ways. The naming of the transport vessel for nuclear waste, Sigyn – not a common name in Sweden –stands out. Like the dragon or snake on the rune stone, Sigyn is a character in Nordic mythology. She is a goddess and

Figure 12.3 — At the Infocenter of Barsebäck nuclear power plant, Sweden, the exhibition features a mock-up of a copper canister for storing spent nuclear fuel. The soft and bright-coloured sitting bench around the metal construction represents an additional protective layer of bentonite clay. (Photograph by Anna Storm).

the wife of Loki who, after many evil actions, was chained and doomed to eternity to get poisonous blood from a snake dripping onto his face. Sigyn, the faithful wife, stays at his side and catches the blood in a bowl, thus hindering the caustic blood to hurt his face. When the bowl is full, and she has to leave temporarily to empty it, Loki shakes his chains in pain so that the Earth tremors.[2]

Also connected to the management of hazardous waste, and partly extending human time perspectives, are the abiotic elements. The sea did not only serve as a necessary coolant when the nuclear plant was in operation, but continues to provide a safe and efficient transport route for the specially built vessel, while the granite bedrock forms a reliable long-term companion together with copper, steel and clay to contain the nuclear power production's unwanted leftovers isolated from humans into eternity. The dismantling of the plant is envisioned to result in a flat mowed green lawn, open and free to be reused for new purposes, a manifestation of the nuclear industry's goal of 'unrestricted release'. In the local municipality, there are elaborate plans to build a fancy residential area on the future denuclearized coastal landscape: "Seaside Barsebäck" (Storm 2014: 72–3). But as the plant owner instead has lately built an interim storage for radioactive waste on site, it will take several decades, if ever, before children will dig and play in sandboxes on the former nuclear territory.

Mutants, monsters and stones

When humans leave nuclear territories, care for radioactive remains is handed over to nonhumans. Nuclear heritage simultaneously disappears by its enclosure into rewilding natures or into controlled biotic and abiotic elements such as green lawns and stones.

The rewilding exemplified in the film about the dismantling of the Yankee Rowe plant and the book about radioactively contaminated abandoned villages in Belarus focusses on large fauna such as bear, bison and wolf. These large animals convey a powerful message about the pristineness of the place, after humans have left, and also drown out more elusive messages carried by microbes and insects. Such megafauna, which were present almost all over the planet prior to the arrival of modern humans, have recently began to generally increase in numbers. Among other things, this is due to the ongoing urbanization leaving large land areas empty of human inhabitation, along with conscious reintroduction schemes (Svenning 2017). Thus, not only does nuclear deterritorialization lead to rewilding, or a post-nuclear 'jungle' (Byshniou 2005), but it forms part of a larger process of human interventions in the production of 'nature' (see Breithoff and Harrison, this volume), contributing to the ways in which rewilding nuclear territories can be conceptualized as normal and under control.

The visualization of an ongoing reinstating of a pre-disturbed status, including reintroduction of species like bison, mirrors traditional ecological restoration work (Kowarik 2005; Jørgensen 2015; Cronon 1995), sometimes with the eventual goal to "erase human history and involvement with the land and flora and fauna" (Jørgensen 2015: 487). In the case of the nuclear plants, the often already existing nature reserves of the security buffer zones also complicate the picture of return and erasure. More recent ecological work in such zones rather emphasizes hybrid landscapes including humans and related social values. This category of a "natureculture" is widely regarded as a "condition of the Anthropocene" (Jørgensen 2017: 849; see Harrison 2015 in relation to Anthropocene heritage) and so forcefully rejects the very idea and possibility of complete human withdrawal from any landscape, radioactive or not. Joseph Masco suggests, however, that "hybrid" might not be the best term to use to scrutinize the complexity of "naturecultures" since this concept formally denotes an infertile offspring, and he suggests that we should use the term "mutant" instead (Masco 2006: 301). In consequence, even if no three-headed snakes were observed in any of the particular places described here, these territories might in fact be understood as mutants, nonetheless.

Pristine or mutant – the megafauna are rich in symbolic connotations. The wolf, for example, in Western imagination appears as a "noble animal, unlike the degraded servant, the dog" (Freccero 2017: M97), which might explain the photographer backing off from his encounter with a wolf in the Chernobyl Zone, as the wolf 'is the master here'. The bison lore, as we saw, brought messages of "survival" and "national martial vigour" and was, in addition, often described as a "monster" of heroic savagery (Schama 1995: 41). Swanson et al. highlight how "monsters… have a double meaning: on one hand, they help us pay attention to ancient chimeric entanglements; on the other, they point us toward the monstrosities of modern Man [sic]" (2017: M2). Is this how we should frame the serpent on the rune stone embracing a world of nuclear waste or if it is a dragon guarding a cursed treasure, or the goddess Sigyn who at some point has to empty the bowl of poisonous substance that will make the Earth tremble?

In times of disaster, we rely on stones. Reliable things are said to be "solid as a rock" and stones "bring a gift of hope, of fortune, of insight, of the possibility of living-with" (Gan et al. 2017: G11). In this way the rune stone with engraved messages from the past to the future, and the bedrock and clay representing deep pasts and deep futures, can be seen as signifying hopes for healing wounds into scars one can live with (Storm, forthcoming). The world's very first nuclear reactor, constructed on the

campus of Chicago University in 1942, was eventually buried underground in a forest outside the city. The site is marked by different stones,
like headstones, one of them with the inscription: 'Caution – Do not dig'
(Storm, Krohn Andersson and Rindzevičiūtė 2019).

What do these suggested characteristics of human withdrawal from
nuclear landscapes tell us about the status of current responsibilities, and
about contemporary human and nonhuman relations to the temporalities of radioactive decay? The nuclear world has long since lent itself to
utopian and dystopian views, and when radioactive waste management
more prominently entered the picture, the nuclear industry found itself in
between a detailed technical discussion and humanity's existential questions of survival and communication with future generations, extending
further into the future than any archaeological record extends into the
past. Perhaps that is why mythology is part of the human imaginations of
radioactive futures, and why the nonhuman is put in charge of guarding
our nuclear heritage.

Human and nonhuman entanglements with radioactive remains

In this chapter, I have argued that humans are now successively leaving
and unmaking nuclear territories. However, this process of deterritorialization is not complete, but partial and in many respects superficial. The
decommissioning and dismantling of nuclear reactors and the evacuation
of people and livestock from contaminated lands are paralleled with a
relocation of nuclear waste to new-built storage sites and a continuous
exchange with the left-behind territories, for example, in terms of bird
migration and human revisitation. Hence, the movement and reconfiguration of nuclear residues not only implies deterritorialization but a reterritorialization of the same places, or of places elsewhere.

Furthermore, such attempts at abandoning and unmaking nuclear territories highlight aspects of human-nonhuman entanglements with consequences for understanding the role of heritage in the Anthropocene.
The hazardous radioactive remains will stay with us into the deep future,
and negotiations about their containment and isolation from living
biota involve questions ranging from the expected corrosion characteristics of copper to possible means of communication with future human

generations. These deliberations could certainly be conceptualized as highly existential heritage processes, although they are seldom articulated as such. Instead, a double objective of making-invisible imbues current nuclear decommissioning activities: first, to let nuclear heritage – in its industrial and social sense – disappear and, second, to move those remnants that cannot disappear – the radioactive waste – principally out of sight. Both these goals are rationalized as unquestionably responsible actions, thus normalizing and removing such processes from critical evaluation.

If unchallenged, this ongoing making-invisible will leave us with an institutionalized nuclear heritage focusing primarily on positive-affirmative and nationalized scientific achievements, narrated partly by the nuclear industry itself, and on difficult-heritage warnings of globalized nuclear disasters and atrocities, narrated primarily within a context of war memorials. The industrial and social aspects of nuclear heritage will to a large extent be lost or, rather, their disappearance transformed into a continuation of the nuclear industry's self-proclaimed identity of cleanness and control. The neat and tidy – responsible – aftermath of a nuclear power plant is in this perspective a green lawn, free and safe to use for any purpose. No signs or markers of previous usage will be relevant as the land is imagined to be completely denuclearized.

However, radioactive remains are not easily rendered invisible, and trigger both technical and existential debate. As a human legacy, or heritage, these remains force humanity to conceptualize responsibilities of care into distant futures but this act of imagination, that is, to safely manage this toxic residue for such long-time horizons is certainly very difficult. In that sense, radioactive remnants may be understood as difficult heritage, akin to, for example, the persistence of ruined and toxic landscapes or memories of the experiences of genocides and war atrocities. But while some difficult heritages have become integrated in public negotiations of reconciliation and the politics of memory, radioactive remains have hitherto eluded such processes, just like the half-lives of radioactive isotopes elude human temporalities. I have suggested in this chapter that one way in which humans have attempted to bypass this fundamental and challenging issue is to hand over the care of radioactive remains to nonhumans, to animals and vegetation, bedrock and clay, sometimes with an added hint of myth. The nonhuman and the mythological both seem to provide a reliable route into ancient pasts as well as into elusive distant futures, and thus may persist even when we have finally left the nuclear territories.

Acknowledgements

The author wants to thank Anders Houltz, Oscar Jacobsson, Lars Kaijser, Tatiana Kasperski, Fredrik Krohn Andersson and Anders Nordebring for their help and valuable comments on an earlier version on this chapter.

Notes

1. There is currently ongoing construction of about 50 new reactors, primarily in East Asia (IAEA 2019).

2. After 30 years in operation, the transport vessel *Sigyn* has recently been taken out of use and replaced with a new ship, named *MS Sigrid*. This name is also related to Nordic mythology and etymologically combines the words 'seger' (victory) and 'frid' (peaceful), while the name itself means 'beautiful'.

References

Blowers, A. 2017. *The Legacy of Nuclear Power*. Abingdon, Oxon: Routledge.

Brown, K. 2013. *Plutopia: Nuclear Families, Atomic Cities, and the Great Soviet and American Plutonium Disasters*. Oxford: Oxford University Press.

Brown, K. 2019. *Manual for Survival: A Chernobyl Guide for the Future*. London: Allen Lane, an imprint of Penguin Books.

Byshniou, I. 2005. Чернобыльские джунгли. 20 лет без человека [Chernobyl Jungle: 20 Years without Humans]. Documentary. https://www.youtube.com/watch?v=zst6O7Skeyc

Cambridge Dictionary. 2019. 'Territory'. https://dictionary.cambridge.org/dictionary/english/territory.

Cronon, W. 1995. The Trouble with Wilderness; or, Getting Back to the Wrong Nature. In *Uncommon Ground: Rethinking the Human Place in Nature*, edited by W. Cronon, 69–90. New York: W. W. Norton & Co.

Dounreay Site Restoration Ltd. 2010. *Dounreay Heritage Strategy: Delivering a cultural legacy through decommissioning*. https://assets.publishing.service.gov.uk/government/uploads/system/uploads/attachment_data/file/718623/Heritage_Strategy_Issue_2_Aug_2010.pdf

Freccero, C. 2017. Wolf, or, Homo homini lupus. In *Arts of Living on a Damaged Planet*, edited by A. L. Tsing, E. Gan and N. Bubandt, M91–M105. Minneapolis: University of Minnesota Press.

Free Dictionary, The. 2019. Territorialize. https://www.thefreedictionary.com/territorialize.

Gan, E., N. Bubandt, A. Tsing and H. Swanson. 2017. Ghosts. Introduction: Haunted Landscapes of the Anthropocene. In *Arts of Living on a Damaged Planet*, edited by A. L. Tsing, E. Gan and N. Bubandt, G1–G14. Minneapolis: University of Minnesota Press.

Harrison, R. 2013. *Heritage: Critical Approaches*. Milton Park, Abingdon: Routledge.

Harrison, R. 2015. Beyond 'Natural' and 'Cultural' Heritage: Toward an Ontological Politics of Heritage in the Age of Anthropocene. *Heritage & Society* 8:1, 24–42.

Harrison, R., C. DeSilvey, C. Holtorf, S. Macdonald, N. Bartolini, E. Breithoff, L.H. Fredheim, A. Lyons, S. May, J. Morgan and S. Penrose. 2020. *Heritage Futures: Comparative approaches to natural and cultural heritage practices*. London: UCL Press.

Holtorf, C. and A. Högberg. 2014. Communicating with Future Generations: What Are the Benefits of Preserving for Future Generations? Nuclear Power and Beyond. *The European Journal of Post-Classical Archaeologies* 4: 343–58.

IAEA. 2019. *Operational & Long-Term Shutdown Reactors; Under Construction Reactors; Permanent Shutdown Reactors*. https://pris.iaea.org/PRIS/WorldStatistics/OperationalReactorsByCountry.aspx.

Jönsson, L.-E. and B. Svensson (eds). 2005. I *industrisamhällets slagskugga: Om problematiska kulturarv*. Stockholm: Carlsson.

Jørgensen, D. 2015. Rethinking Rewilding. *Geoforum* 65: 482–88.

Jørgensen, D. 2017. Competing Ideas of 'Natural' in a Dam Removal Controversy. *Water Alternatives* 10(3): 840–52.

Kowarik, Ingo. 2005. Wild Urban Woodlands: Towards a Conceptual Framework. In *Wild Urban Woodlands: New Perspectives for Urban Forestry*, edited by I. Kowarik and S. Körner, 1–32. Berlin and Heidelberg: Springer-Verlag Berlin Heidelberg.

Krohn Andersson, F. 2012. *Kärnkraftverkets poetik: Begreppsliggöranden av svenska kärnkraftverk 1965–1973*. Stockholm: Stockholms universitet.

Laraia, M. 2018. *Nuclear Decommissioning: Its History, Development, and Current Status*. Cham: Springer International Publishing.

Logan, W. S. and K. Reeves (eds). 2009. *Places of Pain and Shame: Dealing with "Difficult Heritage"*. Abingdon: Routledge.

Macdonald, S. 2009. *Difficult Heritage: Negotiating the Nazi Past in Nuremberg and Beyond*. London: Routledge.

Macdonald, S. 2015. Is 'Difficult Heritage' Still 'Difficult'? *Museum International* 67(1–4): 6–22.

Masco, J. 2006. *The Nuclear Borderlands: The Manhattan Project in post-Cold War New Mexico.* Princeton, NJ: Princeton University Press.

NDA. 2007. *Calder Hall Nuclear Power Station Feasibility Study 2007.* Nuclear Decommissioning Authority. https://tools.nda.gov.uk/publication/nda-calder-hall-nuclear-power-station-feasibility-study-2007/

OECD/NEA. 2015. *Radioactive Waste Management and Constructing Memory for Future Generations: Proceedings of the International Conference and Debate, 15–17 September 2014 Verdun, France.* Paris: Radioactive Waste Management, OECD Publishing. https://doi.org/10.1787/9789264249868-en.

Pratt, M. L. 2017. Coda: Concept and Chronotope. In *Arts of Living on a Damaged Planet,* edited by A. L. Tsing, N. Bubandt, E. Gan and H. A. Swanson, G169–G74. Minneapolis, MN: University of Minnesota Press.

Ramaniuk, D., A. Kliashchuk and I. Byshniou. 2006. *Charnobyl'-Chernobyl-Tschernobyl.* Minsk: Dzianis Ramaniuk.

Schama, S. 1995. *Landscape and Memory.* London: HarperCollins.

Šliavaitė, K. 2005. *From Pioneers to Target Group: Social Change, Ethnicity and Memory in a Lithuanian Nuclear Power Plant Community.* Lund: Lund University.

Storm, A. 2014. *Post-Industrial Landscape Scars.* New York: Palgrave Macmillan.

Storm, A. 2018. Atomic Fish: Sublime and Non-Sublime Nuclear Nature Imaginaries. In *Azimuth: Technology and the Sublime,* edited by G. Rispoli and C. Rosol, VI/12, 59–75.

Storm, A., forthcoming. Scars: Living with Ambiguous Pasts. In *Heritage Ecologies,* edited by Þ. Pétursdóttir and T. Rinke Bangstad. Archaeological Orientations Series. London: Routledge.

Storm, A., F. Krohn Andersson and R. Rindzevičiūtė. 2019. Urban Nuclear Reactors and the Security Theatre: The Making of Atomic Heritage in Chicago, Moscow and Stockholm. In *Urban Heritage: Agents, Access, and Securitization,* edited by H. Oevermann and E. Gantner. New York: Routledge.

Storm, A. and T. Kasperski, forthcoming. Social Contracts of the Mono-Industrial Town: A Proposed Typology of a Historic Phenomenon and Contemporary Challenge. *Industrial Archaeology.*

Svenning, J.-C. 2017. Future Megafaunas: A Historical Perspective on the Scope for a Wilder Anthropocene. In *Arts of Living on a Damaged Planet,* edited by A. L. Tsing, E. Gan and N. Bubandt, G67–G86. Minneapolis: University of Minnesota Press.

Swanson, H., A. Tsing, N. Bubandt and E. Gan. 2017. Monsters. Introduction: Bodies Tumbled into Bodies. In *Arts of Living on a Damaged Planet,* edited by

A. L. Tsing, E. Gan and N. Bubandt, M1–M12. Minneapolis: University of Minnesota Press.

Tafvelin Heldner, M., E. Dahlström Rittsél and P. Lundgren. 2008. *Ågesta: Kärnkraft som kulturarv*. Stockholm: Tekniska museet, Stockholms läns museum, Länsstyrelsen i Stockholms län.

Tunbridge, E. and G. Ashworth. 1996. *Dissonant Heritage: The Management of the Past as a Resource in Conflict*. Chichester: Wiley.

Wendland, A. V. 2015. Inventing the Atomograd: Nuclear Urbanism as a Way of Life in Eastern Europe Before and After Chernobyl. In *The Impact of Disaster: Social and Cultural Approaches to Fukushima and Chernobyl*, edited by T. M. Bohn, T. Feldhoff, L. Gebhardt and A. Graf, 261–87. Berlin: EB Verlag.

Yankee Rowe. 2018. *Demolition Video*. http://www.yankeerowe.com/decommissioning_dismantle.html.

Coda

Chapter 13

The Future is Already Deterritorialized

CLAIRE COLEBROOK

The future is already deterritorialized

The future is already deterritorialized. The present is not simply oriented to its own ongoing existence, but has already been captivated by a future in which humanity ceases to be. One element of the living system that is the Earth has come to regard itself as the point of view from which all futurity ought to be imagined, and this captivation of the future by Anthropos becomes all the more intense as the possibility of various existential threats are played out in the Anthropocenic imaginary. Although there have been many significant and important objections to referring to the Anthropocene (rather than the Capitalocene [Moore 2015] or White Supremocene [Mirzoeff 2018]) it is important to keep Anthropos as a term of consideration, especially when one recognizes the acutely deterritorializing force of Anthropos *and* the intimate relation between the deterritorialized future and existential threats. (I will argue later that Anthropos has *always* been defined by existential threats, that he is set apart from all life in the world by his existential fragility, by the always-present possibility that he may not be. This is a specific mode of deterritorialization where what had formed as a species – homo sapiens in relation to various forces of its milieu – becomes something different by now orienting itself towards its threatened non-existence, and then allowing that idea of non-being to dominate all other relations, especially to those who do not count as Anthropos, do not feel this existential fragility). By repeatedly imagining the destruction of humanity from elsewhere – viral pandemics, zombie or alien invasions, runaway technology, systemic collapse, resource depletion and a supposedly accidental climate catastrophe – the future becomes nothing more than the vanquishing of existential threats.

What the twenty-first century has witnessed is not only the local use of possible catastrophe to contain the present, such that various states of emergency are declared in order to focus policy, law, resources and attention on security concerns; the Anthropocene has produced, or expressed, a species emergency. By imagining humanity's non-being, Anthropos is constituted as that which must be saved, and as that which must destroy all threats. Far from this horizon of non-being chastening the present it has, instead, led to an Anthropocenic imaginary, where the future is oriented to nothing more than the destruction of existential threats. Our current Anthropocenic and post-apocalyptic horizon has been captured by one thing alone: the possibility that we might cease to exist, and that the horizon of our existence must be focused on the destruction of existential threats. We will survive because we must destroy that which threatens our being. Rather than an imaginary that is oriented to the territory, the relations of forces that form us, and rather than be oriented to a higher deterritorialization – relations that cut into the territory and take us elsewhere – we know only Anthropos, and know him only as a being whose existential precarity demands an extensive destruction of all that threatens us.

Of course, the present *does* seem to be one where the species and planet are threatened, and it is ethically imperative not to be glib about the ways in which climate change threatens those who simply cannot afford to be deniers. But this general milieu poses a problem: how does one negotiate the genuine brutality of the present without buying into a fantasy of apocalypse that fetishizes the redemption of humanity by way of a violent destruction of all that renders humanity fragile? How to take the real threat of climate change and systemic collapse seriously without buoying up disaster capitalism and a sense of a global state of emergency that puts all minor sufferings on hold? There is, after all, something worse than the end of the world, and that is the intensification of suffering and barbarism while what calls itself humanity ekes out its last days in existential panic.

It is with this negotiation of the genuinely fragile future that I would mark a distinction between two modes of deterritorialization. The first would be the deterritorialization that enables and characterizes the Anthropocene: humanity emerges as an effect of a series of relations (including colonization), with the future ultimately becoming nothing more than the project of saving humanity from all that threatens their emergent complexity. What forms itself as Anthropos, does not only take itself as the norm and proper end of the whole, but also constantly regards itself as threatened by the present; the future is not the future of the species so much as the flourishing of humanity having fully conquered all that renders it precarious. The second mode would be a decolonized deterritorialization; here the present is captivated, rather than captured, by a quality that creates a cosmic or counter-anthropic stratum. As an example of the first mode of deterritorialization one might think of any number of post-apocalyptic cinematic epics where some threat to humanity – viral, technological, extra-terrestrial, ecological – must be destroyed. The future is given *only* in vanquishing that threat. (We talk of fighting or combating climate change; we talk of working against resource depletion; we predict water wars; we discuss various ways of averting catastrophe: the future is one of being at war with all that renders us fragile.) The second mode of decolonizing deterritorialization would release qualities from Anthropos, no longer defined in terms of survival of the fragile: how might we think of rivers, non-human organisms, sounds and colours if they were not determined in advance from the point of view of Anthropocene humans? The threshold between these two modes of decolonization is difficult if not impossible to fully determine. In Deleuze and Guattari's work, for example, the modes of modernist art that release colours and sounds from everyday human purposiveness at once promise some sense of what is beyond the human at the same time as they strengthen a sense of the modernist subject of artistic sensibility as uniquely threatened and (therefore) worthy.

Cato the Elder was known for ending missives with the imperative that Carthage must be destroyed (Kiernan 2004). I raise this anecdote for two reasons: I have also become prone to a compulsive repetition in every argument I make, referring with horror to Oxford University's Future of Humanity Institute.[1] But just as I find myself akin to Cato the Elder in repeatedly looking with horror at an adversary, the greater similarity is between Cato the Elder and the Institute itself, both intensely focused on destroying existential threats. There is also a global-political resonance between Rome's then civic humanism and Oxford Institute's capacity to see humanity as it happens to be as an impediment to humanity as it ought to be. To destroy Carthage, for Rome, was the only way for a fraction of the species that saw itself as 'humanitas' – what humans ought to be – to become humanity in general. In his 'Letter on Humanism' Heidegger saw the Roman mission of 'humanitas', the cultivation of good citizens through logic, as the loss of the world. *Legien* or speaking about the world became a system unto itself and established the logic of the humanities as that which could and should purvey the world from its lofty but fragile height (Heidegger 1998: 244). What ties that moment of the Roman empire to the present is the possibility of discerning an existential threat, of defining oneself as Anthropos against the barbarians at the gate. This is a specific type of deterritorialization that has its own cogito: there is a threat to my being in the future that must not be, and therefore I am. Deterritorialization takes a territory, a formed set of relations, and generates a new stratum: the 'man' of Western thought and empire increasingly becomes the ordering term of the whole, especially when that man of history is threatened. Dipesh Chakrabarty's step away from postcolonial dispersed peoples towards the now universal humanity threatened by climate change typifies this deterritorialization (Chakrabarty 2009). In the face of non-being we all become members of a humanity whose future must now be focused on life as such. Existential threats are not diminutions of one's being – not catastrophes where there is a massive loss of life – but an event where *who we are* is at risk. For Oxford's Future of Humanity Institute's director Nick Bostrom there is a folkish and fallacious tendency to want to minimize the number of deaths or losses to the species; however, what *should* concern us, he insists, is not how many lives we lose but threats posed to intelligent life as such. It is not rational, according to Bostrom, to focus all or most of our energy on preventing events like the Holocaust or other such large-scale disasters (Bostrom 2013). These occurrences may feel catastrophic but humanity as a whole does recover. One could not say the same if *all* intelligence were annihilated. The difference between a massive loss of life and the loss of a definitive and properly human quality of life makes all the difference; for Bostrom it is *the* difference that ought to guide our relation to the future. The loss of many lives is lamentable, but ultimately humanity as such recovers; the loss of intelligence and the thwarting of technological maturity are truly catastrophic. What is at stake is far more important than reducing suffering, for it is the possibility of *non-being* that really ought to concern us. By imagining the non-existence of a certain type of humanity as utter existential loss, we not only fail to consider different forms of human life, but also fetishize a certain conception of life marked off from non-life, a threshold that (according to Elizabeth Povinelli) characterizes late liberalism (Povinelli 2016).

The concept of existential risk does a great deal of normative, practical and imaginative work. The possibility of the non-being of a fraction of the whole trumps the importance of reducing the barbarism of actuality. '*Existential* risk' marks out a difference between life as such and some quality (such as intelligence) that is the proper potential of life. A fragment justifies and orients the future of the whole. This detached fragment then generates a distinct temporality and modality of the imagination; it is *not* a question of how we ought to live but rather a question of maximizing and intensifying the supposed best of what we already are. 'Intelligence' comes to stand for this definitively human and existential value, extending – without any question – a current techno-scientific trajectory. Where 'humanity' is now the outcome of the development of a certain trait, the future ought to be the maximization of this potentiality. The Future of Humanity Institute's attention to existential risk is not a minor research endeavour but typical of a type of deterritorialization that defines humanity of the Anthropocene and its future. It is oddly consonant with post-apocalyptic culture, where the imminent thought of the non-being of 'humanity' (always represented as Western, urban and predominantly white) generates an imperative and right to life. How many post-apocalyptic end-of-world epics depict a threatened or depleted humanity triumphing over a foe that is humanity in its lesser form? The 'Anthropos' of the Anthropocene is just this self-elevating, threatened and deterritorialized fragment.

Humans as a species branch out into different spatial-temporal directions (the sense of the unity of the species being the effect of a series of movements and relations); what comes to be formed as *homo sapiens* is the effect of centuries of migration and encounter. We might refer to this as territorialization, a relative stability attained through patterns of mobility. One of those lines increasingly develops a certain capacity to change the environment, eventually creating a new stratum: one might think of the imperialist conceptions that one is not simply a being occupying the here and now but is exemplary of humanity in general. This is what Deleuze and Guattari refer to as deterritorialization, where a potentiality of a body takes on a new time and space. One of their best examples refers to the hand as a deterritorialized paw; the body's grasping, clawing, fighting, touching movement starts to become gestural, and then a new layer of sense and a new temporality of inscription are generated (Deleuze and Guattari 1987: 61).

The Anthropocene is also an event of deterritorialization; the global diversity of humans as a species is overtaken by the way in which one species both alters the planet as a living system, *and* generates a deterritorialized future. The future ceases to be multiple and divergent, with 'Anthropos' becoming definitive of a new stratum. This stratum is at once geological (in the narrow sense of the Anthropocene) but also a stratum in Deleuze and Guattari's sense. A whole new mode of relations is formed when the 'humanity' that had defined itself through a specific mode of civilized time comes to stand for that which is threatened, and therefore that which must be saved. The possible non-being of a quality ('humanity') generates the imperative that all must be done to save just that predicate. This event of deterritorialization might seem at first to be a minor scholarly point or even a primarily geological point but it is far more significant than that: the very notion of an existential threat relies on the production of a virtual humanity. A decade of Anthropocene mourning and panic has shown us that it is neither the loss of life as such *nor suffering* that ought to be our concern. Instead, the possible loss of a quality – intelligent life – somehow allows one to define that quality alone as that which gives the species value. If we were to be honest we would admit that human history has been nothing more than loss, suffering and extinction: if one were to weigh suffering and barbarism against what has called itself 'progress' the past would look at least as bleak as the supposed future dystopias we are beginning to imagine. Bostrom, following Parfit, is quite right: it does seem indeed that what is catastrophic for us is the loss of 'intelligence' and not the vast panorama of pain that lies behind us. This is how we should think about 'Anthropos': a capacity – thinking, intelligence, 'civilization' – that outweighs and orients every other value.

Anthropos

Much has already been written about the 'Anthropos' of the Anthropocene, and even more has been written about the 'human' and its aftermath in posthumanism; the various senses and uses are too diverse to enable a single coherent critique. Nevertheless it is just this messy dispersal that forces one to recognize that both 'Anthropos' and 'human' are deterritorializations: more specifically, both are formed through the thought of their non-being, and both move from the possibility of non-being to a virtual form that displaces the actuality of suffering. I have already suggested that a common contemporary manifestation of this deterritorialization occurs in post-apocalyptic disaster cinema: the world of white Western affluence is on the brink of becoming the type of space that the West had always outsourced to its third world, and this very possibility generates an imperative for a truly human and triumphant future. The fact that 'man' is always so threatened with his demise has allowed futurity in the West to focus on the annihilation of risks. In its earliest stages in Plato, it is the horror of *not thinking*, of being enslaved to the shadows and not turning towards the light of reason that defines what we must be. The fact that being enslaved is the founding metaphor of Western politics evidences the extent to which Anthropos is in constant battle with an always threatened non-being. The same applies today when thinkers such as Alain Badiou or Bernard Stiegler warn that humanity is falling back into non-being, seduced by its lesser and enslaved modes. Badiou's *Ethics* sets itself the task of railing against the "violently reactionary" Western order that has become increasingly entrenched, primarily as a result of lazy thinking, a failure to confront the abyss of truth events (Badiou 2001: 5). Stiegler also (but in different ways) is indebted to a Platonism that resists the lure of banality. His pharmacology captures the deterritorialization of the Anthropos and the Anthropocene: we are always at war with stupidity and entropy, have always had to work against falling back into the shadows of non-being that are made possible by the very systems of writing that elevate us. Following Adorno and

Horkheimer and the tendency of enlightenment to enchain itself in ratio-nalization, Stiegler argues:

> Stupidity is *never* foreign to knowledge: knowledge can itself become stupidity par excellence, so to speak. And this is so because knowledge, and in particular theoretical knowledge as passage to the act of reason – or more broadly, *noēsis* – can occur only *intermittently* to a noetic soul that is constantly regressing, and that, as such, is like Sisyphus, perpetually ascending the slope of its own stupidity, given that, as stated by Simonides and cited by Aristotle, "God alone enjoys this privilege," that is, the privilege of being always in actuality, of never being stupid, of never going down the path of disindi-viduation, reification and proletarianization.
>
> This is why not only can knowledge make thought base, but it is essentially a matter of thought's own baseness – ever threatening, ever the threat (Stiegler 2015: 45–6, origi-nal emphasis).

'Humanity' and 'Anthropos' are constantly defined by warding off existential threats: not minor damages, and not suffering. As the Future of Humanity Institute makes clear, one of the main existential threats is failure to reach technological maturity, the failure for *thinking* to maxi-mize itself. Here, again, Bostrom warns against us taking a narrow point of view that would focus on immediate welfare; one needs to consider technology rationally and globally. This is Anthropos in full flight, the diminution of minor harms for the sake of the big *future* picture: "Even from a so-called 'person-affecting' moral perspective, therefore, when assessing whether a flawed realization has occurred, one should focus not on how much value is created just after the attainment of technological maturity but on whether the conditions created are such as to give a good prospect of realizing a large integral of value over the remainder of the universe's lifetime" (Bostrom 2013: 23). It is within this range of consid-eration that Bostrom considers 'humanity's' future path of colonization:

> mature technology would enable a far more efficient use of basic natural resources (such as matter, energy, space, time and negentropy) for the creation of value than is possible

with less advanced technology. And mature technology would allow the harvesting (through space colonization) of far more of these resources than is possible with technology whose reach is limited to Earth and its immediate neighbourhood (Bostrom 2013: 20).

Colonization

Deterritorialization in general accounts for the way in which complex systems generate points of stability that can, in turn, generate new forms of relation. Colonization is a quite specific form of deterritorialization that has always been bound up with a peculiar sense of existential threat. If one thinks of territorialization as the movements that form bodies and space in tandem, one might think of deterritorialization as a movement across that territory that generates a stratum that reorients the whole. One might think here of the way in which white settlers saw the space and movements of Indigenous bodies as non-human, and responded with a hegemonic sense of a virtual humanity that would ideally subsume all bodies. When white Europeans invaded Australia and encountered the Indigenous mode of occupying space that was not their own, they imagined a future in which these seemingly inhuman bodies would no longer threaten their existence. Centuries of genocide, assimilation and displacement were accompanied by an intensified sense of *the human*, an 'Anthropos' that formed itself only by destroying everything that might count as an existential threat. The settler's gaze that demands that 'we' must not amount to this, that 'we' are a humanity of technological maturity, is what has underpinned the Anthropos of the Anthropocene, where 'our' future must not be that of the nomadic, non-state, non-nature-mastering peoples 'we' have destroyed. By imagining the non-being of man – the threat posed to Anthropos – something like the 'future of humanity' is formed and secured. If there is something that counts as an existential threat – technological immaturity, stupidity, loss of intelligence – then a future becomes possible only by way of its destruction. A territory is no longer formed by bodies assembled across a space, but is oriented towards a future where what counts as properly human is maximized to the point where it overtakes the whole.

When Rome set out to destroy Carthage it was not concerned with the threat of a trading partner, an unmanageable province or a power that perhaps might become too strong in allying itself with Numibia. One way to explain Rome's destruction of Carthage is to see it as an example of a deterritorialized future: the potentialities of the globe were no longer played out in a relation among competing powers, but instead overtaken by a sense of *the citizen in general*. What must not be allowed is a future that exists beyond 'Anthropos', beyond what Rome would define as 'humanitas', and what the twenty-first century would elevate in the Anthropocene.

If we think of territorialization as the coming into being of relatively stable bodies by way of the formation of relations, then we can think of deterritorialization as the formation of a different mode or style of relations. In this respect capitalism is at once the most exemplary event of deterritorialization, where relations among bodies and forces become liberated from any intrinsic potentiality, while warding off absolute deterritorialization. What needs to be noted is the specific production of the Earth in the movement of state and then capitalist deterritorialization. The monotheism eventually becomes the single 'man' of subjectivity, the Anthropos who must always shore himself up against the supposedly destructive forces of chaos:

> And that is what is concealed in the two acts of the State: the residence or territoriality of the State inaugurates the great movement of deterritorialization that subordinates all the primitive filiations to the despotic machine (the agrarian problem); the abolition of debts or their accountable transformation initiates the duty of an interminable service to the State that subordinates all the primitive alliances to itself (the problem of debts). The infinite creditor and infinite credit have replaced the blocks of mobile and finite debts. There is always a monotheism on the horizon of despotism: the debt becomes a debt of existence, a debt of the existence of the subjects themselves (Deleuze and Guattari 1987: 215).

Without getting too immersed or bogged down in Deleuze and Guattari's philosophy, I do want to draw attention to the difference between absolute and higher deterritorialization in order to move on to the necessary problem of decolonization. If territorialization amounts to the formation of a space and bodies through the assembling of qualities, and deterritorialization generates a new stratum absolute deterritorialization would be what, for capitalism, *must be destroyed*. The release of qualities and forces from relatively stabilized and relational systems would amount to the end of the world and perhaps the end of embodied and organized/organic life. Deleuze and Guattari refer to *higher deterritorialization* as something different again: if, for example, a bird comes into being by marking out a space of coloured leaves, and producing a refrain that gives this assemblage a marker, and if various human art practices seize upon these colours and rhythms to produce what came to be known as art, then higher deterritorialization would occur when the qualities from which the world was composed could be imagined or intuited *as if for all time and beyond this world*. This conception of higher deterritorialization bears an ambivalent and problematic relation to colonization and futurity. As I have already suggested, white settler culture deterritorializes by way of imagining the future as what must be secured in the face of existential threats. Can one imagine a deterritorialization that does not amount to the dissolution of all lines of assemblage, and that does not rely on the fetishized elevation of the 'Anthropos'? Is it possible to imagine an end of Anthropos that opens out to future worlds?

What this amounts to, or ought to amount to, is a negotiation with the problem of decolonization, especially with the problem of decolonization as a metaphor. If colonization is – insofar as one can think literally – the taking over of a place by a group of bodies from elsewhere, then decolonization would amount to a form of exit. This might be achieved either by imagining the dissolution of what exists, *or* it might more fruitfully be approached as abandoning the deterritorialization of the future that proceeds by way of existential threats. Might there be something other than an Anthropos that has always formed itself by warding off its increasingly fetishized non-being? That is, does deterritorialization have to take the form of Anthropos who can only imagine his continuation as an ongoing war against his dissolution?

It is deterritorialization that enables the Anthropocene, both geologically and conceptually; a potentiality of the species reaches such an intensity that it generates a whole new scale and range of relations. A part overtakes the whole; humanity, man or Anthropos comes to appear as the ground and organizing whole. Intensive agriculture and technoscience gradually create forms of private, high-consumption and entitled individualism that increasingly allow for no future other than that of man as universal, world-transforming subject. There is, then, an intimate relation between the formation of the world as an Anthropos scene, and a future defined by non-being. The more 'the human' understands itself as a fragile emergence capable both of transforming the world but *also* capable of falling back into stupidity, the more the future becomes nothing more than saving who we are. How is it that what comes to understand itself as 'human' as 'Anthropos' is so easily *not* itself, so intimately related to its non-being and disappearance? To be properly human is to be at risk, to be threatened with falling back into being mere life, a part of the world, becoming nothing more than a body without a sense of the globe, man or Anthropos. From Heidegger's conception of *das Fall*, through to Bernard Stiegler's worry that 'we' are becoming stupid by *not* establishing long circuits of attention, to every post-apocalyptic epic that displays a global humanity pushing itself to the brink of annihilation only to save itself: all testify to a humanity that forms itself as that which can always, so easily, be lost. When Heidegger described the shift from a relational *legein* that disclosed the world to a 'logic' that reduced the world and thought to so much standing reserve he was not simply being nostalgic; he was also anticipating Deleuze and Guattari's concept of deterritorialization. An assemblage that brought terms into relation would be subjected to a single logic; a humanity that opened itself and its world through expansion would become subjected to the very complexities it generated. Heidegger's turning back to an original revealing along with his sense of the parasitism, decay and loss that would ensue if the inauthentic potentialities of existence were not destroyed dovetailed tragically with fascism. One might see Heidegger's manoeuvre as a reterritorialization; if *Dasein* is the experience of the world as nothing more than that which is unfolded or cleared by way of time, and if *das Man* is what happens when that event of revealing is taken to be a real thing, then an appeal *back* to thought's origin in order to save the future is a reterritorialization.

To insist that the future be that of Anthropos redeemed and regained is to take the event of deterritorialization – technical 'man' – and posit it as the only future possible, thereby disabling and destroying other futures. Heidegger's all too obvious and predictable fascism ought to render us wary of all those projects – from Rome's destruction of Carthage to the Future of Humanity Institute's fear of not reaching technological maturity – that use an event of deterritorialization to colonize the future. Heidegger advocated returning to the emerging sense of logos, or the 'speaking about' (*legein*) that would draw us closer to the Earth and thereby avoid the utter ungrounding or 'uprooting' of systems that were associated with a fall into detached technologies, the worst of which would be banking/money and idle speech. It was as though for Heidegger the emerging difference of language's coming into being might be regained, restored and renewed. This should at once alert us to the intrinsic possibility of reterritorialization that haunts every event of deterritorialization. 'Humanity' is not the ground or cause of history but is, rather, an effect of historical forces – a deterritorialization; when the future becomes determined as a project of destroying everything that threatens humanity we have reached an intensification that amounts to a reterritorialization. Nothing other than Anthropos is deemed worthy of ongoing life.

There is, however, another possibility that would follow from thinking territorialization not as the original movement of migration that must take over the Earth towards ever greater maturity, but as a staying in place that embraces a certain immaturity. Is territorialization, the creation of spaces through the movement of bodies, the most rigorous way to think about the history of Anthropos and our already territorialized future? Do we need to accept the now common assumption that the Anthropocene has created a tragedy of the commons with nowhere else to go, and no future other than the one we now make and remake for ourselves? Or, do we question the forward path of deterritorialization, where the only future is intensification of the present, and instead open up the radical exit and separation of decolonization? Two paths (at least) open up: technological maturity and the fulfilment of Anthropos against all that threatens his full potentiality, *or* an exit from the war on existential threats, and taking seriously what Deleuze and Guattari referred to as nomadism. This approach would not accept the valorization of emergence, would not assume the prima facie value of a complexity supposedly generated from a less worthy chaos. This, I think, would amount to genuine decolonization – *not* privileging the trajectory of humanity to date, not assuming that enlightenment might still happen someday if we keep on warring with enlightenment's lesser tendencies. Perhaps life without the State would not be the end of the world.

The decolonized future

One way of thinking about the future would be to privilege the ways in which life brings place into being, and this might seem to be especially pertinent today when dreams of globalism and the expansion of liberal affluence have encountered resource depletion and climate change and, in turn, have seemingly generated new forms of enclosed populism and nationalism. One might say that the effect of more and more bodies traversing space has both created an empire of 'man', and then a myth of nationhood and civic birth right. In the wake of putting America first, and making America great again, it might seem necessary to point out that America itself was the result of migration. Any territory, one might want to argue, is already the result of deterritorialization; it is not that there are owned spaces and identified bodies that are then disturbed by movement, but that it is from movement that spaces and ownness come into being. All territories are in part already deterritorialized, disturbed by movements of departure and flight. This would be true not only of the white settlers and slave traders who saw America as a space of freedom, and not only for the subsequent waves of migration, but also for the Indigenous persons whose mode of existence was bound up with the spaces traversed and the forces those journeys brought into being. Even if one thinks of the coming into being of a people and a space as territorialization, this originating move cannot be decoupled from deterritorialization: one becomes who one is not simply by traversing a space but by coupling those formed relations with other migratory patterns. A tribe takes on a relation to a formation of rivers; a city invents itself as the off-spring of the gods; or a nation imagines itself as expressive of humanity in its most liberated or rational form. In all cases a territory, or the formation of bodies through a traversal of space, opens out to another plane that it takes as its ground. But this positing of a transcendental modality is problematic for two reasons. Is it possible or desirable in an era of neoliberalism to think only in terms of movements, forces, and flight? Deleuze and Guattari have persuasively argued for capitalism as deterritorialization; it is perhaps unfair and inaccurate to see their work as buying into a neoliberalism of ongoing self-creation, and yet it is necessary to see the ways in which their work broached the difficult thresholds of deterritorialization. If everyone is ultimately a migrant or nomad, with no essential relation to space, how does one negotiate the violence of displacement and the claims for native title that challenge liberalism's long history of the world as so much appropriable free space? It is telling that one of the common liberal challenges to resurgent white nationalism and anti-immigration rhetoric is that countries such as America and Australia were built on immigration, an appeal to tolerance and openness that fails to register the enslavement and displacement that underpinned these supposed lands of the free.

What might it mean to think of the traversal of space *not* solely as an original territorialization but as a transcendental colonization? This second approach would engender a less flat conception of territorialization. This possibility is already implicit in Deleuze and Guattari's account of space, life and *de*territorialization: how do certain elements in a field of relations create a new plane of relations that captures the first? What one thinks of as colonization in the quite literal and extreme sense – the history of white settlement and displacement – has a prehistory in all those moments of historical capture and empire that subject space to a body that leaps outside the terrain and takes itself as the origin of the whole. Slavery and colonization do not begin with the European conquering of the globe. One has to go back a long way, *though it is possible*, to think of genuinely nomadic groupings of bodies that have not yet been subjected to the State form, have not yet allowed somebody to stand in for the sense of the whole. This potentiality of stateless nomadism is not, and should not be a metaphor; it is important to think just how precarious and fleeting life without state subjection has been, while also insisting that it nevertheless opens the thought of a higher deterritorialization beyond capitalism. One might think here of James Scott's *Against the*

Grain where the emergence of state forms is *not* an unquestioned flourishing of maturity, or many of the recent articulations of Indigenous modes of existence where being a self is *not* defined by a sense of a universal Anthropos, but instead by one's relation with multiple and inhuman others (Scott 2017; Viveiros de Castro 2015). John Protevi's recent *Edges of the State* (2019) identifies moments where states breakdown and 'nonstate peoples' offer other modes of existence; David Wengrow and David Graeber write about non-state forms of existence that bear a different mode of complexity and dynamism that is at odds with the conception that the present and future are, and should be, thought of as mature versions of humanity's childhood (Wengrow and Graeber 2015).

Deterritorialization is always an ex post facto futural form. A potentiality appears to be the ground and origin of the whole, and thereby dictates the proper form of the future. One might think of this in many ways, one of which would be Lee Edelman's concept of reproductive humanism, where the child stands for an innocent promise of humanity to come (Edelman 2004). When this is bound up with environmental imperatives the future becomes that which must be saved for our children, the future thereby becoming the fulfilment of a past that never was (Sheldon 2016). This structure is evident both in post-apocalyptic culture and Anthropocene studies. Actual humanity may have come near to destroying itself and the planet, but the very feature that made humanity possible – techno-scientific indomitability – is what will save the future. This sense that who we *really are* might save us is given both in the forms of geo-constructivism criticized by Frederic Neyrat, who ties both environmental and managerial forms of saving the Earth to a common notion of the Earth as object of our making (Neyrat 2019), and in fantasies that the colonizing spirit that formed America will also generate our cosmic future. The future is at once deterritorialized by no longer being bound to humanity as it happens to be, and reterritorialized by being the fulfilment of what we ought to be. Think of post-apocalyptic epics, such as Christopher Nolan's 2014 *Interstellar* where a rapacious corporate capitalism has not only destroyed the planet, but has limited the future to be nothing more than mere living on. The future is, however, saved (by a former astronaut) for the sake of the yet-to-be-corrupted child, who promises humanity as it ought to be. Similarly, the Future of Humanity

Institute argues for 'technological maturity' – not the survival of who we are, nor some conception of humanity that might exist otherwise, but the elimination of existential threats that would hamper the flourishing of who we imagine we ought to be.

Deterritorialization and territorialization

Without even turning to the work of Deleuze and Guattari, the proper names attached most often to conceptions of deterritorialization, it is possible to think of two possible ways in which an ethics of territorialization (and deterritorialization) might be posed. The first would be to think of territorialization as a form of unnatural or elicit takeover, the occupation of a space or the organization of a space by a life or force not proper to the space. Territorialization might be something akin to colonization, which not only installs an order but *identifies* each element that falls subject to such ordering. If this were so then deterritorialization would be something of a liberating movement, allowing forces to be released from their capture. This conception of territorialization would need to assume that there are natural bodies, whose proper nature precedes order, and that there are spaces that exist prior to transit. Perhaps the story of the first migration from Africa would enable this understanding, of an original home and dwelling *from which* everything else is theft and impropriety.

Two problems present themselves here: do we really want to grant all migrations the same status? Do we really want to say that we are all migrants really, thereby downplaying the violence of colonization? Second: *all* bodies – human, animal, political, planetary – merge from movement and space. Even those first humans emerged from movement and the traversal of space, evolving in relation to a milieu that was also composed from movements and encounters among forces. Territorialization and deterritorialization would not be before and after terms, and not amount to a binary of good and evil, but would always be in a contested relation to each, with territorialization referring to forces entering into relation, and deterritorialization referring to elements that branch off to form new strata. This is the key significance of thinking about life *not* from the point of view of bodies but from the point of forces. What this would amount to is an abandonment of a proper space and embodiment that gets overtaken by an external power, and instead a potentiality in any body for a 'line of flight', where a potentiality generates a new plane and strata of relations.

Territorialization and deterritorialization are dynamic and relational concepts, generating a different way of thinking about politics, ethics and futurity. All futures and all spaces are, *to some extent*, always already deterritorialized; the human bodies that form a tribe, by marking out a space, and taking on the sounds, figures and movements of their milieu are already effects of an entering into relation of pre-human forces. When a tribe is then subjected to the power of the chieftain, and the Earth appears as a divine ground from which authority emerges, a new event of deterritorialization opens a new system of relations, and a new future. As Deleuze and Guattari trace the history of bodies and state forms in *Anti-Oedipus* and *A Thousand Plateaus*, it is *deterritorialization* that stratifies and allows bodies to appear to be indebted to a higher body, ranging from the chieftain, to the despot, to 'the subject'. From a territory, a relation among forces that produces an assemblage of bodies, one moves to what appears to be a space, polity, life or law that offers the ground and reason of any part of the whole:

> It is overcoding that impoverishes the Earth for the benefit of the deterritorialized full body, and that on this full body renders the movement of debt infinite. It is a measure of Nietzsche's force to have stressed the importance of such a movement that begins with the founders of States, these artists with a look of bronze, creating 'an oppressive and remorseless machine,' erecting before any perspective of liberation an ironclad impossibility (Deleuze and Guattari 1983: 199).

How, then, might one negotiate a politics in this dynamic movement of territorialization and deterritorialization? One path that is not viable is to see territorialization as bad and deterritorialization as good (or vice versa). Moving beyond good and evil one might start intuitively: how do we balance the theoretical insight that there is no such thing as a proper or original body (with everything beginning as movement, migration and encounters among forces) with late capital's overtaking of all relations and spaces and subsuming them beneath the logic of exchange? I think the answer to this question is to add another concept into the mix: decolonization. The problem with colonization is, after all, neither of territorialization – the formation of a space – nor deterritorialization (the generation of a new stratum that captures the relations of the whole). Colonization operates not with the dynamism of territorialization but with the fixity of a single event of deterritorialization – the formation of the figure of man as Anthropos, and the reduction of all territories to variants of the human. Colonizers more often than not did not see themselves as thieves but as bearers of enlightenment, humanism and progress. Colonization deterritorializes – generating the global network of late capitalism – and then reterritorializes on the figure of Anthropos: 'we' are now all human, all facing the same precarious future. Precarity, here, is a loss of the conditions of *man:* the terror that we might fall back into the conditions that we surveyed in the 'third world' and recognized as immature and inhuman.

One might start to think of the difference between the postcolonial and decolonization in terms of deterritorialization. Once the colonizer has physically left the territory the colony can still be subjected to the figure of man, with the very terms of liberation, sovereignty, rights and ownership relying on the Anthropocenic logic that subjected the world to a single system of exchange. (The Anthropocenic differs from the anthropocentric in this respect: man is no longer at the centre of nature viewed as a natural resource, but is rather the purveyor of the Earth as a single living system with all life and humans now bound to the same precarious future.) Decolonization might, however, be thought of neither as a return to the territory, nor as another event of deterritorialization where some other stratification exists alongside Anthropos. As Anthropocene discourse has made clear, the future is deterritorialized, and there is no other horizon than that of precarious humanity and its seemingly prima facie right to life.

It might be better, then, to see territorialization as an immanent movement, as the coming into relation of forces *from which bodies emerge.* This would then mark a difference between territorialization and colonization: the former would be the co-formation of bodies and spaces, and this would explain why Deleuze and Guattari make so much of nomadism, or the traversal of a space that brings bodies and the relations of a territory into being. If territorialization is quasi-originary, not reducible to life itself but better thought of as the movements that allow something like 'the lived' to come into being, then *deterritorialization* would perhaps be closer to the potentiality for colonization that might nevertheless generate an *uncivil* future (a world in which the borders of the civic have not yet subjected terms to the sense of 'the human') (Muecke 2019). To think of territorialization *not* as the coming into relation of bodies, but as the formation of bodies through forces that take on a quality by way of their encounter allows us to think of a quite distinct relation among the terms colonization, decolonization, territorialization and deterritorialization.

Territorialization is not something that happens to bodies, but brings bodies into being. When Deleuze and Guattari write about the hand being the deterritorialized paw, they do so within a *geology* of morals, where layers and sedimentations allow for further formations, but also – as the current discourse on the Anthropocene has made evident – where later formations can create a new composition of the whole. For the most part humans have adapted to their climate, but their technological formations have altered the climate in turn, with the Anthropocene signalling a point where the Earth as a living system is altered by the technological changes that various ecosystems had made possible. To deterritorialize is not the dissolution of territorial relations; it is better thought of as the ongoing creation of further relations of possibility. The forces that allow relations to form relatively stable wholes may take on a life of their own and destabilize but also reconfigure the initial territory. Deterritorialization is double-edged, releasing the forces that formed territories into new relations, but also generating new stabilities.

What is the value of thinking about decolonization alongside deterritorialization? I want to suggest that the answer to this question resides in the difference between theories of territorialization and theories of emergence. If one follows theories of emergence, social wholes and organs are complex assemblages that are generated from less complex elements. If that is so then any event of decolonization would be the overthrow of one state form, leaving a vacuum or absence. Even in theories that are not strictly or explicitly tied to emergence, the assumption that wholes and bodies are *more* complex and differentiated than their elements characterizes a great deal of cultural theory. One might go back as far as psychoanalysis and its premise that without some subjection to the symbolic order one falls into psychosis, or liberal theory's minimal requirement that one imagine one's own decisions as universalizable for humanity in general, or Judith Butler's account of recognition, where some minimal performance of normativity is essential to being a subject (Butler 2010). In all these cases the assembled whole is more complex than whatever precedes, especially if the prior self is effected through subjection. One way to read Franz Fanon would be to see his work as a continuation of the psychoanalytic and existential assumption that who I am is effected through the assumption of the recognition of the other. Here his work would be in line with a familiar argument in postcolonial theory where the territory after colonization has no language or sense of self other than that of the colonizer. If one takes away the governing order one is left with an absence or void. However, there is another and more fruitful way to read Fanon, where the destruction and exit from what counts as man – and *not* admittance to the human – is the path to the decolonized future:

> The social revolution cannot draw its poetry from the past, but only from the future. It cannot begin with itself before it has stripped itself of all its superstitions concerning the past. Earlier revolutions relied on memories out of world history in order to drug themselves against their own content. In order to find their own content the revolutions of the nineteenth century have had to let the dead bury their dead. Before, the expression exceeded the content; now, the content exceeds the expression (Fanon 2008: 199).

The idea of humanity, or the sense of the social whole, does not elevate and render worthy and complex the bodies it assembles; it *reduces* the complex relations, forces and potentialities that make up any body. This is the doubled-edged sword of deterritorialization, as both creation and annihilation. One might think here of all the desires and interests that are created by the production of social strata; deterritorialization is neither the negation nor creation of complexity but the formation of different strata. The value of thinking in terms of strata or plateaus – of thinking rhizomatically – is that one can account both for relations of subjection and for spaces that exist alongside each other. One can think of the deterritorialization of 'man' as a subsuming stratification, allowing every territory to be seen as a variant of a single Anthropos. One can also think of what Deleuze and Guattari refer to as higher deterritorialization where a fragment is detached from the assembled strata of relations and intuited from the point of view of the cosmos (Deleuze and Guattari 1994: 197). This would grant us a fully decolonized and affirmative sense of the end of the world; detached from the formed world of sense, survival, man, history, technological maturity and the lived. One might see a predicate as out of time, as that from which formed worlds might be possible.

This conception of higher deterritorialization yielding an affirmative and decolonized end of the world is possible because the concept of territory is not grounded in a theory of emergence, but its opposite: it is not that complexity and formations come into being from simple units, but rather that the complexity and intensity of matters are reduced when matters are formed into relations. Light, for example, could be so much more than the colours perceived by the human eye; and the human eye could be more than the eye of the sensory-motor organism that perceives lines and colours in terms of its own world rather than as forces that open up the cosmos. To deterritorialize is therefore not a linear movement away from ordering, just as territorialization is not reducible to ordering; one might even see territorialization not as order from disorder, but the formation of relations from a chaos that has far more complexity than emergent and subsequent wholes.

I want to conclude by exploring two questions that are raised when one thinks about deterritorialization as a concept in Deleuze and Guattari's sense, as a way in which thinking orients and configures itself (with 'thinking' *not* being reducible to intellect). First: part of the motor of the concept is its attempt to think of political movements and bodies as contractions and stabilizations of forces that are more complex and differentiated than the strata that emerge. This raises the question of vitalism, and whether a theory of life generates an ethics. Does higher or absolute deterritorialization – a release of forces from any specific world – constitute an ethics that is beyond good and evil? One would have to think about forces *as such* and not insofar as they serve the history and world of Anthropos. This would generate an ethics at a cosmic level, a thought of what might be beyond the way the world happens to be for us. If this does count as a form of vitalism it is one in which actual life is itself a territorialization of the prevital and cosmic (Deleuze and Guattari 1987: 43). If this is so there seems to be an imperative to pursue ethical inquiry beyond the level of the actual, in order to think deterritorialization in terms of an inhuman and pre-vital potentiality. Does this amount to a naturalized or vital politics? Would we say, as Deleuze and Guattari seem to do, that embodied life has a prior condition of intensive complexity that offers the potential for a future beyond the lived body? If so, deterritorialization would be a movement from which organic life emerges that might nevertheless provide a way for life to move beyond itself. The *prevital* or inhuman conditions for life would be life's 'proper' end. This might at first appear to be a repetition of a Western and high modernist fetish for transcendental elevation, where one negates the world as it is for the sake of what might be thought. That is, indeed, one way in which deterritorialization might be negotiated, and this would lead to a second question: if a radical theory of deterriorialization takes us back to life, how do we think about life? Is life best thought of at the level of the organism, with Anthropos as the exemplary rational animal who is the steward of the planet and the future, *or* might life be thought beyond organic forms – the life that exceeds animality? Such a life has often been meditated upon by various Indigenous peoples oriented towards life's cosmic and planetary dimensions.

Rather than saving the world as it is by elevating a fragment (Anthropos who becomes the raison d'être of the world), one would exit from the current stratification. One might think of the ways in which various forms of Indigenous culture hint at this type of ethics, where the current world is seen as a contraction of a grander world of non-human spirits and places, and where this world is multiple, capable of generating any number of encounters with spirit-others. Deterritorialization cannot, then, be opposed to territorialization as a simple move of freedom. Rather, what needs to be negotiated – beyond good and evil – are the various ways in which deterritorialization either captures all other strata within the range of Anthropos, or enables a thought of life beyond any single stratum

To move towards a conclusion, we might say that territorialization is the coming into being of relations through reductions of intensive difference, allowing bodies to emerge as formed matters. Accompanying this movement there are always various lines of deterritorialization: the rivers, birds, sands and undulations of one's mapped space might be experienced as forces of the cosmos. This sense of the cosmos might generate a higher deterritorialization of expressive matters as such – the potentiality from which worlds emerge – or, it might be drawn back to the vision and range of Anthropocene man. It is not a question of whether or not the future is deterritorialized but whether that deterritorialization is thought in line with logics of emergence or of decolonization. If social orders are thought of as complexities that have emerged from an anarchic chaos, then it follows that when they are swept away what remains is disorder. This leaves the classic postcolonial critique and predicament: we are left with the tools of the master's house, and all we can do is continue to speak ironically. This is one way to read minor literature and deterritorialization. One speaks one's own language as a foreigner, creating murmurs and destabilizations. However, if chaos is thought of as *more* differentiated and complex than social orders, then the exit of decolonization would amount to a creative opening, allowing for new worlds well beyond the Anthropos who always saw the 'new world' as requiring his care. This yields a different conception of the minor, not as a deviation or deflection, but the cutting into the sounds and perceptions that have composed the whole to release new worlds. If Franz Fanon can be read as a theorist insisting on the ways in which colonization installs its figure of normative whiteness in the psyche of the colonized, thereby indicating a deterritorialization that subjects us all to Anthropos, he can also be read as offering a different or higher deterritorialization.

It was through Fanon that Deleuze and Guattari reversed the oedipality of colonization. The colonized subject is not territorialized in the family, and then further subjected to the colonizer; it is precisely at the scene of colonization that one recognizes that the figure of familial, individual Anthropos is the effect – not the ground – of empire. It is decolonization – the recognition of those who have no desire for Anthropos – that enables higher deterritorialization:

When Frantz Fanon encounters a case of persecution psychosis linked to the death of the mother, he first asks himself if he has 'to deal with an unconscious guilt complex following on the death of the mother, as Freud had described in *Mourning and Melancholia*.' But he soon learns that the mother has been killed by a French soldier, and that the subject himself has murdered the wife of a colonist whose disembowelled ghost perpetually appears before him, carrying along with it and tearing apart the memory of the mother. It could always be said that these extreme situations of war trauma, of colonization, of dire poverty and so on, are unfavourable to the construction of the Oedipal apparatus – and that it is precisely because of this that these situations favour a psychotic development or explosion – but we have a strong feeling that the problem lies elsewhere. Apart from the fact that a certain degree of comfort found in the bourgeois family is admittedly necessary to turn out oedipalized subjects, the question of knowing what is actually invested in the comfortable conditions of a supposedly normal or normative Oedipus is pushed still further into the background.

The revolutionary is the first to have the right to say: "Oedipus? Never heard of it." (Deleuze and Guattari 1983: 96).

Through Fanon and decolonization, Deleuze and Guattari reverse the temporality of deterritorialization; rather than begin with the simplicity of the subject and the family that then generates complex social wholes to arrive at the universality of humanity, they see the original scene as multiple and political, with what comes to be known as Anthropos and the family as a contraction and impoverishment of rich and complex differences. The logic of colonization has always relied on something akin to a philosophy of emergence, allowing settler culture to see itself as bringing into being the proper potentiality or true complexity of the territory which – on this account – has less order than the imposed state. From the point of view of emergence, the task of life would be to reach its greatest level of achievement in terms of the intricate, complex and ever-more sophisticated systems of order. *Decolonization*, by contrast, might then be thought of – if we follow the territorializing/deterritorializing potential – as the articulation of intensities that are cosmic (existing beyond the orders they have brought into being). Life should not be thought of in terms of the forms which have emerged; the Anthropos who arrives at the end of history and recognizes himself as the author of a new era of the Anthropocene can think of no future other than his own end.

The nature of life

One way of thinking about modern theory, and one that is dominant in what is often referred to as new materialism, is that post-Kantian thought (and therefore liberalism) precluded a transition from a theory of life to a politics or ethics. Quentin Meillassoux (2008) has criticized this as the assumption of correlationism, that we only know the world as it is given to us, with no possibility of stepping outside of subjectivity. Meillassoux is not the first to criticize the ethics of liberal subjectivism; prior objections from neo-Aristotelian ethics and phenomenology had insisted that even if one cannot know life in itself, this should not preclude the formation of ethics by way of the lived. The twentieth century was marked by a split between forms of post-Kantianism that insisted that 'life' was yet one more symptom of metaphysical over-reaching that needed to be bracketed or suspended, and various vitalisms and quasi-vitalisms that grounded the ethical and political on life, including theories of emergence. Deterritorialization, when coupled with decolonization, provides a conceptual exit from the impasse between Kantian anti-foundationalism and new materialisms, which are often also new literalisms.

From Kant to Rawls, liberalism has insisted that the social order cannot be based on what one claims to know, as if one could possess expertise of life that would yield imperatives for us all. In the absence of such expertise, and with the explicit and self-conscious refusal to make an elevated exception of oneself, the liberal subject imagines the social order as one that might be accepted by any possible rational subject regardless of their position in the social whole (Rawls 1971). Liberalism and its aftermath rely upon an event of futural deterritorialization: the present is lived *as if* it were to be judged by any subject whatever. One imagines oneself as a fragment of humanity in general. This mode of deterritorialization, the deterritorialization of Anthropos, of humanity in general, may have been forged in liberalism and the suspension of nature in itself, but it continues into the twentieth century. Deconstruction, or what came to be known as 'theory', intensified the bracketing or suspension of life: not only can we not know life with any degree that might enable us to prescribe a future, the very conditions of knowing anything at all – that we write, think and speak within a differential system – are essentially open and self-ungrounding. For deconstruction, the condition for life – the knowledge of the lived in its stable predicable form – is death: one can only know life through some system that remains stable through time (Derrida 1978). Not a repetition of that which preserves, sustains and enhances itself, but a tracing movement that each time incurs loss, nonbeing and corruption. The seeming opposition between life in itself that would provide a foundation for ethics, and an attention to the systems through which life is known, reaches its limit in decolonized deterritorialization. Rather than turning back to life itself (as new materialisms and theories of emergence tend to do) and rather than bracketing life within human or inscriptive systems, one might think of decolonization as the creation of a higher deterritorialization, a break with the stratum of Anthropos.

This is where decolonization needs to be taken seriously. Not only the Anthropocene, but various forms of seeming posthumanism allow a new term of 'life' to create a new moral stratum that organizes the present. Everything we do ought to be oriented to saving life, where life is either intelligent life, the life of *our* planet or a life that forms the basis for a general theory of emergence. I want to suggest that such seemingly radical turns to life are continuations of Anthropos and its mode of all-subsuming *relative* deterritorialization. To turn back to life, especially in theories of emergence, is to produce a point of view that accounts for all other strata and forms of relations. This is the case even when life is no longer seen as foundational. Indeed, it is perhaps recent anti-foundational accounts of life that enable us to mark a clear distinction between those modes of deterritorialization that have enabled an all-encompassing Anthropos, and decolonizing deterritorializations that cut into the world as it is to generate some type of exit.

Two recent theories, both of them post-deconstructive, follow this cue. Timothy Morton's 'Queer Ecology' uses evolutionary theory to argue that biology is already anti-essentialist, and that queer theory is the natural ethic for environmentalism:

> Just read Darwin. Evolution means that life-forms are made of other life-forms. Entities are mutually determining: they exist in relation to each other and derive from each other. Nothing exists independently, and nothing comes from nothing. At the DNA level, it's impossible to tell a 'genuine' code sequence from a viral code insertion...
>
> In a sense molecular biology confronts issues of authenticity similar to those in textual studies. Just as deconstruction showed that, at a certain level at any rate, no text is totally authentic, biology shows us that there is no authentic life-form. This is good news for a queer theory of ecology, which would suppose a multiplication of differences at as many levels and on as many scales as possible (Morton 2010: 275).

Using physics and the life sciences, Karen Barad has also argued for the queerness of life and, like Morton though in different ways, for the blurring of identities because of the relational nature of life. Writing about slime moulds, and the moralizing rhetoric of the reporting on their behaviour, Barad concludes that queer performativity is not the privilege of drag queens or even gender-conforming humans, but occurs at the level of pre-intentional amoebic life:

> [S]lime moulds [of which social amoebas or cellular slime moulds (*Dictyosteliida*) are classified as one kind] are 'no more than a bag of amoebae encased in a thin slime sheath, yet they manage to have various behaviours that are equal to those of animals who possess muscles and nerves with ganglia – that is, simple brains.' What is or isn't an 'individual' is not a clear and distinct matter, and that seems to be precisely the scientific sticking point: the question of the nature of identity is ripe here – it's what's so spectacularly exciting from a scientific point of view. No wonder that social amoebas are taken to be model organisms in molecular biology and genetics for studying communication and cell differentiation. Social amoebas queer the nature of identity, calling into question the

individual/group binary. In fact, when it comes to queering identity, the social amoeba enjoys multiple indeterminacies, and has managed to hood-wink scientists' ongoing attempts to nail down its taxonomy, its species-being defying not only classification by phylum but also by kingdom (Barad 2012: 26).

Barad neither warns against, nor shies away from anthropomorphism, but rather questions that there is such a thing as a bounded and autonomous Anthropos who could or should attribute agency to non-humans. Instead, the human – like other queer compositional entities that act, *but do so always and only in relation* – is queer like everything else. It is not one's sexuality that makes one queer: one can only be an individual and then claim something like 'a' sexuality if there have already been all sorts of strange and never fully intended couplings. This is not true only of life, but also of time and – if we want to think at the atomic level – the very composition of anything at all:

> Where and when do quantum leaps happen? If the nature of causality is troubled to such a degree that effect does not simply follow cause end over end in an unfolding of existence through time, how is it possible to orient oneself in space or in time? Can we even continue to presume that space and time are still 'there'?
>
> This queer causality entails the disruption of dis/continuity, a disruption so destabilizing, so downright dizzying, that it is difficult to believe that it is that which makes for the stability of existence itself. Or rather, to put it a bit more precisely, if the indeterminate nature of existence by its nature teeters on the cusp of stability and instability, of possibility and impossibility, then the dynamic relationality between continuity and discontinuity is crucial to the open-ended becoming of the world which resists acausality as much as determinism (Barad 2010: 240).

To argue that life is queer, as both Barad and Morton do (though in different ways), and then to tie this queerness to futurity and the Anthropocene, generates two different forms of posthuman futurity both of which nevertheless rely on some ethic or post-ethic of emergence. In this, both models challenge conventional understandings of heritage as embedded in the genealogical relations of people, politics, history and place (Sterling and Harrison, this volume). Barad, writing about post-nuclear landscapes in the time of the Anthropocene, argues that rather than a planet, or a species humanity, that goes through time, past events are bound up with each other in multiple temporalities – not by erasing the human and replacing it with some non-human natural innocence, but by reading the space of the Earth and its events as already wounding, scarring, killing and preserving itself differentially across the species and its landscape. Barad concludes:

> In these troubling times, how can we not trouble time? Nothing less than the nature of and possibilities for change and conceptions of history, memory, causality, politics and justice are conditioned by it. At the very core of QFT are questions of time and being. The indeterminacy of time-being opens up the nature of matter to a dynamism of the play of being and nothingness. Is there something about the nature of this dynamism that might lend some insight into what the practice of the politically committed work of mourning attuned to justice might look like? Or that would make it possible to trace the practices of historical erasure and political a-*void*-ance, to hear the silent cries, the murmuring silence of the void in its materiality and potentiality? What are the conditions of im/possibilities of living-dying in voids produced by technoscientific research and development, projects entangled with the military-industrial complex and other forms of colonial conquest? (Barad 2018: 215).

Both Barad and Morton stress the relational nature of life, and life's impurity and non-identity, to argue that the future ought to be neither the erasure nor fulfilment of the human, but a world in which the queerness of life amounts to a deflation of human exceptionalism. There is no pure nature to which we can return, nor is there a human nature that ought to realize its full potential and maximize its self-recognition for the future. The Anthropocene is not an opportunity to step up and save the day for us all – a common motif of reterritorialization that one finds in post-apocalyptic culture, where humanity in its better form is the only force that may save the world. Rather, the Anthropocene is a moment of recognizing that the pristine nature that we lost never existed (Morton 2009) and that the future of humanity is not human, but genuinely collective. A dominant motif of the Anthropocene is reterritorialization, such that humanity finds itself at the end of history and (however guilty *or by way of being guilty*) finds its proper self for the future. Against this, both Morton and Barad (and a series of other major theorists of the Anthropocene, including Bruno Latour [2017] and Donna Haraway [2016]) argue for something like a levelling out of humanity by way of relationality and non-identity. By recognizing that there has never been a self-present or self-determining humanity for which nature is a mere backdrop, one might allow for a future of fully Earth-bound kin.

Such arguments for a posthuman future are at odds both with some scholarly attempts to think of a new sense of species unity and redemption (Ellis 2016), and the post-apocalyptic imaginary, where the sense of imminent annihilation is precisely what prompts a regrouping of humanity, and an imperative for survival. One might contrast two projections for a future of the Anthropocene, one with a focus on historical emergence and self-recognition – the supposed 'good' Anthropocene of geoengineering – and another with a sense of vital emergence and a sense of dissolution. It is in the spirit of saving what humanity ought to be that Naomi Klein has insisted that it is capitalism, and not human nature, that has led to catastrophic climate change; if this is so then it is still possible for humanity to diagnose and overcome the historical trajectory that has damaged the planet. Her sense of a proper humanity at the end of history that might redeem itself and meekly inherit the Earth is typical of the popular post-apocalyptic imaginary in which it is *not* human nature, not anything that one might recognize as a species universal, but capitalism that captures relations among humans and deflects living bodies from their own desires. Following up on her 2014 book, *This Changes Everything*, Klein disputed an extended piece in the *New York Times Magazine* where Nathaniel Rich argued that inaction on climate change was due to a human inability to make short-term sacrifices:

> If an inability to sacrifice in the short term for a shot at health
> and safety in the future is baked into our collective DNA, then

we have no hope of turning things around in time to avert truly catastrophic warming.

If, on the other hand, we humans really were on the brink of saving ourselves in the '80s, but were swamped by a tide of elite, free-market fanaticism – one that was opposed by millions of people around the world – then there is something quite concrete we can do about it. We can confront that economic order and try to replace it with something that is rooted in both human and planetary security, one that does not place the quest for growth and profit at all costs at its centre.

And the good news – and, yes, there is some – is that today, unlike in 1989, a young and growing movement of green democratic socialists is advancing in the United States with precisely that vision. And that represents more than just an electoral alternative – it's our one and only planetary lifeline…

There is nothing essential about humans living under capitalism; we humans are capable of organizing ourselves into all kinds of different social orders, including societies with much longer time horizons and far more respect for natural life-support systems. Indeed, humans have lived that way for the vast majority of our history and many Indigenous cultures keep earth-centred cosmologies alive to this day. Capitalism is a tiny blip in the collective story of our species (Klein 2018).

Klein is quite right to note that an appeal to 'human nature' not only forecloses any critical account of the different ways in which life has generated social formations, but also narrows the range of what futures might be imagined. Even so, what her critical account leaves open is the threshold at which one refuses nature in one's account of life, emergence and futurity. That is, it seems reasonable and necessary to say that there is nothing natural (and therefore nothing inevitable) about capitalism, but to what extent does a claim for the non-naturalness of the present – because of other possibilities of life – allow for a dissolution of the present? This is where I think that the concept of deterritorialization can expand the range of the problem of the human: rather than refuse any concept of human nature and insist that we can always become other than what 'we' are, we might see 'human nature' as a formation that exists in ways that are far more complex (and embedded in the Earth) than capitalism.

Futurity

As I have already argued, accounts of life and emergence have allowed for a certain dissolution of the present. If there is no such thing as humanity that simply goes through history in order to recognize itself, and there is instead an always intertwined mesh that is nevertheless not fully in command of itself, then perhaps the present is innocent after all. This conclusion is taken up by Timothy Morton, and it is one that ought to prompt us to question a simple ethics of deterritorialization:

> Every aspect of hyperobjects reinforces our particular lameness with regard to them. The viscosity that glues us to the hyperobject forces us to acknowledge that we are oozing, suppurating with nonhuman beings: mercury, radioactive particles, hydrocarbons, mutagenic cells, future beings unrelated to us who also live in the shadow of hyperobjects. The nonlocality of hyperobjects scoops out the foreground–background manifolds that constitute human worlds. The undulating temporality that hyperobjects emit bathes us in a spatiotemporal vortex that is radically different from human-scale time. The phasing of hyperobjects forcibly reminds us that we are not the measure of all things, as Protagoras and correlationism promise. And like a wafting theatre curtain, inter-objectivity floats in front of objects, a demonic zone of threatening illusion, a symptom of the Rift between essence and appearance (Morton 2013: 197).

Morton's three concepts of hyperobject, lameness and Rift all work to demote humans from self-mastery and world-mastery. Hyperobjects can only be known in their effects, never as they are in themselves. We see wildfires, flooding, tick and mosquito borne viruses, sea-level rises, heatwaves and deforestation but never an object that might be known as climate change, or even climate. For this reason the very world and relationality that composes our being is never something we can grasp;

what makes us who we are is all too *real*, and therefore distant. We are not self-composing, for the rift between what we know and what there really is renders us lame. Like Barad, Morton concludes with a demotion of human supremacy and autonomy.

Here, again, we are brought up against two modes of deterritorialization, both operating with different inflections of emergence. First, one might think of the ways in which 'humanity' emerges from all the disparate modes of existence of *homo sapiens*. Through empire, colonization, slavery and global capitalism a figure of the average white man comes to stand for the whole. A body formed within the territory is no longer a body among others but comes to regard itself as the essence and end of the whole; everybody in the territory is now seen as an expression of man, an example of a humanity that will find itself at the end of history. If one accepts this notion of deterritorialization, where a body emerges and becomes a force in its own right, one can intensify this movement and demand a full actualization and redemption of the human. We have found ourselves at the end of history, and once recognizing our emergence know that we too can turn back, accept our role as planetary guardians and allow human life to flourish. This could amount either to a new or queered humanity of self-recognition and lameness, or a hyperhumanity – where the very technoscience and mastery that has allowed for altering the Earth at a geological level can lead to a heightened control of 'our' milieu. To recognize emergence would be to take the trajectory of complexity as a moral imperative, as a justification for 'technological maturity' and the saving of what has come to know itself as humanity at all costs. This is what I mean when I suggest that the future is already deterritorialized: what thinks of itself as humanity no longer claims to be some divine essence, and not even a universal norm, but a procedure of maximization of potentiality. This can occur at a micro-level, where the twenty-first century is no longer a soul-searching journey of discovering who one is, but a day-to-day self-management, intensifying what one might become.

If the history of human technology has unfortunately rendered the planet uninhabitable then we can embark on interstellar travel, reverse extinction, create various forms of carbon sinks, find other forms of energy or become some form of superintelligence that leaves those requirements and limits behind. This conception of deterritorialization has a long history in philosophy, and contemporary purchase in both pop and high culture. The liberal imperative (perhaps stated most forcefully by Kant) is that if it is possible to think or desire another state of affairs – such as imagining humanity as *not* bound by the causal order of nature – then the very thought of that elevation creates an imperative; we should always be a humanity to come. Today it is the very imagination of planetary destruction that allows humanity to free itself from causality and the past. Because climate change is a global catastrophe, humanity must emerge as a futural force, retrieving itself from the wreckage by imagining itself not as what it actually is, but demanding the emergence of a higher potentiality. If life can become other than what it is, and if human life to date has been destructive, then the future becomes one in which the evil potentialities are cast off and set into the past, while the future becomes the moment to be born again.

The second mode of deterritorialized futurity would see emergence neither from the point of view of the creation of worthy complexity from nightmarish chaos, *nor* as an ethical imperative of ungrounded becoming. Here it might be worth thinking about the future and deterritorialization in a more radically linear manner. It is all too easy and common to dismiss linearity, but a certain non-linearity has enabled a neoliberal and hyperhumanist form of deterritorialization. Whatever we happen to be now, and whatever we happen to have become, this does not preclude demanding a future in which renewal, redemption and renovation allow for a planet of tomorrow. Rather than accepting that who we are now is the result of a long history of repetition without difference, and that what has come to call itself humanity is bound up with capitalism and anthropogenic climate change, we imagine all sorts of redemptive turns and breaks. We think that we might refer to something like the Capitalocene and thereby allow humanity to be a virtuous and virtual remainder that might rise, like a phoenix from the ashes, having broken from its grubby past. Because we are committed to a certain logic of emergence we define humanity as nothing other than a capacity for becoming-other. What might happen if deterritorialization were not thought of as a godlike emergence ex nihilo, as ongoing self-creation and renewal, but as always bound up with forces allowing for highly singular lines of flight? Here I go back to what it might be to think beyond a theory of emergence and its morality of complexity from chaos. If complexity has *more* difference than the stable forms that emerge, then a future that is not bound up with humanity's self-narrativizing and self-preserving line of redemption would require a dramatic reversal in order to generate a line of flight. One might think of various forms of what we too often look at as pre-history or the uncivilized and inhuman as offering a 'higher deterritorialization'. In this case one might deploy one's theory of life not as a mode of normativity, where whatever is is right, but as a mode of exit: how might an account of life open up to the unlived? The unlived would amount to all those forms of existence that hitherto have appeared as the end of the world, but that might after all offer 'a' future, even if it is not ours.

Notes

1.   https://www.fhi.ox.ac.uk

References

Badiou, A. 2001. *Ethics.* Trans. Peter Hallward. London: Verso.

Barad, K. 2010. Quantum Entanglements and Hauntological Relations of Inheritance: Dis/continuities, SpaceTime Enfoldings, and Justice-to-Come. *Derrida Today* 3(2): 240–68.

Barad, K. 2012. Nature's Queer Performativity. *Kvinder, Køn & Forskning* 1-2: 25–53.

Barad, K. 2018. Troubling Time/s and Ecologies of Nothingness: Re-turning, Re-membering, and Facing the Incalculable. In *Eco-Deconstruction: Derrida and Environmental Philosophy*, edited by M. Fritsch, P. Lynes and D. Wood, 206–248. New York: Fordham University Press.

Bostrom, N. 2013. Existential Risk Prevention as Global Priority. *Global Policy* 4(1): 15–31.

Butler, J. 2010. Performative Agency. *Journal of Cultural Economy* 3(2): 147–61.

Chakrabarty, D. 2009. The Climate of History: Four Theses. *Critical Inquiry* 35(2): 197–222.

Deleuze, G. and F. Guattari. 1983. *Anti-Oedipus: Capitalism and Schizophrenia.* Trans. Brian Massumi. Minneapolis: University of Minnesota Press.

Deleuze, G. and F. Guattari. 1987. *A Thousand Plateaus: Capitalism and Schizophrenia.* Trans. Brian Massumi. Minneapolis: University of Minnesota Press.

Deleuze, G. and F. Guattari. 1994. *What is Philosophy?* Trans. Hugh Tomlinson and Graham Burchell. New York: Columbia.

Derrida, J. 1978. *Writing and Difference.* Trans. Alan Bass. London: Routledge.

Edelman, L. 2004. *No Future: Queer Theory and the Death Drive.* Durham, NC: Duke University Press.

Ellis, E. 2016. Evolving Toward a Better Anthropocene. https://thebreakthrough. org/issues/conservation/evolving-toward-a-better-anthropocene.

Fanon, F. 2008. *Black Skin, White Masks.* Trans. Richard Philcox. New York: Grove Press.

Haraway, D. J. 2016. *Staying with the Trouble: Making Kin in the Chthulucene.* Durham, NC: Duke University Press.

Heidegger, M. 1998. Letter on Humanism. In *Pathmarks,* edited by W. McNeill, 239–76. Cambridge: Cambridge University Press.

Kiernan, B. 2004. The First Genocide: Carthage, 146 BC. *Diogenes* 51(3): 27–39.

Klein, N. 2018. Capitalism Killed our Climate Momentum, Not "Human Nature". https://theintercept.com/2018/08/03/climate-change-new-york-times-magazine/.

Latour, B. 2017. *Down to Earth: Politics in the New Climatic Regime.* Cambridge: Polity Press.

Meillassoux, Q. 2008. *After Finitude: An Essay on the Necessity of Contingency.* Trans. Ray Brassier. London: Continuum.

Mirzoeff, N. 2018. It's Not the Anthropocene, It's the White Supremacy Scene; or, The Geological Color Line. In *After Extinction,* edited by R. Grusin, 123–50. Minneapolis: University of Minnesota Press.

Moore, J. 2015. *Capitalism in the Web of Life: Ecology and the Accumulation of Capital.* London: Verso.

Morton, T. 2009. *Ecology Without Nature.* Cambridge: Harvard University Press.

Morton, T. 2010. Queer Ecology. *PMLA* 125(2): 73–282.

Morton, T. 2013. *Hyperobjects.* Minneapolis: University of Minnesota Press.

Muecke, S. 2019. Rebooting the Idea of 'Civilisation' for Australian Soil. https:// theconversation.com/friday-essay-rebooting-the-idea-of-civilisation-for-australian-soil-112682?fbclid=IwAR3oV7ORI4BQK0yTiKhtpZYASJTLyqfXlq3y vlmP67HlTZkScieoAEQ0dtU.

Neyrat, F. 2019. *The Unconstructable Earth: An Ecology of Separation.* Trans. Drew S. Burk. New York: Fordham University Press.

Povinelli, E. 2016. *Geontologies: A Requiem to Late Liberalism.* Durham: Duke University Press.

Protevi, J. 2019. *Edges of the State.* Minneapolis: University of Minnesota Press.

Rawls, J. 1971. *A Theory of Justice.* Cambridge, MA: Harvard University Press.

Scott, J. C. 2017. *Against the Grain: A Deep History of the Earliest States.* New Haven, CT: Yale University Press.

Sheldon, R. 2016. *The Child to Come: Life After the Human Catastrophe.* Minneapolis: University of Minnesota Press.

Stiegler, B. 2015. *States of Shock: Stupidity and Knowledge in the 21st Century.* Cambridge: Polity.

Viveiros de Castro, E. 2015. *The Relative Native.* Chicago, IL: Hau Books.

Wengrow, D. and D. Graeber. 2015. Farewell to the 'childhood of man': Ritual, Seasonality, and the Origins of Inequality. *Journal of the Royal Anthropological Institute.* 21(3): 597–619.

About the Contributors

Cecilia Åsberg is Guest Professor of Science and Technology Studies, Gender and Environment at KTH Royal Institute of Technology Stockholm, and Professor of Gender, Nature, Culture at Linköping University. She is Founding Director of the Posthumanities Hub, and the Seed Box: An Environmental Humanities Collaboratory; and associate editor of the journal *Environmental Humanities*. Åsberg has published extensively (in Swedish, Dutch, English); taught gender studies, Environmental Humanities, STS and posthumanities to BA, MA and PhD students in various research, managing and teaching positions at KTH, Linkoping University and Utrecht University (NL), and been guest professor at several German and Nordic universities (Rachel Carson Centre, LMU). Recent publications include 'Planetary Speculation: Cultivating More-Than-Human Arts', in *Cosmological Arrows*. Stockholm: Art And Theory, 2019; 'Feminist Posthumanities in the Anthropocene: Forays into the Postnatural' in *Journal of Posthuman Studies* (2018); 'Toxic Embodiment' *Environmental Humanities* 11, no. 2 (2019, special issue ed. with Olga Cielemęcka); and *A Feminist Companion to the Posthumanities* (2018, ed. with Rosi Braidotti).

Anna Bohlin is Associate Professor in Social Anthropology at the School of Global Studies, University of Gothenburg, Sweden. She has previously researched issues of memory and temporality in relation to place, in particular in connection with various state projects such as democratization (South Africa) or river restoration (Sweden). Current research interests include people's relations to everyday domestic items; waste; and second-hand and reuse as embodied and alternative forms of heritage. She is part of the leadership group of the Centre for Critical Heritage Studies (University of Gothenburg/University College London) and the Research Cluster Making Global Heritage Futures, where she explores the intersection between things/materials, temporality and sustainability, most recently within the project 'Staying (with) Things: Alternatives to Circular Living and Consuming', Swedish Research Council 2020-2024.

Esther Breithoff is a UKRI Future Leaders Fellow and Lecturer in Contemporary Archaeology and Heritage at Birkbeck, University of London. She has held postdoctoral research positions at UiT The Arctic University of Norway and University College London. Her research spans the fields of Contemporary Archaeology and Critical Heritage Studies and traces a common set of interests in the relationships between conflicts, resources, recycling and rights across the human/non-human divide in the Anthropocene. Esther is also Affiliated Researcher on the Heritage Futures research programme, based at UCL, and the Review Editor for the *Journal of Contemporary Archaeology*.

Denis Byrne is a Senior Research Fellow at the Institute for Culture and Society, Western Sydney University, where he is convenor of the Heritage & Environment research programme. He is an archaeologist whose work has mostly been in the fields of Indigenous and migrant heritage in Australia as well as in the cultural politics of heritage conservation in Southeast Asia. His current research is centred on the ARC-funded China-Australia Heritage Corridor project which, focussing on the period from the mid-1800s to mid-1900s, investigates connections between the built environment of migrants from Zhongshan (Guangdong) in Australia and that of their ancestral villages in China. He is also researching the history of coastal reclamations in the Asia-Pacific as examples of Anthropocene heritage. His books *Surface Collection* (2007) and *Counterheritage* (2014) challenge western-derived heritage practices in Asia and explore new approaches to the writing of archaeology and heritage.

Claire Colebrook is Edwin Erle Sparks Professor of English, Philosophy and Women's Gender and Sexuality Studies at Penn State. She has published numerous works on Gilles Deleuze, visual art, poetry, queer theory, film studies, contemporary literature, theory, cultural studies and visual culture. She is the editor (with Tom Cohen) of the Critical Climate Change and CCC2 Irreversibility Book Series for Open Humanities Press.

Caitlin DeSilvey is Professor of Cultural Geography, based at the University of Exeter's Cornwall campus, where she is Associate Director for Transdisciplinary Research in the Environment and Sustainability Institute. Her research into the cultural significance of material change has involved extensive collaboration with heritage practitioners, archaeologists, ecologists, artists and others. From 2015-2019 she was co-investigator on the Heritage Futures project (funded by the Arts and Humanities Research Council), and in 2016-17 she was a fellow at the Centre for Advanced Study, Olso. Her publications include *Anticipatory History* (2011, with Simon Naylor and Colin Sackett) and *Visible Mending* (2013, with Steven Bond and James R Ryan). *Curated Decay: Heritage Beyond Saving* (2017) received the University of Mary Washington's Center for Historic Preservation 2018 Book Prize, awarded each year to an author whose book has a positive impact on preservation in the United States.

Christina Fredengren is an Associate Professor in the Department of Archaeology and Classical Studies, Stockholm University. She is a founding member of the Stockholm University Environmental Humanities network together with Claudia Egerer and Karin Dirke. As Scientific Leader of the Deep Time research cluster at the SeedBox at Linköping University and affiliated researcher at the Posthumanities hub, her research deals with archaeology, heritage and matters of intra-generational ethics and care. She has particular research interests in human-animal relations, sacrifice and water; these are reflected in her current research projects, including the VR funded project 'Waters of the Time' and the Formas funded project 'Checking-in-with-Deep Time'.

Rodney Harrison is Professor of Heritage Studies at the UCL Institute of Archaeology and Arts and Humanities Research Council (AHRC) Heritage Priority Area Leadership Fellow (2017-2020). He has experience working in, teaching and researching natural and cultural heritage conservation, management and preservation in the UK, Europe, Australia, North America and South America. He is the (co) author or (co) editor of 17 books and guest edited journal volumes and over 80 peer reviewed journal articles and book chapters and is the founding editor of the *Journal of Contemporary Archaeology*. Between 2015 and 2019 he was principal investigator on the AHRC funded Heritage Futures research programme www.heritage-futures.org. His research has been funded by AHRC, GCRF/UKRI, British Academy, Wenner-Gren Foundation, Australian Research Council, Australian Institute of Aboriginal and Torres Strait Islander Studies and the European Commission.

Colin Sterling is an AHRC Early Career Leadership Fellow at UCL Institute of Archaeology. His research investigates the ideas and practices of heritage from a range of theoretical and historical perspectives, with a core focus on critical-creative approaches to heritage making. He is currently writing a book with Rodney Harrison on more-than-human heritage in the Anthropocene, which aims to expand the framework of critical heritage studies to better address the urgent problems of a warming world. Colin was previously a Project Curator at the Royal Institute of British Architects and has worked as a heritage consultant internationally, specializing in curatorial planning, audience research and interpretation. His first monograph *Heritage, Photography, and the Affective Past* was published by Routledge in 2019. He has a long-standing interest in the relationship between art and heritage, and is currently working on a new project investigating the impact of experiential and immersive design across the heritage sector.

Anna Storm is Professor of Technology and Social Change at Linköping University in Sweden. Her research interests are centred on industrial and post-industrial landscapes and their transformation, in physical as well as imaginary sense, and comprising both cultural and 'natural' environments. Such landscapes not only trigger perspectives of power relations but challenge our understandings of ecology, aesthetics, memory and heritage. Recent work more specifically deals with nuclear nature imaginaries, deployed through human relations to animals in nuclear settings, such as fish, snakes, bison and wild boar. Anna Storm is currently leading the multidisciplinary project 'Atomic Heritage goes Critical: Waste, Community and Nuclear Imaginaries', and is senior researcher in 'Cold War Coasts: The Transnational Co-Production of Militarized Landscapes' and 'Nuclearwaters: Putting Water at the Centre of

Nuclear Energy History'. She is the author of *Post-Industrial Landscape Scars* (Palgrave Macmillan 2014), which was shortlisted for the Turku Book Award 2015.

J. Kelechi Ugwuanyi teaches in the Department of Archaeology and Tourism, University of Nigeria. He recently completed a PhD in heritage studies in the Department of Archaeology, University of York. His dissertation explored the negotiation between the global heritage discourse and the existing beliefs and value systems in the context of the village arena (or 'square') among the Igbo of Nigeria. The work bridges the gap between authorized heritage institutions and Indigenous/local communities to encourage inclusion/sustainability in heritage management. John has a diploma in tourism and museum studies, as well as a BA and an MA (both cum laude) in archaeology and tourism from the University of Nigeria. Currently, he is the assistant editor of the *Journal of African Cultural Heritage Studies*. He has published locally and internationally, and has his interests focused mainly on heritage studies, museum, tourism, Indigenous knowledge systems and contemporary archaeology.

Adrian Van Allen is an Associate Researcher in the Department of Anthropology, Smithsonian National Museum of Natural History and California Academy of Sciences. Adrian is a cultural anthropologist who studies museums as technologies for constructing natural-cultural worlds. Currently an Abe Fellow of the Social Science Research Council, she is conducting a comparative study of biobanking policy and practice in natural history museums in the U.S.A., France and Japan. She is also working on a book that examines the material culture of wildlife biobanking in the face of mass biodiversity loss, focused on an ethnography of the Smithsonian's Global Genome Initiative. Through exploring museum specimens as mobile and transformative of a variety of scales of time, her research traces the multiplicity not only between but also within objects as they negotiate reconstructed pasts and imagined futures.

Anatolijs Venovcevs is a PhD candidate at UiT: The Arctic University of Norway. His research focuses on the spatially and temporally dispersed heritage of twentieth century single-industry mining towns in Labrador,

Canada and Murmansk Region, Russia. He received his Masters at the Memorial University of Newfoundland in St. John's, Canada and has previously worked as the GIS Technician for the Town of Happy Valley-Goose Bay, Labrador. Besides his expertise in contemporary archaeology, GIS and industrial heritage, he is interested in presenting knowledge in new and unconventional ways.

Joanna Zylinska is Professor of New Media and Communications at Goldsmiths, University of London. The author of a number of books including *The End of Man: A Feminist Counterapocalypse* (University of Minnesota Press, 2018), *Nonhuman Photography* (MIT Press, 2017) and *Minimal Ethics for the Anthropocene* (Open Humanities Press, 2014), she is also a translator of Stanislaw Lem's philosophical treatise, *Summa Technologiae* (University of Minnesota Press, 2013). Zylinska combines her philosophical writings with image-based art practice and curatorial work. Her current research involves an exploration of the relationship between artificial intelligence and Anthropocene stupidity.

www.ingramcontent.com/pod-product-compliance
Lightning Source LLC
Chambersburg PA
CBHW020653270326
41928CB00005B/104